Laser Ultrasonics

Techniques and Applications

Laser Ultrasonics
Techniques and Applications

C B Scruby and L E Drain

National Nondestructive Testing Centre, AEA Technology,
Harwell Laboratory

Published in 1990 by
Taylor & Francis Group
270 Madison Avenue
New York, NY 10016

Published in Great Britain by
Taylor & Francis Group
2 Park Square
Milton Park, Abingdon
Oxon OX14 4RN

No claim to original U.S. Government works
Printed in the United States of America on acid-free paper
10 9 8 7 6 5 4 3 2

International Standard Book Number-10: 0-7503-0050-7 (Hardcover)
International Standard Book Number-13: 978-0-7503-0050-6 (Hardcover)
Library of Congress catalog number: 90-42609
Consultant Editor: **A E de Barr**

This book contains information obtained from authentic and highly regarded sources. Reprinted material is quoted with permission, and sources are indicated. A wide variety of references are listed. Reasonable efforts have been made to publish reliable data and information, but the author and the publisher cannot assume responsibility for the validity of all materials or for the consequences of their use.

Library of Congress Cataloging-in-Publication Data

Catalog record is available from the Library of Congress

Taylor & Francis Group
is the Academic Division of Informa plc.

Visit the Taylor & Francis Web site at
http://www.taylorandfrancis.com

Contents

Preface

Ultrasonics has made a major impact in many areas, including medical diagnosis and industrial non-destructive testing. As a result of some pioneering work in the 1960s and early 1970s, followed by more systematic studies in the late 1970s and 1980s, laser techniques have now been established as a viable non-contact alternative to piezoelectric transducers for generating and receiving ultrasound. Substantial research has quantified the interactions between the optical and ultrasonic fields, and various potential applications have been demonstrated, mostly under laboratory conditions.

Meanwhile there has been a steady growth of interest in diagnostic techniques across all areas of industry. Process monitoring, on-line quality control, in-service inspection, non-destructive testing, etc, all need sensors. In the case of measurements during manufacturing and fabrication, the desire is for non-contact or, better still, remote sensors that do not disturb the process under investigation. Optical techniques are therefore the obvious candidate. However, only ultrasonics (and to a lesser extent radiography) can probe to significant depths within engineering materials, which is where some of the measurements must be made to assure material quality. Consequently, the ideal sensor ought to use optics outside the specimen and ultrasonics inside. Laser ultrasonics offers just such a combination. On these grounds this subject should therefore be poised ready for exploitation by industry.

The purpose of this book is to review the principles underlying both generation and reception processes, in preparation for a discussion of likely applications. Thus the first chapter will briefly introduce the two main subjects of this book: ultrasonics and lasers. Techniques based on the diffraction of light are discussed in Chapter 2. While the main thrust of the book is towards the use of lasers for ultrasonic measurement, there are some important uses of ultrasonic devices in laser optics and a brief review of these is included. In Chapter 3 we shall explore various types of laser interferometry in depth, before considering applications of interferometry to ultrasonics in Chapter 4. Chapter 5 takes the form of a comprehensive study of the generation by laser of ultrasonic waves, whilst Chapter 6 reviews a very wide range of applications for laser-generated ultrasound (mostly, but not entirely, in combination with laser reception). The reader will find only a very few references to medical ultrasonics. This is because contact probes are so well suited to most aspects of medical diagnosis and imaging. The final chapter will, after a summary, attempt to assess the future of this new technology.

As the first book solely on laser techniques in ultrasonics, it is written for as broad a readership as possible. Our desire is that researchers and industrialists alike will find parts, and hopefully all, of the book useful. We hope that it will foster a wider understanding of the potentialities and limitations of the various techniques, and also suggest profitable applications.

Acknowledgments

We are indebted to many colleagues at Harwell and elsewhere for their encouragement, advice and support, especially A E Hughes, A M Stoneham FRS, R S Sharpe, F A Wedgwood, H N G Wadley, S B Palmer and R J Dewhurst.

We wish to thank B C Moss for providing some unpublished technical information and experimental data, and also F K Brocklehurst and K J Davies for assisting in checking the manuscript and page proofs.

We would like to acknowledge our thanks to AEA Technology, who have encouraged the publication of this book. We are especially grateful to the Harwell Tracing Office, and to Mrs C Davis in particular, for overseeing the preparation of the line diagrams used to illustrate the text.

One of us (CBS) wishes to offer very special thanks to his wife and children, for their tolerance and patience during the preparation of the manuscript.

We are grateful to the following for granting permission to reproduce figures included in this book.

The authors of all figures not originated by ourselves.
S Hirzel Verlag for figures 2.4 and 3.29(a).
The Japanese Journal of Applied Physics for figure 2.7.
The IEEE for figures 2.12, 3.26 and 4.34.
The American Institute of Physics for figures 2.16, 2.20, 2.25, 3.39, 4.17–4.21, 4.44–4.46, 5.12, 5.13, 5.20, 5.22, 5.23, 5.52, 6.16, 6.25, 6.28 and 6.29.
The American Society for Non-Destructive Testing for figures 2.26, 4.41–4.43, 6.2 and 6.3.
Dantec Elektronic for figure 3.27.
The Optical Society of America for figures 4.8, 4.12, 4.22 and 4.23.
B and W Loudspeakers for figure 4.9(a).
MIRA for figure 4.10.
The IEE for figure 4.16.
Academic Press for figure 5.7.
The British Institute of Non-Destructive Testing for figure 5.43.
The Institute of Acoustics for figure 5.50.
The American Society of Mechanical Engineers for figure 5.53.
Harry Diamond Laboratories for figures 6.4–6.6.
The Royal Society for figures 6.14, 6.15 and 6.30.
Butterworth and Co (Publishers) Ltd for figure 6.17.
Gordon and Breach Science Publishers for figures 6.22 and 6.23.
University of California Lawrence Livermore Laboratory for figure 6.27.

Glossary of Symbols

Other symbols are defined where they occur and may have several uses.

a, a_n	Amplitude of sinusoidal normal displacement of a surface.
a_t	Amplitude of transverse displacement of a surface.
δa_N	Amplitude of sinusoidal normal surface displacement corresponding to a signal to noise ratio of unity.
A	Area.
A_0, A_1, A_2	Amplitudes of incident and reflected longitudinal waves.
B_0, B_1, B_2	Amplitudes of incident and reflected shear waves.
B	Bulk modulus of elasticity.
c	Velocity of light *in vacuo* $= 2.9979 \times 10^8 \text{ m s}^{-1}$.
c_1, c_L	Compression wave velocity.
c_2, c_T	Shear wave velocity.
c_R	Rayleigh wave velocity.
C	Specific thermal capacity.
C	Electrical capacity.
d	Diameter of laser beam or illuminated spot.
D	Diameter of receiving optics.
D_{subs}	Force dipole strengths.
E	Energy.
E	Electric field.
f	Ultrasonic frequency.
Δf	Bandwidth.
F	Force.
F	Focal length.
$\mathscr{F}$	Finesse of Fabry–Perot interferometer.
G_{subs}	Green's functions.
h	Thickness or separation.
h	Planck's constant $= 6.626 \times 10^{-34} \text{ J s}$.
$H(t)$	Heaviside function.
i	Detector current.
δi_N	RMS detector noise current.
I	Light intensity or electromagnetic power density per unit area.
$J_n(\ \ )$	Bessel function of nth order.
k	Ratio of compression to shear wave velocity.
k	Boltzmann's constant $= 1.381 \times 10^{-23} \text{ J K}^{-1}$.
K	Thermal conductivity.
l	Length.

L	Difference in optical path length in an interferometer.
L_c	Optical path length around a laser cavity.
L	Latent heat of vaporization.
$\mathscr{L}$	Optical beam length through an acoustic field.
M	Mass.
M_2	Figure of merit of an acoustic-optic material.
n	Order of diffraction or interference.
p	Pressure.
p	Photoelastic constant.
Q	Opto-acoustic diffraction parameter/quality factor of laser cavity.
r	Spherical polar or cylindrical polar coordinate.
r_0	Radius of the waist of a laser beam to $1/e^2$ intensity points.
r_M	Radius of mirror (aperture) in confocal Fabry–Perot interferometer.
r_z	Radius of laser beam at distance z from the waist.
R	Optical reflectivity.
R_0	Radius of laser beam to $1/e^2$ point.
R_M	Radius of curvature of spherical mirrors in confocal interferometer.
R	Electrical resistance.
t	Time.
T	Temperature.
$\boldsymbol{u}$	Vector displacement of a point.
u	Scalar displacement of a point.
u_n	Normal displacement of a surface.
u_t	Transverse displacement of a surface.
u	Velocity of surface or object.
v	Generalized velocity of an acoustic wave.
v_s	Velocity of a surface wave.
v	Raman–Nath parameter.
v_0	Normalized Raman–Nath parameter.
V	Voltage.
V_0	Amplitude of (voltage) interference signal.
V	Volume.
W	Light power.
W_0	Laser power output.
W_I	Light power into interferometer.
δW_N	RMS noise equivalent light power fluctuation.
δx_N	RMS noise equivalent displacement.
α	Coefficient of linear thermal expansion.
γ	Absorption coefficient.
$\gamma_1, \gamma_2, \gamma_3$	Direction cosines.
δ	Skin depth.

$\delta(\ \ )$	Delta function.
δ_{ij}	Dirac delta ($= 1$ if $i = j$; $= 0$ if $i \neq j$)
ε_{subs}	Elastic strain.
η	Quantum efficiency.
θ	Angle of diffraction or scattering.
θ, θ_0	Angle of incidence of ultrasonic waves at a surface.
θ_1, θ_2	Angles of reflection of compression and shear waves at a surface.
κ	Thermal diffusivity.
λ	Optical wavelength.
λ_0	Optical wavelength *in vacuo*.
λ	Lamé elastic constant.
Λ	Ultrasonic wavelength.
Λ_d	Diffusion length.
μ	Refractive index.
μ^*	Effective refractive index.
μ_0	Permeability of free space $= 4\pi \times 10^{-7}\ \mathrm{H\,m^{-1}}$.
μ_r	Relative permeability.
μ	Lamé elastic constant.
v	Light frequency.
ν	Poisson's ratio.
ρ	Density.
ρ	Radius of detector aperture.
σ_{subs}	Elastic stress.
σ	Electrical conductivity.
τ	Time constant or period.
τ_D	Delay time in interferometer.
τ_D^*	Effective delay time in interferometer.
ϕ	Phase (optical or acoustic).
Φ	Scalar potential.
Ψ	Vector potential.
Ω	Solid angle.

1 Introduction

Since their invention in the early 1960s lasers have found applications in most areas of science and technology. Laser techniques are thus used in such diverse fields as spectroscopy, metrology, inspection, communications, fusion research, weapons systems and surgery. As a source of light the laser has a number of unique characteristics, which together make it a very versatile device. These include its ability to generate a well-collimated, intense beam of coherent light at a precisely defined wavelength. Lasers can be readily constructed to operate in continuous or pulsed mode (with pulse lengths from milliseconds down to picoseconds), and at wavelengths from infrared to ultraviolet. Not all the applications take advantage of all the features of the laser: some (e.g. laser welding) mainly require a high-intensity, well-collimated beam that can be accurately focused, while others (e.g. interferometry) make use of the accurate wavelength and high coherence.

While the greatest initial impact of lasers was undoubtedly on optical techniques, they have more recently begun to make a significant contribution in fields such as ultrasonics. Ultrasonics has now become a broad and mature technology. Ultrasonic inspection methods are routinely used for the non-destructive examination of engineering materials and structures, ultrasonic diagnosis is routinely used in medicine, while ultrasonic measurements of displacement and dimension are widely used throughout industry. The combination of laser techniques with ultrasonics has led to some exciting new discoveries and applications. Before proceeding to consider the combination of lasers with ultrasonics, we briefly introduce the fields of ultrasonics and laser physics separately.

1.1 ULTRASONICS

Energy can be propagated through solids, liquids and gases as acoustic waves. In fluids there can in general only be one propagation mode, of longitudinal waves (i.e. displacements parallel to the direction of propagation) which take the form of alternate compressions and rarefactions; thus they are also known as compression(al) waves. Acoustic waves propagate with a velocity which is characteristic of the fluid medium. In elastic solids, which include most metals, the situation is more complicated. The medium can now accommodate transverse waves (displacements perpendicular to the direction of propagation) in addition to longitudinal waves (figure 1.1). The particle motions in what are known as elastic waves can in general be resolved into three perpendicular components, two transverse and one longitudinal. Each of the

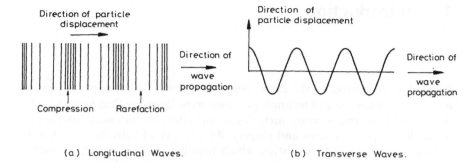

Figure 1.1 Schematic of (*a*) longitudinal and (*b*) transverse waves. In the latter case there is a plane of polarization defined in the plane of the page.

three modes has its own characteristic velocity, although in isotropic solids the two transverse velocities are equal (and also smaller than the longitudinal velocity).

Acoustic waves (whether naturally or artificially generated) cover an extremely wide frequency spectrum. The audible range extends from about 50 Hz to a frequency between 12 and 20 kHz depending upon the age, etc, of the individual. Below this range lies infrasound, including most of the energy in earthquakes, for instance. Acoustic frequencies above about 20 kHz are known as ultrasound. Most commercial ultrasonics takes place within the range 50 kHz–20 MHz, although higher frequency measurements are increasingly being made, to ~ 100 MHz. The more recent technique of acoustic (or ultrasonic) microscopy uses even higher frequencies, i.e. 200 MHz–2 GHz. Gigahertz frequencies are currently the upper limit for ultrasonics. Above these frequencies there are quantized lattice vibrations (phonons). Phonon frequencies in solids typically extend up to 10^{13} Hz. For present purposes we shall however restrict ourselves to a range below 10^9 Hz, which can in practice be treated by classical mechanics.

In common with all wave motions, the velocity of ultrasound is equal to the product of its wavelength and frequency. Thus for a typical longitudinal sound velocity in a solid of 5000 m s^{-1}, the audible range extends over the wavelength range 0.25–100 m. At the lower end of the ultrasonic spectrum, the wavelengths in most solids are fractions of a metre, which still correspond to relatively large-scale vibrations. However, in the 1–5 MHz range (one of the most widely used in medical and industrial ultrasonics) the wavelengths are a few millimetres in metals and just less than a millimetre in water. Thus the waves are far more localized, and can probe the medium with much greater resolution. At a frequency of 2 GHz in the acoustic microscope the wavelength (in water) is 750 nm, comparable with that of visible light. Thermally excited phonon wavelengths may be as small as several interatomic spacings.

Ultrasound can be made to propagate relatively large distances in many solids and liquids. However, in addition to geometrical spreading, which takes the form of the inverse square law in three dimensions, there are a number of medium-specific loss mechanisms, such as absorption by the medium and scattering from discontinuities. All of these cause the signal to be attenuated. Although attenuation can have various origins, it generally increases as the frequency of the ultrasound increases. Thus, while ultrasound at tens or hundreds of kilohertz will propagate many metres in metals, these distances are typically reduced to centimetres much above 10 MHz. As gigahertz frequencies are approached, the waves will only propagate small fractions of a millimetre. Absorption and scattering similarly restrict the use of high-frequency ultrasound in liquids. In gases such as air the attenuation is extremely high and propagation paths of only a few centimetres may be attainable once the frequency rises much above 1 MHz.

In common with other wave motions, ultrasound undergoes various changes at boundaries, whether internal or external. When a beam of ultrasound impinges upon a boundary between two different fluid media, some of the energy will in general be transmitted through the boundary into the second medium, although refraction will occur. The rest of the energy will be reflected back from the boundary into the first medium. The proportions of ultrasonic energy transmitted and reflected can be calculated if the properties of the media are known. If the second medium is a gas such as air, then transmission into it can often be neglected because it is so small.

However, if either or both of the media are elastic solids, the situation becomes more complicated since solids can support three propagation modes. The conditions at the boundary ensure that in most cases at least two of the propagation modes couple together, so that energy can be transferred from compression to shear waves or vice versa. Mode conversion, as this is known, can be a benefit since potentially it enables more information to be extracted about the specimen under inspection. However, it also adds to the complexity of ultrasonic data and may lead to difficulties in interpretation.

The multiple reflections and mode conversions that occur at the boundaries of the specimen interfere with and reinforce one another to transfer energy into the normal modes of the specimen. These normal modes (specimen resonances) build up particularly quickly when the specimen is a regular solid such as a rectangular plate. These normal modes are generally avoided in ultrasonic measurements by methods such as time-gating (i.e. selecting a short time window) and band-pass filtering (the normal modes generally occur at lower frequencies).

Surfaces and interfaces are also of great interest for another reason: they are able to support various types of surface and interface wave. Surface waves are often detected with relatively large amplitude: this is because they only propagate over a two-dimensional surface rather than throughout three-dimensional space. Waves on rods and strips suffer even less attenuation due to geometrical spreading since they propagate in only one dimension. The most important surface wave is the Rayleigh wave. It causes points on the surface to move in an elliptical motion with (predominantly) perpendicular and parallel components. Also important are Lamb and other waves that couple together the motions on surfaces in close proximity. Energy can readily be coupled into interfacial waves from bulk waves (and vice versa) at boundaries.

From a theoretical viewpoint it is usually convenient to consider two main types of wave: plane waves and spherical waves. In the former case the source of the ultrasound is considered to be at infinity so that the wavefronts are parallel planes. Under these conditions the calculation of ultrasonic field strengths is usually simplified. The second most convenient ideal source is an infinitesimally small point. This generates spherical wavefronts. Whereas in the former case there is no attenuation due to geometrical spreading, in the latter the inverse square law applies, so that the intensity decays with the square of the distance from the point source. Regrettably the ultrasound from practical sources can only be approximated to either of these two cases.

The simplest model of a more realistic ultrasonic source is a vibrating piston of finite radius. Under continuous excitation the region immediately in front of the source receives waves from all points across the piston but after slightly different time delays. These waves thus interfere with one another to produce a complex pattern of interference fringes, with maxima and minima in intensity. This region is known as the 'near field' of the source. It is considered to be the least satisfactory for making ultrasonic measurements, and is usually treated as a 'dead zone'. At greater distances, the waves from different parts of the piston arrive almost simultaneously so that there is negligible interference. This region (extending to infinity) is known as the 'far field' of the source, and is the one mostly used in ultrasonic applications. In the far field, the intensity obeys the inverse square law, so that the ultrasound behaves as if it has emanated from a point. Approximately plane waves can be obtained either by working at a reasonably large range using a small diameter source, or else by using the source in conjunction with an ultrasonic lens.

1.1.1 Generation and detection of ultrasound

There are a number of ways to generate and receive ultrasound, but the most common is by means of a piezoelectric transducer (or probe), a diagram of which is shown in figure 1.2(a). If a piezoelectric material is subjected to stress or pressure then its faces become charged electrostatically, generating a potential difference. Conversely, if a potential difference is applied between two of its faces, then a stress and/or strain is induced in the material. In these two modes the piezoelectric material can be used as a receiver and generator of sound waves, respectively. The most common piezoelectric

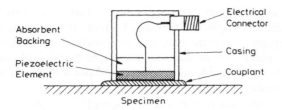

(a) Schematic of Ultrasonic Transducer.

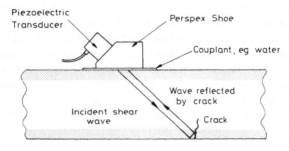

(b) Schematic of Defect Detection by Ultrasonics.

Figure 1.2 Schematics showing (a) ultrasonic transducer and (b) its use in non-destructive testing to detect a defect. The diagram shows a compression wave probe attached to an angled perspex shoe which causes mode conversion to shear wave in the specimen.

materials are ceramics such as lead zirconate titanate (abbreviated to PZT). Crystalline quartz, one of the first piezoelectric materials to be discovered, is still used for transducer construction. Above approximately 300 °C high-temperature materials such as lead metaniobate and lithium niobate are employed. More recently, piezoelectric polymers such as polyvinylidene fluoride (PVF_2 or PVDF) have been discovered.

There must be some medium to transmit (or 'couple') the ultrasound from the transducer to the sample and vice versa, hence the term 'contact' transducer. For measurements that are carried out in water or another liquid, the fluid itself acts as the couplant, which is extremely convenient. Indeed, many biological materials behave very similarly to water from an ultrasonic point of view, so that water can readily be used as the couplant. Except at very high frequencies, water is a very good liquid for ultrasonic propagation, with low absorption. For ultrasonic measurements in solids, there is a choice between 'dry contact' of transducer to sample, minimal fluid couplant in the form of a thin layer of liquid, oil or grease, and full immersion of the sample in a tank of couplant liquid.

The couplant between transducer and sample is frequently the weakest and least well characterized link in the ultrasonic measurement. Elimination of couplant, i.e. dry contact, rarely gives satisfactory results: it is subject to great variations in sensitivity among other things, and is generally only employed when fluid couplants are prohibited. Sometimes a solid couplant is used, such as a soft metal. Full immersion in, for example, water does however guarantee reasonably consistent results. It is less convenient for shear-wave generation in the solid, because mode conversion at the surface must be relied upon. It is clearly more suitable for small than large samples, unless large immersion tanks are to hand. The use of a thin layer of couplant between probe and sample is undoubtedly the most convenient method from a practical point of view, since compression and shear waves can readily be generated over a wide range of incident angles. However, great care is needed if sensitivity fluctuations caused by minute variations in couplant thickness are to be avoided. Fluid couplants are mostly restricted to a limited temperature range around room temperature, although there are some silicone couplants for use above 100 °C.

1.1.2   Non-contact transducers

Because of the problems associated with contact transducers, there has been a steady effort to develop 'non-contact' ultrasonic transducers that are couplant free. The best known are electromagnetic acoustic transducers (EMATs for short). In the EMAT transmitter a current is induced to flow in the surface of the sample by means of a coil close to the surface. A solenoid (or permanent magnet) at the surface generates a magnetic field perpendicular to the flow of current, so that a force is exerted on the material carrying the current. This localized stress field acts as a source of ultrasound. In the EMAT receiver the magnetic field is arranged so that movement of the surface causes an EMF to be induced in the sensing coil. EMATs are thus restricted to conducting specimens, or those with a conducting surface layer.

Other non-contact ultrasonic transducers include magnetostrictive devices, which are suitable for ferromagnetic materials, and capacitive transducers. The latter require locally polished surfaces and are mainly limited in their application to the reception of ultrasound in the laboratory. Optical methods are the other option for non-contact ultrasonics, and these are the subject of this book.

1.1.3 Shortcomings of conventional transducers

Laser techniques have been developed for ultrasonic application for two reasons: either some shortcoming in conventional ultrasonic techniques has been identified (e.g. difficulties with high-temperature application), i.e. 'a problem looking for a solution', or else some feature of the laser has been identified (e.g. ease of displacement calibration against the wavelength of light), i.e. 'a solution looking for a problem'. We remind readers that the latter was a comment made about lasers themselves in the 1960s! We shall therefore consider briefly some of the shortcomings of the more conventional methods for generating and receiving ultrasonic energy, followed by some of the benefits offered by laser ultrasonic systems.

As already mentioned, most of the problems with contact transducers originate in the couplant. For ultrasonic purposes, the couplant layer is a source of considerable variability in sensitivity and also in bandwidth. At the one extreme there may be insufficient couplant so that there are dry areas leading to poor transmission while, at the other extreme, too thick a layer of couplant can lead to significant losses in the couplant and/or the development of resonances. Furthermore, the transducer performs differently if loaded by fluid couplant rather than by a solid. These problems of variability are exacerbated when the probe is scanned at any speed across the surface, especially if it is rough. The need to maintain good coupling therefore tends to restrict scanning speeds. This problem, together with any sensitivity variations, can be ameliorated by the use of an immersion inspection technique if practicable. Although it is rarely very significant, it is also noted that contact transducers load the surface of the specimen and must therefore influence the ultrasonic propagation.

A further important shortcoming of most ultrasonic couplants is that they cannot be used over a sufficiently wide temperature range for all applications. Thus there is no suitable fluid couplant for use above 500°C. Many fluids for ultrasonic coupling can also adversely affect the surface of the specimen under study. Thus even water may be excluded for some applications because it causes corrosion, while there may be concern about possible chemical attack from, for instance, trace nitrates and chlorides in commercial couplants. For some specialized materials no liquid may be permitted to contact the surface, because of possible degradation. The EMAT does not suffer from these couplant problems, since it is a non-contact technique. It is still susceptible to sensitivity variations due to changes in 'lift-off' (i.e. separation between probe and surface) caused by roughness and surface deposits.

The piezoelectric element in a transducer acts as a resonant system. The most common resonant mode is a simple thickness resonance between the two parallel faces of a thin disc. The element thus has a high Q (low bandwidth). While this is clearly a benefit if a high-sensitivity, single-frequency system is called for, it is far from ideal otherwise. Various methods are available for increasing bandwidth and for lowering the Q of the system, mainly by the judicious use of absorbent backing material. The resulting frequency response is then that of a damped harmonic oscillator, which is still some way from the desired flat response. A further complication is that radial (and coupled thickness–radial) modes may also be excited in the piezoelectric disc, especially when the Q of the thickness resonance is reduced. These will appear as additional peaks in the frequency spectrum.

EMATS also suffer from resonance and bandwidth problems. Because their sensitivity is inherently much lower than a piezoelectric system, some frequency tuning may be necessary to obtain an acceptable signal to noise ratio. Furthermore, because the transmitter and receiver coils are inductive, their impedance increases with frequency, making them more difficult to use much above a few megahertz.

Another drawback for some applications is the size of most contact piezoelectric probes and EMATS. They may also fail to meet spatial resolution requirements. It is difficult to construct these probes to have acceptable efficiency and a spatial resolution better than a few millimetres, although it is noted that focusing and synthetic aperture techniques can often be used to give superior resolution. The size of the probes may also limit their use in restricted spaces, or on very small specimens. Finally the diameter of the probe defines the extent of the near field, so that in a contact mode of operation, the material immediately below the surface of the specimen may fall into a form of 'dead zone'.

As we shall see in the following chapters, laser techniques offer solutions, or at least partial solutions, to these problems. Thus they are entirely couplant free, non-contact and remote from the specimen. They are also capable of flat spectral response over large bandwidths. Light beams can be made extremely small to give good access in confined spaces and to give high spatial resolution, and can also be scanned readily across the specimen. An additional benefit of laser techniques is that they are capable of a high degree of absolute accuracy, since measurements can in principle be calibrated against the wavelength of light. There may be sensitivity penalties for using lasers rather than piezoelectric probes, and certainly they are likely to be more costly and complex to use. Nevertheless they are beginning to make a small but significant impact in a limited number of applications where their benefits over other probes outweight their disadvantages.

1.1.4 Applications of ultrasonics

Although ultrasonics has been studied since the times of such men as Lord Rayleigh, it is only since the Second World War that the discipline has begun to make a significant impact as a practical tool. It would require a substantial book to list and discuss the significance of all these applications. All that the authors propose to do in this chapter is to remind the reader of some of the most important current application areas of ultrasound and comment briefly on whether laser techniques could offer any significant advance in these areas.

It is in the medical area that the non-specialist has become most aware of the power of ultrasound to visualize the interior of optically opaque objects. Thus ultrasonic imaging systems are now routinely in use in diagnostic medicine because they are non-invasive and virtually harmless. Such has been the impact of the technique that most pregnant women in the UK are given an ultrasonic scan to check the size and health of their unborn children. The remarkable quality and resolution of the image and the speed with which it is obtained are usually surprising to the patient. There are many other medical applications of ultrasound in diagnosis and treatment. It is not easy, however, to foresee many medical applications of laser ultrasound, since contact techniques are so satisfactory, and the use of lasers would raise questions about safety. However, medical transducers do need calibration, both for sensitivity and to ensure their maximum power does not exceed the accepted safe limit, and laser techniques are potentially of importance here.

Ultrasonics has made an important impact in the manufacture of microwave electronic devices. In order to generate controlled time delays, a SAW (surface acoustic wave, i.e. Rayleigh wave) at high-megahertz or gigahertz frequencies is generated on a substrate and received a short distance away. The resulting SAW device can then be used as a tapped delay line and transversal filter for on-line signal processing. Applications for laser ultrasonics here would probably be limited to investigating SAW propagation in the laboratory.

There are increasing uses of high-power ultrasound for a variety of applications. A very common use is in the ultrasonic cleaner, where large particle amplitudes at frequencies just above the audible range cause contamination on the surface of the specimen to be 'shaken' off by processes such as cavitation in the cleaning fluid. High-power ultrasound is used plastically to deform metals and weld various materials. As a final example, there is a growing interest in sonochemistry, where high-power ultrasound is used to catalyse and accelerate otherwise rather inefficient chemical reactions. A further increasing industrial use of ultrasound is in non-invasive plant measurements. Ultrasonic sensors can be temporarily or permanently attached to process plant to measure such parameters as fluid flow, fluid level, and product composition and concentration. it is envisaged that they will eventually become as much a part of plant monitoring as anemometers, pressure gauges and thermocouples. No applications for laser ultrasound are as yet reported for this area.

The final industrial application area for review here is the field known broadly as non-destructive testing (or evaluation), abbreviated to NDT.

Although ultrasonic NDT techniques have been developed more recently, they now vie with radiography, visual inspection and magnetic techniques as the most reliable method for assuring the integrity of a wide range of engineering structures, from aircraft, to bridges, oil rigs and nuclear reactors. A common arrangement for ultrasonic inspection is shown in figure 1.2(*b*). The reason for the growth of interest in ultrasonics for defect detection is that it enables images to be formed of the interior of materials such as steel, which are opaque otherwise. For instance, as part of the inspection of the pressure vessel of a pressurized water reactor, steel sections as thick as 250 mm are routinely inspected by ultrasonic techniques. Defects no greater than a few millimetres can now be detected and sized reliably.

In addition to the detection of bulk and surface defects in a wide range of materials, including metals, concrete, ceramics and composites, ultrasound is also used to characterize the materials themselves in terms of their ultrasonic velocity and attenuation. There is also a growing interest in being able to carry out these measurements on the production line, so as to 'close the loop' and initiate process control.

It is without doubt this industrial NDT area where laser ultrasonics will make the largest impact in the medium term. It is in industrial ultrasonics where non-contact methods are being called for, either because of high ambient temperatures, or because the material cannot be contaminated with couplant. It is also in industrial ultrasonic measurements where high bandwidths and spatial resolution are occasionally called for, or where access is too restricted for contact probes to be efficiently used.

The bibliography at the end of this chapter lists a small number of standard texts on acoustic and ultrasonic wave propagation, together with texts on the industrial applications of ultrasonics.

1.2 LASERS

The application of optical methods to ultrasonics predates the laser in such techniques as Schlieren photography for visualizing ultrasonic fields. However, the advent of the laser has made such a large contribution to the subject that optical measurements using conventional light sources are now of much less significance. We shall therefore concentrate entirely on the application of laser optics to ultrasonics for the purposes of this book.

Most of the optics that is needed to understand the subject matter of this book is likely to be well known. However, because the laser itself may be for some readers still a relatively recent invention, and because its unique properties are essential to what follows, we shall briefly introduce the basic principles of laser operation and the main properties of laser radiation. A more detailed description of the various aspects of optics and laser science can be found in a number of standard texts, recommended to the reader in the bibliography at the end of this chapter.

1.2.1 Electromagnetic radiation

The laser radiation used in ultrasonic experiments is a form of electromagnetic radiation, the full spectrum of which extends from less than 10^4 Hz for very long radio waves to greater than 10^{21} Hz for high-energy gamma rays. The velocity of these waves in vacuum is 2.9979×10^8 m s^{-1}. This universal constant will be denoted by c. The frequency range of interest in this book corresponds to wavelengths (*in vacuo*) from approximately 200 nm to 10 μm, which includes the visible range, normally taken to extend from 400 to 700 nm. Until the advent of lasers, it was not possible to generate strictly monochromatic (i.e. single-frequency) electromagnetic radiation in this range. At lower frequencies, e.g. radiowaves and microwaves, single-frequency radiation is generated as a matter of course.

Electromagnetic radiation is a propagating wave motion in which energy is alternately transferred between electric and magnetic fields. These fields are normally perpendicular to each other and to the direction of propagation. The radiation can thus show polarization effects characteristic of transverse wave motion (somewhat analogous to shear waves in elastic solids). In plane-polarized radiation, the electric field has a constant direction: the plane containing this direction and the direction of propagation is referred to as the plane of polarization. Light can also be circularly polarized, when the electric field (and hence also magnetic field) is constant in amplitude, but rotates at the frequency of the radiation. Elliptical polarization is the general form of polarized light.

It is important to match polarizations in interference experiments. Furthermore, polarization properties may be used with advantage to eliminate background and provide a simple method for directional discrimination. Whilst polarization has little influence on laser generation, where only a high-energy pulse is required, it usually plays an important part in the laser reception optics, since many optical components (such as beam splitters) are influenced by polarization direction.

It is predominantly the electric field which interacts with matter in non-magnetic materials. Although we shall also be concerned with magnetic materials such as steel, for simplicity all descriptions of light will be made in terms of the electric vector. The electric polarizability of a medium leads to a reduction in wave velocity by a factor μ, known as the refractive index. The velocity of light in that medium is given by c/μ. Consequently, the wavelength is also reduced by this factor μ in the medium.

The concept of optical path length will be used in discussion of the propagation of light, particularly in interferometry. This is a measure of the time taken for light to travel between two points, and in the case of single-frequency radiation, of the phase difference. Optical path length may be defined as the distance the light would have travelled *in vacuo* during the time taken on the actual path.

1.2.2 Quantization of light

A consequence of the quantum theory is that electromagnetic radiation behaves in many ways as if composed of discrete packets, or 'photons', whose energy is hv, where h is Planck's constant and v is the frequency of the radiation. Whilst the treatment of most of the phenomena discussed in this book does not need to involve the quantization of the electromagnetic field, quantum theory is needed to understand the physical operation of a laser. The photon structure of light also imposes a fundamental limit on the sensitivity of techniques that depend upon the recording of very small changes in light level (see §1.2.5).

Atomic and molecular systems have a number of discrete energy states or levels in which they can exist. There is a ground state and some excited states. Each system can undergo a transition from a lower to a higher state by the absorption of energy in the form of incident radiation of the appropriate frequency. In the reverse process, known as emission, the system loses energy by a transition from a higher to a lower state. Absorption and emission occur at discrete frequencies which manifest themselves by characteristic 'lines' in the spectrum of the atom or molecule. Net absorption occurs when the population of the higher states is smaller than the population of the lower states, as in normal thermal equilibrium. Emission may occur spontaneously or in response to incident radiation of the same frequency in which case it is known as 'stimulated emission'.

Suppose the energy of the ground state is E_0, and the energy of one excited state is E_1. Then, according to quantum theory, the system can absorb or emit a quantum of energy in the form of electromagnetic radiation of frequency v, such that

$$h v = E_1 - E_0. \tag{1.1}$$

1.2.3 Basic principles of laser operation

The laser is a device which amplifies the intensity of light by means of a quantum process known as stimulated emission. Indeed the name LASER is an acronym standing for light amplification by stimulated emission of radiation. A practical laser needs a means of amplification (i.e. stimulated emission, as referred to in the previous section) and a means for feeding the energy back into the system to build up sustained oscillation. The operation of the simplest type of laser can most readily be understood in terms of a quantum mechanical model as follows.

Imagine a given atom (or molecule) has just three energy levels, E_0, E_1 and E_2, such that $E_2 > E_1 > E_0$ (figure 1.3). There will in reality be more

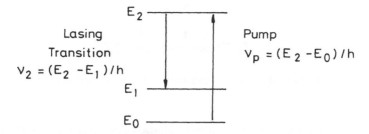

Figure 1.3 Simplified energy level diagram for a laser medium.

than three levels, but for present purposes we can neglect other levels. The 'ground-state' E_0 is well populated, whereas the intermediate and upper states are more sparsely populated. Suppose now that the atom absorbs a quantum of incident radiation such that the atom is excited into the upper state. From quantum theory, the radiation must have a frequency ν_p such that

$$h\nu_p = E_2 - E_0 \qquad (1.2)$$

where h is Planck's constant. In laser terminology this process of absorption is known as 'pumping', so that ν_p is the 'pumping frequency'.

Pumping tends to equalize the population of two states so that E_2 becomes well populated. This reverses the normal occupancy of E_2 and E_1 and is known as population inversion. Emission (i.e. 'stimulated emission') can now occur in response to incident radiation, at a frequency ν given by

$$h\nu = E_2 - E_1. \qquad (1.3)$$

We note that necessarily $\nu < \nu_p$, so that the pumping frequency must always be higher than that of the radiation to be amplified.

There are two main techniques for pumping a laser system. Either the laser medium is illuminated with an intense light source such as a discharge tube, or an electrical discharge is passed through a gaseous medium. In most laser systems the sequence of transitions is more complicated than the simple three-state model described above. A discussion of such systems is beyond the scope of this book. However, because it is the most widely used, we should note that the helium–neon system involves the interaction of two energy levels in the helium with two levels in the neon. The neon atoms are excited by collisions with helium atoms of almost identical energy, so that although the radiation is characteristic of the neon, helium gas is also required to make a laser.

In order to obtain light of sufficient intensity for practical use, there has to be some mechanism for feeding the energy back into the laser system and thereby building up the amplitude of oscillations in a resonant system. The usual way of obtaining sustained oscillations is to site a high-performance mirror at each end of the lasing medium (figure 1.4). In the simplest system

Pump

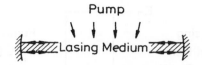

Figure 1.4   Basic arrangement for a laser oscillator.

both mirrors are plane, and accurately aligned perpendicular to the axis of the laser. Thus the light can be reflected backwards and forwards through the lasing medium. On each pass it stimulates further emission from the medium and is thus amplified in intensity. This is known as a resonant cavity. In order to 'tap off' some of the energy into a continuous beam for external use, one of the mirrors is manufactured to transmit a small fraction of the light. Figure 1.5 shows two designs of gas laser. The first has mirrors external to the lasing medium, with the light passing through the windows inclined at the Brewster angle to eliminate reflective losses for polarization in the plane of incidence to the windows. The second, with integral mirrors, is favoured for small, mass-produced lasers.

The simplified practical arrangement for a laser introduced above should be suitable for a continuously running laser provided pumping is continuously applied. In many systems the lasing threshold is high, so that a large excitation power is required. There can then be practical problems associated

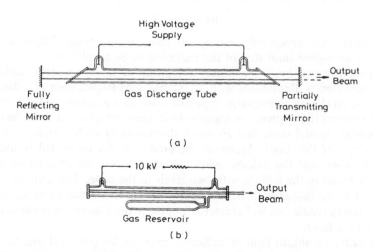

Figure 1.5   Illustration of gas laser construction: (*a*) Conventional arrangement, with independent mirrors and Brewster angle windows into the gas discharge medium. (*b*) Laser with integral mirrors, a design commonly used for small mass-produced lasers (e.g. helium–neon).

with the need to dissipate the heat that is produced as a by-product. Unless temperature stability is maintained, the operation of the laser will be unsatisfactory.

Because many continuous-wave (cw) gas lasers are used in applications such as interferometry, where wavelength purity and coherence are important, these lasers are usually built with specially designed optical components. The main problem is that, in a simpler laser, a number of longitudinal and transverse cavity modes will be excited. The result is that energy is amplified over a narrow range of frequencies instead of the desired single frequency, giving rise in turn to inter-mode beat frequencies, and there is a variable distribution of energy across the beam. The generally preferred transverse mode (usually the lowest-order mode with circular symmetry, TEM_{00}) can be selected, and higher-order modes suppressed, by the use of at least one curved (concave) mirror at the end of the cavity and/or a suitable aperture within the cavity.

Multiple longitudinal wavelengths can be a significant problem in interferometry. Depending on the wavelength and cavity length, there may be as many as 10^6 wavelengths in the cavity, so that the frequencies of different modes differ by as little as one part in 10^6. Thus several modes may come within the width of the laser spectral line. These modes beat together to give frequencies that may interfere with the signal being measured in addition to raising the background noise level. For instance, the inter-mode beats from a multimode helium–neon laser can be particularly problematical if measurements are made with bandwidths extending to greater than 100 MHz, and at lower frequencies in some circumstances (§3.5.2). Inter-mode beats are extremely sensitive to the precise length of the cavity, so that minute temperature fluctuations can cause the beats to intensify or be attenuated. Single-mode lasers have recently become more readily available. The usual approach is to incorporate a Fabry–Perot étalon (see §3.11) into the optical path which is tuned to one longitudinal mode. However, the temperature of the cavity must be carefully stabilized to ensure that the longitudinal mode to which the étalon is tuned remains excited. The disadvantage of selecting a single mode is that the available energy is reduced, although an improvement in signal/noise is usually claimed. A satisfactory alternative is to employ a multimode laser and then both use a balanced detection system and adjust the path difference between the two arms of the interferometer to minimize the effects of inter-mode beats if necessary.

The study of laser ultrasonics is somewhat unusual in that it often requires both a pulsed laser to generate the ultrasound, and a continuously running (often called continuous wave) laser to receive the ultrasound. Many laser systems (e.g. carbon dioxide) can be made to run in both continuous and pulsed modes, depending whether a sufficient population inversion can be permanently maintained at low enough input power levels for the laser to be adequately cooled. However, some systems tend to be best used as cw lasers (e.g. helium–neon) and others (e.g. ruby) are only practicable as pulsed lasers.

1.2.4 Pulsed lasers

The simple laser arrangement discussed above will operate in a pulsed mode if it is pumped for example by a pulsed flash-tube. Depending on the type of laser, pulses of duration typically $100\,\mu s$–$1\,ms$ can be obtained. In this 'normal' mode of operation the pulses tend not to be uniform, but consist of many 'spikes' of microsecond duration due to relaxation oscillations (figure 1.6(a)). Although high-energy pulses can be produced in this way, the normal mode is not particularly useful for laser ultrasonics because the pulse duration is too large. An additional technique, known as Q-switching or Q-spoiling, is needed to obtain pulses in the required 1–$100\,ns$ range (figure 1.6(b)). The Q (i.e. quality) factor of a cavity resonator is the energy stored in the cavity divided by the energy lost from the cavity per round trip of the light within the cavity. Thus if Q is low the cavity oscillations are suppressed and stored energy builds up within the lasing medium. When the Q is high, the

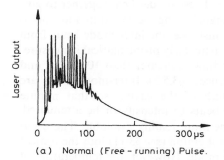

(a)   Normal (Free – running) Pulse.

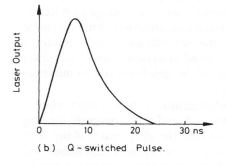

(b)   Q – switched Pulse.

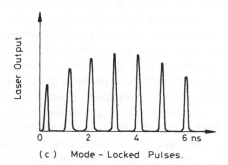

(c)   Mode – Locked Pulses.

Figure 1.6   Schematic to illustrate the differences between (a) 'normal' (i.e. free-running), (b) Q-switched and (c) mode-locked pulses from a solid state laser. Note that the shapes and durations can vary considerably in practice.

cavity can support oscillations into which energy is supplied from the medium. Thus switching from low to high Q results in the rapid extraction of power from the laser cavity.

Practical Q switches take the form of elements with variable absorption that are inserted between the mirrors. Two commonly used Q switches are the Pockels cell and bleachable (saturable) dye. In the Pockels cell a potassium dihydrogen phosphate (KDP) crystal is employed which rotates the plane of polarization of the light when an electric field is applied to it. This is used in combination with a polarizer to form an extremely rapid optical switch (figure 1.7). The saturable dye is a simpler system: it is a small cell containing a dye which is opaque to the laser light at low intensities but transparent at high intensities (the molecules are excited to a higher, non-absorbing state). However, the switching operation in the Pockels cell, unlike the bleachable dye, can be precisely timed, which is an important advantage for many applications. The technique used to produce the highest power, shortest duration Q-switched pulses is to set a low Q while the laser medium is excited by the pump. When the maximum attainable population inversion is reached, the Q is switched to a high value so that there is an unimpeded optical path between the mirrors. Oscillations now build up extremely rapidly as energy is dumped into the cavity mode from the highly populated excited state in the laser medium. The power is thus able to reach its maximum in a matter of nanoseconds, to give an asymmetrical pulse with a slightly longer fall than rise time, and a duration of 10–50 ns typically. By convention, pulse duration is mostly defined as the full width of the pulse at half height (FWHH).

The question of cavity modes was discussed in the context of CW lasers. Transverse and longitudinal modes also play an important part in pulsed lasers. The laser pulse is normally multimode. The pattern of transverse modes tends to vary over a period of time, making it difficult to focus to a

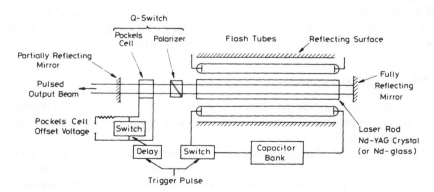

Figure 1.7 Schematic diagram of a pulsed neodymium YAG (or glass) laser with provision for Q-switching.

very small area, and can give rise to hot spots in the laser ultrasonic source. For many laser ultrasonic applications these effects are relatively unimportant so that the multimode pulse can be used, which means that the maximum output energy of the laser is available. If the spatial profile of the pulse is important for a given application, then the single TEM_{00} mode can be selected by, for instance, inserting a special aperture into the cavity to suppress higher order modes. In a typical Q-switched pulse a number of different longitudinal resonant modes of the cavity are simultaneously excited, which may interfere to give inter-mode beats which appear as a higher frequency ripple on the pulse. When the Q switch is a Pockels cell the ripple amplitude is small, so that for most laser ultrasonic applications the effect can be ignored.

Yet shorter pulses (e.g. 30 ps) are beginning to be explored in laser ultrasonic studies. These are obtained by exploiting multiple longitudinal modes described in the previous paragraph, in a phenomenon known as 'mode locking'. If, for instance, a saturable dye is used as the Q switch, the non-linear behaviour of the dye tends to mix together different cavity modes, coupling or 'locking' them together to give constant phase relationships. This has the effect of producing a regular train of mode-locked pulses (figure 1.6(c)), which physically correspond to a single pulse circulating in the cavity. There are various techniques which are beyond the scope of this book for isolating individual mode-locked pulses. The application of acousto-optic cells to mode locking is described in §2.5.4.

Also of potential importance to non-contact ultrasonics are lasers that can be pulsed repetitively. Although repetition rates up to a kilohertz have been obtained from solid state lasers, adequate cooling becomes a serious problem, especially as laser rods tend to be thermal insulators. Gas (in particular metal-vapour and excimer) lasers present less of a heat transfer problem and repetition rates as high as 20 kHz are available. In contrast, however, solid state lasers can store higher individual pulse energies than gas medium lasers. In order to maintain a manageable average power the energy per pulse drops off as the rate is increased. Conventional ultrasonic inspection often uses repetition rates as high as $1-10$ kHz for signal averaging purposes or for speed. Thus pulsed lasers are available of comparable rates.

1.2.5 Sensitivity limits

In §1.2.2 it was stated that the photon structure of light imposes a fundamental limit on the sensitivity of techniques that depend upon the recording of very small changes in light level.

The response of photodetectors arises in the high-resolution limit from discrete events, e.g., the emission of a photoelectron from a photomultiplier cathode or the production of an electron–hole pair in a semiconductor. These events occur randomly in response to the absorption of energy from the electromagnetic field. Although the mean rate of occurrence may be fixed when the illumination is constant, the irregularities in timing lead to short-term fluctuations, or noise, in the output from the detector against which the signal must be detected.

Quantum theory dictates that the smallest package of light energy that can be detected is a single photon. Thus there cannot be more events generated in the detector than there are photons absorbed, and the detector may be regarded as responding to the absorption of individual photons. The perfect detector produces an event for every photon unit of energy absorbed. In practice, detectors are less efficient than this, the mean number of events in time τ being

$$\bar{n} = \frac{\eta W \tau}{h\nu} \tag{1.4}$$

where W is the total power of the radiation falling on the detector (in watts). η is the quantum efficiency which must be less than unity. In practice, quantum efficiencies in the optical region can be quite high, e.g., over 80% for silicon diode detectors in the near infrared, 20% or more for photomultipliers in the blue region of the spectrum. The variation of quantum efficiency with wavelength for typical detectors of these types is shown in figure 1.8.

Statistical theory for randomly occurring events gives for the mean square deviation in the number

$$\langle (n - \bar{n})^2 \rangle = \bar{n} = \frac{\eta W \tau}{h\nu}. \tag{1.5}$$

Thus if i_0 is the mean output current of the detector, and i_N the fluctuating noise component, then

$$\frac{\langle i_N^2 \rangle}{i_0^2} = \frac{h\nu}{\eta W \tau}. \tag{1.6}$$

The mean square noise current is thus $\propto 1/\tau$, and hence to the bandwidth selected. It is otherwise independent of frequency, i.e. the noise is 'white'. In terms of the bandwidth Δf, the formula for the noise intensity is as follows (Ross 1966):

$$\frac{\langle i_N^2 \rangle}{i_0^2} = \frac{2h\nu\Delta f}{\eta W}. \tag{1.7}$$

Expressed in another way, we may say that the RMS apparent fluctuation of the power level due to photon noise, the 'noise equivalent power', is given by

$$\delta W_N = (\langle W_N^2 \rangle)^{1/2} = (2h\nu W\Delta f/\eta)^{1/2}. \tag{1.8}$$

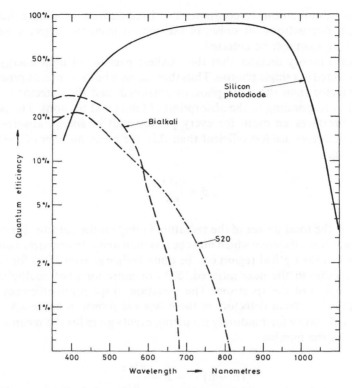

Figure 1.8 Variation of quantum efficiency with wavelength for common light detectors. Curves are shown for a typical silicon photodiode and for photomultipliers with bialkali (KCsSb) and S20 multialkali (NaKCsSb) cathodes.

1.3 MAIN CHARACTERISTICS OF LASER LIGHT

Lasers are characterized by a number of key optical properties, most of which play an important role in the interaction with ultrasonic fields. Their four major optical properties are: monochromaticity, coherence, directionality and high intensity, which are inter-related. For instance, coherence length is inversely proportional to the number of cavity modes excited, and thus to the effective bandwidth of the laser, so that a perfectly monochromatic laser should have an infinite coherence length. We consider each property in turn with particular reference to the use of lasers in ultrasonics.

1.3.1 Monochromaticity

Light from sources other than lasers covers a range of frequencies. True single-frequency operation of a gas laser can be achieved by careful design. Although it is more difficult, such a laser can be constructed to have a frequency stability at constant temperature better than one part in 10^{10}. Multimode operation of course reduces the monochromaticity of a laser. A small helium–neon laser generally has three or four longitudinal modes excited with a spacing of a few hundred megahertz, depending on the cavity length. This still gives a bandwidth of a few parts in 10^6. Solid state lasers tend to have rather larger frequency spreads.

Monochromaticity is important for some ultrasonic applications, in particular the interferometric measurement of ultrasonic fields. Firstly it is necessary in order to obtain high coherence, and secondly it enables accurate absolute measurements of ultrasonic displacement to be made, since calibration is against the wavelength of the light. Monochromaticity is of less importance for the generation of ultrasound by laser. It does, however, enable narrow-band optical filters to be used to separate source and receiver beams of light.

1.3.2 Coherence

Coherence is an important property when it comes to building an optical system to detect ultrasonic waves, and therefore we shall define it carefully. In simple terms the word coherence is used to describe how well a wave disturbance at one point in space or time correlates with the disturbance at another point. If there is a well-defined phase relationship between the light at two different points in space (i.e. two light beams), or at two different times (i.e. one beam split into two with a delay between the parts, as is typical in an interferometer) then the two light disturbances can be brought together to produce a predictable interference pattern. If the light is completely incoherent so that there is no predictable relationship (i.e. random phase) between the two disturbances, no interference fringes will be formed. On the other hand, perfectly coherent light will produce excellent fringes with 100% visibility.

If $\mathbf{E}_1(t)$ and $\mathbf{E}_2(t + \tau)$ represent the electric field vectors of the light at one point x_1 and time t, and a second point x_2 and time $t + \tau$, respectively, then the correlation between the electric field at these two points and times can be expressed in terms of an integral over a number of cycles of the periodic field, thus:

$$\mathbf{G}_{12}(\tau) = \langle \mathbf{E}_1(t) \cdot \mathbf{E}_2^*(t + \tau) \rangle \qquad (1.9)$$

where $\mathbf{E}^*$ is the complex conjugate of $\mathbf{E}$.

The coherence function (complex degree of coherence of the light) is obtained by normalizing $\mathbf{G}_{12}$ thus:

$$g_{12}(\tau) = \mathbf{G}_{12}(\tau)/(\mathbf{G}_{11}(0) \cdot \mathbf{G}_{22}(0))^{1/2}. \qquad (1.10)$$

The modulus of $g_{12}(\tau)$ lies between 0 and 1. The function covers both spatial and temporal coherence. Thus, if $g_{12}(0) = 0$, the light at points 1 and 2 is spatially incoherent. If $g_{12}(0) = 1$, there is complete coherence so that perfect interference fringes could be formed. Complete temporal coherence between the light at time t and time $t + \tau$ at one point is satisfied by $g_{11}(\tau) = 1$, whereas temporal incoherence is represented by $g_{11}(\tau) = 0$.

Since coherence is determined by how well the wavefronts can remain 'in step' it is clearly related to bandwidth. For instance, light containing either a spread of wavelengths or a discrete number of incommensurate wavelengths will have an amplitude and phase pattern that is never reproduced exactly. An approximate expression for the distance travelled by a light ray before it loses coherence, known as the coherence length, is given by

$$L_{coh} = c/\delta v$$
$$= \lambda^2/\delta\lambda \tag{1.11}$$

where λ is the wavelength and $\delta\lambda$ and δv are the spread in wavelength and frequency. This is discussed further in Chapter 3.

Thus for a filament lamp where $\delta\lambda$ is of the order of λ, the coherence length is very small—of the order of the wavelength. For nominally monochromatic sources such as sodium vapour lamps, the coherence length is of the order of a millimetre. However, for a single-mode laser with a frequency spread of one in 10^{10}, the coherence length must be of the order of 10^{10} wavelengths, i.e. 6 km. A multimode helium–neon laser with a frequency spread of the order of 3 in 10^6 will have a coherence length of 20 cm. One important difference between a multimode gas laser and a conventional 'monochromatic' light source is that the pattern of coherence and incoherence is repeated with the periodicity of the round-trip laser cavity length (L_c). Thus the light from a multimode laser with a mirror separation of 30 cm will regain coherence around 60, 120, 180 cm, etc (cf §3.5.1). This is because the modes that cause the incoherence have commensurate wavelengths. In general coherence is regained in an interferometer with optical path length difference L, when (if n is an integer)

$$L = nL_c. \tag{1.12}$$

Coherence is important to the reception of ultrasonic waves by laser. Most techniques involve some form of interferometer, in which good coherence between probe and reference beam is essential. Conventional 'monochromatic' light sources can be used for interferometry, but because their coherence length is only of the order of a few millimetres, they must be set up with the probe and reference beams equal in length to this accuracy. Beam collimation loses so much light that they become too insensitive. This imposes enormous practical restrictions on the use of non-laser light sources in interferometry. The use of a gas laser on the other hand enables a much longer probe than reference beam to be used. This means that a compact instrument (incorporating all except the probe beam) can be built, and that the distance to the sample is not critical. If a multimode laser is used, care must be taken to ensure that one is working at a coherence repeat distance, but this still gives considerable flexibility.

Temporal coherence is unimportant and spatial coherence relatively unimportant to the generation of ultrasound by laser, that is unless a particularly highly collimated beam or very sharp focus is required. Otherwise, incoherent light will generate thermal stresses of the same magnitude as coherent light. This is fortunate since the coherence length of pulsed solid state lasers operating in multimode may only be of the order of centimetres.

1.3.3   Directionality

Directionality is a function of the spatial coherence of the beam of light. The radiation produced by a laser is confined to a narrow cone of angles. The beam divergence for a typical gas laser is of the order of 1 milliradian, although some commercial helium–neon lasers are available with divergences of a few tenths of a milliradian. Comparable beam divergences are now available from pulsed solid state laser systems. It is only possible to attain such low beam divergence from conventional light sources by severe collimation which leads to extremely low intensities. The diameter of the beam is typically about 1 mm for a gas laser such as helium–neon, and in the range 1–20 mm for a pulsed solid state laser.

The light from gas and solid state lasers thus forms a highly collimated beam which is extremely valuable for laser ultrasonics since it enables the beam to be focused to a very small spot. This means not only high spatial resolution, but also, in the case of ultrasonic displacement measurement by interferometer, the ability to collect a larger fraction of the scattered light from a rough surface, thereby increasing sensitivity. For laser generation it means that very high incident power densities are attainable. Low beam divergence also means that the beam can travel distances of the order of several metres from laser to specimen without appreciable spreading and losses. Thus both laser generation and reception of ultrasound can be made genuinely remote techniques.

1.3.4   Gaussian beam optics

The output from a laser oscillating in the lowest transverse TEM_{00} mode is a beam which is as parallel as possible consistent with diffraction (i.e. spreading due to the wave nature of electromagnetic radiation). A beam of width d will gradually increase in width, with a divergence of approximately λ/d. Typically a laser beam has a diameter of about a millimetre, leading to a divergence of approximately 1 milliradian (a few minutes of arc) at visible wavelengths. Another consequence of diffraction is that a beam cannot be focused to a true point, thus placing a limitation on the power density that it is possible to obtain from a given laser. With an ideal focusing lens of focal length F, the minimum diameter at the 'waist' of the beam is $\sim \lambda F/d$, where d is now the diameter of the beam at the lens.

Calculation of beam propagation taking account of diffraction is particularly simple for the Gaussian profile characteristic of the TEM$_{00}$ mode, since the beam retains its Gaussian form at all cross sections. All other beam-profile shapes change as the beam converges to or diverges from the waist. For a Gaussian beam, the width and convergence at any cross section may be calculated using the general beam shape shown in figure 1.9. The intensity at a distance r from the axis of the beam and at a distance z from the waist is given by

$$I(r, z) = I(0, z)\exp(-2r^2/r_z^2) \tag{1.13}$$

where r_z is the radius of the beam cross section measured to the points at which the intensity falls to $1/e^2$ of its maximum value. This is the most commonly used convention in specifying beam width. The relationship between r_z and r_0 is as follows:

$$r_z^2 = \frac{\lambda^2 z^2}{\pi^2 r_0^2} + r_0^2. \tag{1.14}$$

The beam convergence or divergence angle φ (to the $1/e^2$ intensity points) at a distance z is given by

$$\varphi = \left(\frac{2\lambda}{\pi r_0}\right)\left(1 + \frac{\pi^2 r_0^4}{\lambda^2 z^2}\right)^{-1/2}. \tag{1.15}$$

Using these equations, the propagation of Gaussian beams through optical systems may be calculated. Changes in beam convergence introduced by lenses normally placed where beam diameters are largest are computed by normal geometrical optics.

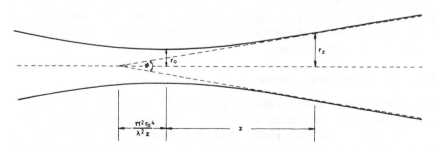

Figure 1.9 Diagram illustrating the propagation of a generalized Gaussian beam. The full curve indicates the edge of the beam measured to the $1/e^2$ intensity points.

1.3.5   High intensity

This is perhaps the property for which lasers are best known outside the field of optics. Although the optical power output from a small helium–neon laser may only be say 2 mW, a beam diameter of 0.5 mm leads to a power density of about 1 W cm^{-2}. Such a beam can readily be focused by a simple lens to a spot of diameter 0.05 mm because it is monochromatic and coherent. The incident power density is then 100 W cm^{-2}. High-power carbon dioxide lasers can be obtained which deliver a substantial fraction of 1 kW cm^{-2} without focusing. Focusing optics produces a beam of sufficient intensity to melt, cut or weld structural materials.

Intensity is a most important property for laser reception of ultrasound, since it will be shown in Chapter 3 that the sensitivity of a single-mode laser interferometer system (defined as signal to noise ratio for a fixed bandwidth) increases with the square root of light intensity, provided other conditions remain constant. The limiting factor becomes the intensity at which the specimen is damaged or otherwise adversely affected by intense irradiation. Most work to date has been carried out with helium–neon lasers of power in the range 1–10 mW. Some studies have been carried out with argon ion lasers whose power typically extends to several watts, but increased power frequently brings the penalty of additional noise and multimode interference.

Intensity is a crucial factor for the generation of ultrasound by laser since incident power intensities typically in the range 10^4–10^6 W cm^{-2} are needed to act as a thermoelastic source of ultrasound. Plasma and ablation (a strong ultrasonic source as will be discussed in Chapter 4) require incident power densities in excess of 10 MW cm^{-2}. Such high-power densities are impossible to attain by conventional light sources. Pulsed lasers which incorporate a Q switch routinely produce beams of these intensities and higher for periods of a few tens of nanoseconds. By way of example, neodymium:YAG lasers (wavelength 1.06 μm) are commonly sold to deliver 0.1 J per 15 ns pulse in a beam of diameter 5 mm. A simple calculation shows that the unfocused beam has an approximate peak power of 6 MW and thus a power density in excess of 10^7 W cm^{-2}. In fact, because the laser pulse is so short, we shall see in Chapter 4 that it is the energy per pulse that largely controls the ultrasonic amplitude in the thermoelastic regime. The limiting factor is not the power that is available from commercial lasers, but rather the threshold for damage in the irradiated sample. Indeed, many pulsed lasers are too powerful for ultrasonic applications, and must be used at reduced power if the technique is not to be destructive.

It should be mentioned here that the typical pulse length of a Q-switched pulse of 10–50 ns is by coincidence virtually ideal for ultrasonic generation. If there is no spreading of the pulse in the sample by for instance thermal diffusion, then 10 and 50 ns pulses have bandwidths (as conventionally defined) of 30 and 6 MHz. Much ultrasonic non-destructive testing is carried out between 1 and 5 MHz, while ultrasonic materials characterization requires a somewhat greater bandwidth. If higher bandwidths (> 100 MHz) are required then mode-locked laser pulses are available.

1.4 LASERS FOR ULTRASONICS

1.4.1 Common laser systems

A field such as laser technology develops so rapidly that information soon becomes out of date. In this section we propose to list and discuss briefly the most common laser systems at the time of writing that might be of interest to researchers and users of laser ultrasonics. We shall thus limit ourselves to gas lasers, solid state lasers and semiconductor lasers. We shall omit other types of laser such as the dye laser because as yet no clear laser ultrasonic application has been identified for these.

(a) Gas lasers

Gas lasers were until recently the most common type of laser, and find their application in many fields. The helium–neon system, among the most popular, is the work-horse of laser interferometry, being the main method for remote ultrasonic reception. Other gas laser systems with higher power, such as the argon ion system, are also used for interferometry, although much less widely. Gas lasers have also been used in pulsed mode for the generation of ultrasound, both the carbon dioxide system and the family of excimer lasers.

Although grouped together under the one heading, gas lasers embrace a number of different types of laser system. The argon system is based on electronic transitions in the argon ion. There are thus many different laser lines available within the argon spectrum. The helium–neon system on the other hand depends on the juxtaposition of energy levels in both the helium and the neon. The carbon dioxide system is different again, because it is based upon transitions between molecular states (rotational) of the carbon dioxide molecule. Excimer laser systems are different from other gas laser systems in that the ground state is unstable, and the inert gas halide molecules (e.g. KrF) are only stable in the excited state. Tables 1.1 and 1.2 list the commercially available gas (including metal vapour) laser systems, with some of their properties.

(b) Solid state lasers

Solid state lasers have been used both commercially and in research for some time, although not to the extent of gas lasers. The first solid state laser was the ruby laser: it is still used to deliver high-energy pulses of visible red light.

Table 1.1 Commercially available CW gas laser systems with possible applications for interferometry and ultrasonic reception. All are pumped by electrical discharge.

Laser system	Principal wavelengths (μm)	Other wavelengths (μm)	Output	Comments
Argon	0.488, 0.514	Many lines between 0.351 and 1.093	<20 W	Single-mode power less, suitable for interferometry, more power than helium–neon
Carbon dioxide	10.6	Many lines between 9.0 and 11.0	<10 kW	Very high powers attainable, 10 μm too long for most interferometry
Carbon monoxide	Many lines between 5.0 and 6.4		<20 W	Wavelength too long for most interferometry
Helium–cadmium	0.442	0.325	<75 mW	No recorded applications yet to interferometry, reputation for high noise
Helium–neon	0.633	Six other lines between 0.543 and 3.39	<50 mW	Multimode; single mode stabilized <3 mW ideal for interferometry
Krypton	0.647	A number of lines between 0.351 and 0.752	<6 W	No recorded applications yet to interferometry, less efficient substitute for argon

The most popular solid state systems are based on transitions in neodymium. The laser medium is either neodymium-doped glass or neodymium-doped YAG (yttrium–aluminium–garnet). The fundamental wavelength for the neodymium system is in the infrared but visible light can be obtained by frequency doubling techniques.

Solid state lasers are more generally associated with short pulse operation, and are the most common laser type used for the generation of ultrasound. However, the Nd:YAG system in particular is increasingly being used in CW operation. A recent development, for instance, is the laser-diode pumped Nd:YAG CW laser which is extremely compact, and dispenses with the need for high voltages to power a discharge tube. These lasers are therefore increasingly an option for the reception of ultrasound in interferometers. Tables 1.3 and 1.4 list the most common commercially available systems.

(c)   Semiconductor lasers

The third main group of lasers of interest in laser ultrasonics are those based on a semiconductor system such as gallium aluminium arsenide (GaAlAs). They tend to have shorter coherence lengths, less well-collimated beams and lower power than gas or solid state lasers. However, they are rapidly being improved and are therefore making a major impact on optics. A further drawback for much interferometry is that they all emit in the infrared, making alignment difficult.

A large number of diode lasers using gallium arsenide are available commercially. They emit in the wavelength range 700–860 nm. Some laser diodes at the lower end of this range emit radiation which is sufficiently visible for instrument alignment purposes. The general-purpose types (much used in communication) usually have a spectral line width of about 1–2 nm, corresponding to a coherence length of a fraction of a millimetre. However, single-mode lasers with coherence lengths up to a metre can be obtained and even some low-cost types may have coherence lengths in excess of 30 mm. Output powers are generally in the range 1–50 mW, but outputs up to 1 W are possible. Diode lasers based on the semiconductor InGaAsP are also available, emitting in the range 1170–1585 nm (mostly at 1300 nm). Although these have advantages for fibre-optic communication, they are not suitable for interferometry since their wavelength is too long for many common optical materials. In addition, they mostly have less power, a shorter coherence length and are more expensive than GaAlAs types. There are also several specialized low-power semiconductor lasers emitting at longer wavelengths in the infrared, which are of no interest in the context of this book.

1.4.2   Choice of laser systems for ultrasonic measurements

As the above tables show, there is a wide choice of laser systems available: indeed the choice continues to increase as a result of research and development.

Table 1.2 Commercially available pulsed gas lasers with possible application to ultrasonic generation. Note that it is not possible to obtain maximum output, minimum pulse length and maximum rate simultaneously in one laser.

Gas	Wavelength (μm)	Output (J)	Pulse length (ns)	Repetition rate (Hz)	Comments
Carbon dioxide	10.6	<2200	6 ns–1100 μs	<10^4	IR, possible for ultrasonic generation
Copper vapour	0.51, 0.578	<0.03	10–60	<2×10^4	Mostly only suitable for high repetition rate
Argon fluoride	0.193	<0.6	5.5–25	<500	Excimer, UV, possible for ultrasonic generation
Nitrogen	0.337, 0.428	<0.035	50 ps–10 ns	<100	UV, possible for ultrasonic generation
Fluorine	0.157	<0.03	10 ns	<500	Excimer, UV
Krypton fluoride	0.248	<1.5	30 ps–200 ns	<500	Excimer, UV, possible for ultrasonic generation
Xenon chloride	0.308	<400	30 ps–80 μs	<500	Excimer, UV, possible for ultrasonic generation
Xenon fluoride	0.351	<10	8 ns–1 μs	<500	Excimer, UV, possible for ultrasonic generation

Table 1.3 Commercially available solid state cw lasers, indicating suitability for the reception of ultrasound (* indicates frequency multiplication).

System	Wavelength (μm)	Output	Comments
Nd:YAG (quartz iodine pumped)	0.355*, 0.53*, 1.06, 1.32	< 1 kW	Single mode (lower power) suitable for interferometry in infrared or visible
Nd:YAG (laser-diode pumped)	0.53*, 1.06	< 1 mW	Rapidly improving, likely to be very suitable for interferometry

Table 1.4 Commercially available solid state pulsed lasers (flash-tube pumped) with applications to the generation of ultrasound. Note that it is not possible to obtain maximum output, minimum pulse length and maximum rate simultaneously. (* indicates frequency multiplication.)

Medium	Wavelength (μm)	Output (J)	Pulse length	Repetition rate	Comments
Nd:YAG	0.266*, 0.355*, 0.532*, 1.06, 1.32	< 150	30 ps–20 ms	< 50 kHz	Ideal for ultrasonic generation over a range of wavelengths
Nd:glass	0.266*, 0.355*, 0.532*, 1.06	< 10^4	100 ps–2 ms	< 120 Hz	Similar to Nd:YAG, less compact
Ruby	0.694	< 400	15 ns–3 ms	< 120 Hz	Good for ultrasonic generation in visible

However, most of the laboratory demonstrations of laser ultrasonics have been carried out with a limited number of laser systems, which we shall now briefly note.

Laser generation of ultrasound does not impose very stringent conditions on the source laser. All that is required is the ability to deliver a reasonably high pulsed energy density to a small area of the specimen, where the pulse length is in the range 1–100 ns. As already mentioned, wavelength is not critical, nor is monochromaticity or coherence. Most of the studies in the literature have employed a solid state laser, either ruby (for some of the early studies) or Nd:YAG (which has dominated the most recent studies). The Nd:YAG laser has proved itself to be a versatile system under Q-switched conditions. Adequate pulse energy can be obtained from Nd:YAG lasers of modest size, while the pulse lengths are close to ideal for ultrasonic studies in metals. However, the fundamental wavelength is in the near infrared (1.064 μm) which makes alignment, etc, more difficult than with a visible laser, in addition to increasing the risk of accidental eye damage. The Nd:YAG laser is available with frequency doubling, etc, to give several harmonics of the fundamental frequency. The frequency doubled wavelength of 0.532 μm is well placed in the middle of the visible, while the harmonics of 0.266 μm and 0.355 μm are in the ultraviolet. As a later chapter will show, ultrasonic generation is more efficient as the optical wavelength is reduced.

Pulsed gas lasers have been used for laboratory studies, mainly carbon dioxide in the infrared (10.64 μm) and, more recently, nitrogen and excimer lasers in the ultraviolet. Although carbon dioxide lasers are capable of high power, absorption at a metal surface tends to be somewhat inefficient at such a long wavelength; in non-metals this is not a problem. Furthermore, 10.64 μm is sufficiently far into the infrared to require special materials for optical components such as lenses.

The reception of ultrasound by laser on the other hand is dependent upon all the fundamental properties of the laser: monochromaticity, coherence, directionality and high power density. Because of its ready availability and excellent optical characteristics, notably monochromaticity and coherence, the helium–neon laser has dominated much of the work. This system is however somewhat limited with regard to maximum power (which controls sensitivity, as will be shown in a later chapter). Thus other more powerful laser systems are actively being explored. The obvious choice is the argon ion laser, which not only can deliver higher power, but is also inherently more sensitive because of its shorter wavelength. Argon ion lasers are bulky and notoriously 'noisy' (although they are becoming less so). The CW Nd:YAG laser is also available as a solid state option, although it suffers from a number of drawbacks including bulk. An interesting new option is the laser-diode pumped Nd:YAG laser referred to above.

1.4.3   Safety considerations

All lasers must be treated with great respect from the point of view of safety. The greatest danger is of accidental irradiation of the eye leading to retinal damage. However, there is also the possibility of skin damage with the more powerful lasers, with the formation of skin lesions. The main reason for these safety hazards is the extremely high power densities that even quite small lasers can deliver. Thus a typical small helium–neon laser delivering 1 mW in a 1 mm diameter unfocused beam gives rise to a local power density of 1.3 m W mm^{-2}. This is well in excess of the threshold for retinal damage, formally designated the maximum permissible exposure (MPE), from continuous irradiation, i.e. 25 μW mm^{-2} for visible light. This value assumes a flicker response time of 0.25 s for the eye.

Table 1.5

	Classification					
Requirements	Class 1 (except high-powered class 1)	Class 1 (high-powered)	Class 2	Class 3A	Class 3B	Class 4
Laser responsible officer	No	Yes	No	Yes	Yes	Yes
Registration	No	Yes	No	Yes	Yes	Yes
Standing orders	No	Yes	No	Yes	Yes	Yes
Hazard label on laser	No	Yes	No	Yes	Yes	Yes
Supplementary label(s) on laser	Yes	Yes	Yes	Yes	Yes	Yes
Use only in designated laser area	No	No (laser is enclosed in protective housing)	No	Yes (unless LRO considers unnecessary)	Yes	Yes
Warning sign at entrance to DLA	No	No	No	Yes (unless LRO considers unnecessary)	Yes	Yes
Warning lights at entrance to DLA	No	No	No	No	Yes	Yes
Interlocked access to DLA	No	No (protective housing is interlocked)	No	No	Yes	Yes
Beam stops or attenuators	No	Yes (inherent in design)	No	No	Yes	Yes
Enclosure of beam paths	No	Yes (interlocked)	No	No	Yes (where practicable)	Yes
Care needed to prevent specular reflections	No	Yes (design must prevent these)	No	No	Yes	Yes

Table 1.5 (*cont.*)

Requirements	Classification					
	Class 1 (except high-powered class 1)	Class 1 (high-powered)	Class 2	Class 3A	Class 3B	Class 4
Requirements concerning optical viewing aids	No	Aids to be filtered and secured to prevent exposure above MPE	No	Aids to be approved by OLRO	Aids to be interlocked or filtered to prevent exposure above MPE	Yes
Key of key control to be removed when laser not in use	No	No	No	No	Yes	Yes
Training	Yes	Yes	Yes	Yes	Yes	Yes
Laser goggles to be provided	No	No	No	No	Yes (where OLRO considers necessary)	Yes
Protective clothing to be worn	No	No	No	No	Yes (where necessary)	Yes (where necessary)

The near infrared is more hazardous since the radiation is still focused onto the retina by the lens, and yet is invisible so that there is no automatic closing of the eye. Thus a very long exposure must be assumed, so that the threshold for radiation from a CW Nd:YAG laser is only $16 \, \mu W \, mm^{-2}$. Lasers delivering CW radiation outside the part of the electromagnetic spectrum where the retina is most sensitive, or where there is a relatively high absorption by the water and tissue, i.e. further into the infrared ($> 2.5 \, \mu m$ wavelength), have a correspondingly higher MPE of $1 \, mW \, mm^{-2}$.

Q-switched pulsed lasers are intrinsically more hazardous because they can deliver instantaneous power densities well in excess of $1 \, MW \, cm^{-2}$. The average energy density in a typical 100 mJ laser beam of diameter 5 mm is approximately $5 \, mJ \, mm^{-2}$. The threshold for retinal damage at wavelengths $0.4 - 1.4 \, \mu m$ is $5 \, nJ \, mm^{-2}$, assuming a typical Q-switched pulse length in the range 1 nanosecond – several microseconds, and low repetition rate. Thus direct ocular exposure to Q-switched laser beams in the visible or near infrared is virtually guaranteed to cause damage to the retina.

It is outside the scope of this book to cover all types of laser and their use. For a more detailed and authoritative treatment of laser hazards, the reader should consult, for instance, one of the following publications: BS 4803, ANSI Z.136.1, CLM-SC17.

It is clear from the above that various safety precautions are essential when using high-power lasers for ultrasonic studies, as for any other application. It is clear, for instance, that there need to be registrations and defined responsibilities to control their use. There need also to be clear instructions (standing orders). There need to be various types of warning signs, labels and lights, and special areas designated for laser use. There need to be various precautions to prevent those not directly involved with a laser experiment from being harmed. Thus entry to laser areas needs to be restricted and interlocks fitted so that the laser is made safe on entry. High-power lasers also require key operation. Then there are precautions for the laser operators themselves, including protective goggles and other clothing, beam stops, beam path enclosure, care regarding specular beam reflection and special arrangements for optical viewing aids. Finally there needs to be specific training for all operators.

Not all lasers present the same potential hazard, so that the requisite safety precautions may also vary according to the type of laser. Because of this it was decided to classify lasers in a number of categories for safety purposes, depending upon wavelength, CW power and energy per pulse. The classes are defined as follows (CLM-SC17).

Class 1. Inherently safe lasers, whose output cannot exceed the MPE. Potentially more powerful laser systems can be placed in this class if they are 'boxed' or otherwise enclosed so that high-power radiation cannot escape to the outside.

Class 2. Mostly low-powered cw lasers of output < 1 mW operating at a visible wavelength. Although these are not inherently safe, the natural blink reflex normally affords sufficient protection to the eye.

Class 3A. Lasers emitting < 5 m W in the visible spectrum.

Class 3B. Lasers emitting visible and/or invisible radiation, if cw of power < 0.5 W, or if pulsed giving a radiant exposure < 0.1 J mm^{-2} per pulse.

Class 4. High-power cw or pulsed lasers whose output powers or radiant exposures exceed those specified for class 3B systems. They should only be used with extreme caution.

Table 1.5 summarizes the control measures appropriate for each class of laser.

This book does not set out to be an authority on safety. The purpose of this section is to draw the reader's attention to potential hazards, and encourage him to consult an authoritative document on laser safety such as BS 4803, ANSI Z.136.1 or CLM-SC17. We would finally like to point out (in case the reader is intimidated by the complexity of safety precautions) that they can be implemented without too much disruption to the experiment, especially if the comment about 'boxed' systems in the definition of class 1 is noted. Many workers have rapidly learned how to work effectively within the confines of these safety codes.

BIBLIOGRAPHY

Ultrasonics and non-destructive testing

Achenbach J D 1973 *Wave Propagation in Elastic Solids* (Amsterdam: North-Holland)

Auld B A 1973 *Acoustic Fields and Waves in Solids* (New York: Wiley)

Farley M F and Nichols R W (eds) 1988 *Non-destructive Testing* vols 1–4 (Oxford: Pergamon)

Graff K F 1975 *Wave Motion in Elastic Solids* (Oxford: Clarendon)

Krautkramer J and Krautkramer H 1977 *Ultrasonic Testing of Materials* 2nd edn (Berlin: Springer)

Lamb Sir H 1931 *The Dynamical Theory of Sound* (London: E Arnold)

Mason W P and Thurston R N *Physical Acoustics* vols I–XX (New York: Academic)

Matthews H 1977 *Surface Wave Filters* (New York: Wiley)

Rayleigh Lord 1926 *Theory of Sound* (London: Macmillan)

Sharpe R S *Research Techniques in Nondestructive Testing* vols 1–9 (New York: Academic)

Sharpe R S, Cole H A and West J 1973 *Quality Technology Handbook* 3rd edn (Guildford: IPC)

Silk M G 1984 *Ultrasonic Transducers for Nondestructive Testing* (Bristol: Adam Hilger)

Szilard J 1982 *Ultrasonic Testing* (New York: Wiley)

Optics and lasers

Allen L and Jones D G C 1967 *Principles of Gas Lasers* (London: Butterworths)

Arecchi F T and Schulz-Dubois E O 1972 *Laser Handbook* (Amsterdam: North-Holland)

Bertolotti M 1983 *Masers and Lasers* (Bristol: Adam Hilger)

Bloom A L 1968 *Gas lasers* (New York: Wiley)

Born M and Wolf E 1980 *Principles of Optics* 6th edn (Oxford: Pergamon)

Brown R 1968 *Lasers* (London: Aldus)

Heavens O S 1971 *Lasers* (London: Duckworth)

Hecht J 1986 *The Laser Guidebook* (New York: McGraw-Hill)

Jenkins F A and White H E 1976 *Fundamentals of Optics* 4th edn (New York: McGraw-Hill)

Lengyel B A 1971 *Lasers* 2nd edn (New York: Wiley)

O'Shea D C and Callen W R 1977 *Introduction to Lasers and their Applications* (Reading, MA: Addison-Wesley)

Pressley R J (ed) 1971 *Handbook of Lasers* (Boca Raton, FL: CRC Press)

Ross M 1966 *Laser Receivers* (New York: Wiley)

Siegman A E 1986 *Lasers* (Oxford: Oxford University Press)

Svelto O 1982 *Principles of Lasers* 2nd edn (trans. D C Hanna) (New York: Plenum)

Weber M J (ed) 1986 *Handbook of Laser Science and Technology* (Boca Raton, FL: CRC Press)

Young M 1984 *Optics and Lasers* 2nd edn (Berlin: Springer)

2 Acousto-optic Interactions

2.1 THE ACOUSTO-OPTIC EFFECT

2.1.1 Introduction

The effects of acoustic waves on light transmission through transparent media have been known for a long time (Lucas and Biquard 1932, Debye and Sears 1932). They arise from refractive index variations associated with the pressure fluctuations of sound waves. These variations may be detected optically by refraction, diffraction or interference effects. The advent of lasers has made the phenomena easier to observe and has generated many applications. Soon after lasers became available, the potential for their use in acousto-optics was appreciated (Quate *et al* 1965, Korpel *et al* 1966, Whitman and Korpel 1969).

Although the acousto-optic effect is widely used for the study and measurement of ultrasonic waves, the principal interest is in acousto-optic devices for deflecting, modulating or frequency-shifting light beams and in signal processing. This is now a well-developed subject (Korpel 1988) not always directly relevant to ultrasonic studies, the main theme of this book, and will therefore not be comprehensively discussed.

2.1.2 The diffraction of light by ultrasonic waves

A plane acoustic wave travelling in a transparent medium produces a periodic variation of refractive index which can act as a diffraction grating. This is illustrated in figure 2.1 where a laser beam is shown incident transversely on an ultrasonic wave travelling through a liquid. Diffracted beams are

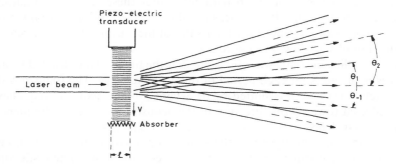

Figure 2.1 The diffraction of a laser beam by an ultrasonic wave.

produced as from a conventional diffraction grating at angles θ_n from the original direction given by

$$\Lambda \sin \theta_n = n\lambda \qquad (2.1)$$

where λ and Λ are the wavelengths of the light and acoustic waves, respectively, and n the order of diffraction. $n = 0$ corresponds to the undeflected beam. Since no appreciable absorption or blocking of light takes place, the total energy in the diffracted beams (including $n = 0$) is equal to that in the incident beam.

If the width of the beam, $\mathscr{L}$, is sufficiently small, the fluctuations produced by the wave may be treated as a simple phase grating across which the phase of the light varies sinusoidally,

$$\phi = \phi_0 + \frac{2\pi \mathscr{L} \Delta \mu}{\lambda_0} \cos\left(\frac{2\pi x}{\Lambda}\right) \qquad (2.2)$$

where $\Delta \mu$ is the peak fluctuation in refractive index and λ_0 the wavelength of the light in vacuum. x is the coordinate in the direction of propagation of the ultrasonic waves. The light intensity in the various orders may then be readily calculated. The intensity in the nth order is

$$I_n = I_1 (J_n(v))^2 \qquad (2.3)$$

where J_n is the Bessel function of the first kind of order n. I_1 is the incident light intensity and v is the amplitude of the phase fluctuation,

$$v = \frac{2\pi \Delta p \mathscr{L}}{\lambda_0} \left(\frac{\partial \mu}{\partial p}\right)_S. \qquad (2.4)$$

Δp is the peak pressure fluctuation of the acoustic wave and $(\partial \mu / \partial p)_S$ is the coefficient of variation of refractive index with pressure, assumed to be linear over the small range usually involved. This is required under adiabatic conditions (indicated by the subscript S) as is usual in sound propagation, as there is no time for heat transfer during an acoustic period. The quantity v is known as the Raman–Nath parameter after these early investigators (Raman and Nath 1935). It is a measure of the optical diffraction power of the ultrasonic beam. A plot of the intensities of the various diffracted beams as a function of v is shown in figure 2.2.

For a travelling acoustic wave, we note that the phase of a diffracted beam advances by n cycles for each period of the acoustic wave. The frequency of the diffracted beam is thus shifted by nf where f is the acoustic frequency. Light deflected towards the direction of the acoustic wave has its frequency increased; that deflected away has a decreased frequency.

Diffraction angles produced by the acousto-optic effect are usually rather small. This is a consequence of the large discrepancy between acoustic and optical wavelengths. For example, in water an ultrasonic wave of frequency

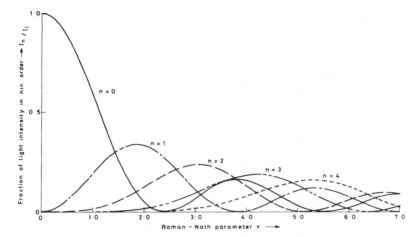

Figure 2.2 Light intensities in beams diffracted by a narrow ultrasonic beam (Raman–Nath regime) as a function of ultrasonic excitation. n is the order of diffraction.

10 MHz has a wavelength of 150 μm compared with optical wavelengths typically around 0.5 μm, giving a separation of diffracted orders of 3.3 milliradians or 11 minutes of arc. This may be compared with a typical laser divergence angle of around a milliradian. Thus the diffracted beams may not always be easy to resolve and expansion of the laser beam may be desirable to reduce its divergence.

A laser beam of Gaussian profile with radius r_0 to the $1/e^2$ intensity points gives a divergence with an angular Gaussian distribution of width $2\lambda/\pi r_0$ (also to $1/e^2$ intensity points); see equation (1.15). We may thus deduce that the ratio of the separation of the diffracted orders to the divergence of the laser beam is $\frac{1}{4}\pi N$ where N is the number of acoustic waves within the width of the laser beam, $2r_0$. At least three acoustic cycles within the laser beam are required for the satisfactory resolution of the acousto-optic diffraction effect, assuming that the beam has a Gaussian profile and diffraction-limited divergence. More cycles may be necessary in less favourable circumstances.

If diffraction orders do overlap, the time-averaged diffraction pattern may be obtained by superposition of the diffracted beams, each broadened by the divergence characteristic of the incident beam. However, in overlapping regions, modulations at the ultrasonic frequency will occur (see §2.2.2).

2.1.3 The piezo-optic coefficient

To obtain quantitative measurements of acoustic or ultrasonic fields from diffraction experiments, it is necessary to know the refractive index variation

Table 2.1 Adiabatic piezo-optic coefficients for liquids. S, adiabatic values, L, by Lorentz theory. Measurements were taken close to 23 °C. RV, Raman and Venkateraman (1939), RE, Reisler and Eisenberg (1965).

Liquid	Wavelength (nm)	Refractive index	$(\partial\mu/\partial p)_S$ $(10^{-10}$ Pa$^{-1})$	$\rho(\partial\mu/\partial\rho)_S$	$\rho(d\mu/d\rho)_L$	Reference
H_2O	589	1.333	1.447	0.321	0.367	RV
	546	1.334	1.431	0.317	0.369	RE
D_2O	546	1.330	1.443	0.309	0.363	RE
CH_3OH	589	1.328	3.376	0.328	0.361	RV
	546	1.330	3.383	0.328	0.363	RE
C_6H_6	589	1.501	3.533	0.530	0.592	RV
CS_2	589	1.626	4.332	0.717	0.782	RV

with pressure or density of the medium. The required adiabatic coefficients $(\partial\mu/\partial p)_S$ and $(\partial\mu/\partial\rho)_S$ may be measured directly or deduced with theoretical assumptions. The difference between isothermal and adiabatic values is in fact usually quite small but is significant for accurate measurements. Piezo-optic coefficients have been measured directly for several liquids (Raman and Venkateraman 1939, Reisler and Eisenberg 1965) and some values are given in table 2.1.

It seems plausible to assume that the refractive index depends only on the density ρ of the liquid, there being no direct dependence on temperature or pressure. In this case, the adiabatic and isothermal coefficients of refractive index with respect to density are the same. The required adiabatic coefficient with respect to pressure may then be deduced from isothermal data and the adiabatic bulk modulus B_S which may be readily deduced from the velocity of sound and the density. This assumption is quite accurate enough for most purposes: see, for example, the values for water given in table 2.2. It may be noted, however, that a small intrinsic dependence of refractive index on temperature has been observed for some liquids (Waxler and Weir 1963).

An estimate of $\partial\mu/\partial\rho$ may be obtained purely theoretically using a formula due to Lorentz (1880). The electric polarizability of the atoms or molecules of the medium is assumed to be constant, unaffected by the distance between them. The electric field they experience from the light wave may be shown to be that inside a spherical cavity in the medium. This leads to the result

$$\frac{\mu^2 - 1}{\mu^2 - 2} \propto \rho. \tag{2.5}$$

Hence

$$\frac{d\mu}{d\rho} = \frac{(\mu^2 - 1)(\mu^2 + 2)}{6\mu\rho}. \tag{2.6}$$

These values are shown in table 2.1 for comparison.

Table 2.2 Piezo-optic coefficients for water wavelength $\lambda = 633$ nm.

Temperature (°C)	Isothermal coefficients		Adiabatic coefficients	
	$(\partial\mu/\partial p)_T$ $(\times 10^{-10}\ \mathrm{Pa}^{-1})$	$\rho(\partial\mu/\partial\rho)_T$	$(\partial\mu/\partial p)_S$ $(\times 10^{-10}\ \mathrm{Pa}^{-1})$	$\rho(\partial\mu/\partial\rho)_S$
15	1.515	0.324	1.508	0.324
20	1.486	0.324	1.473	0.323
25	1.463	0.323	1.444	0.323

The most commonly used liquid for ultrasonic propagation, water, has been the subject of a number of studies (Raman and Venkateraman 1939, Waxler and Weir 1963, Waxler *et al* 1964, Reisler and Eisenberg 1965). The results have been summarized by an equation for the dependence of refractive index on temperature and density proposed by Eisenberg (1965), and this has been used, together with thermodynamic data, to calculate the values of the piezo-optic constants given in table 2.2. These are for a wavelength of 633 nm (He–Ne laser). They should be increased by 1.0% or 1.3% for the argon ion laser wavelengths of 514 and 488 nm, respectively.

2.2   THE MEASUREMENT OF ULTRASONIC FIELDS

2.2.1   Direct measurements of the diffracted beams

A typical arrangement for the measurement of the intensity of an ultrasonic wave by the acousto-optic effect is shown in figure 2.3. Travelling ultrasonic waves are produced in a water tank by a piezoelectric transducer. Their path is terminated by a sound-absorbing medium to prevent the formation of standing waves by reflection from the walls of the tank. An expanded laser beam is transmitted transversely to the sound direction through windows in the tank. The far-field (Fraunhofer) diffraction pattern is formed by focusing

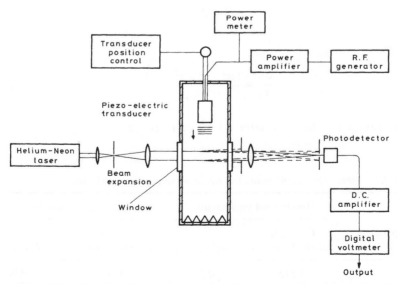

Figure 2.3   The measurement of ultrasonic beam power by optical diffraction.

the emergent beams by a lens. The photodetector in the focal plane may be positioned to receive any of the diffraction orders. The ratio of the power received to that in the transmitted beam with no ultrasonic waves present is determined. These ratios may usually be interpreted in terms of the elementary diffraction theory for a phase grating given in §2.1.2. The Raman–Nath parameter v may be obtained from the reduction in the directly transmitted beam (zeroth order), I_0, via the relation

$$\frac{I_0}{I_1} = (J_0(v))^2. \qquad (2.7)$$

The intensity of this beam decreases to zero as v increases to 2.4, as will be seen from figure 2.2. Alternatively, the intensities in the various diffraction orders may be used (Haron *et al* 1975). Deviations from the distribution of intensities given by equation (2.3) may arise from the presence of harmonics produced by the transducer or by non-linear propagation effects.

Careful measurements of the power in the acoustic field from various transducers have been made by this method and compared with measurements based on radiation pressure (Reibold 1976/77, Haron *et al* 1975). Good agreement within experimental error of a few per cent is obtained in the frequency range 1–10 MHz.

An important consideration in the interpretation of these results is the structure of the sound field normally associated with the near-field region of an acoustic transducer, which results in variations of the peak sound pressure with distance from the transducer. There is a consequent variation in the optical effect. The Raman–Nath parameter observed is the result of integrating the effect along the optical path. Assuming the motion to be piston like, Ingenito and Cook (1969) have calculated the form of this integrated effect and their theory may be used for correcting experimental results. An example of the variation of the Raman–Nath parameter with distance z from the face of a circular transducer of radius r_t is shown in figure 2.4. v is normalized to v_0, its value for true plane-wave interaction across the diameter of the transducer. The extent of the near-field region or 'Fresnel zone' is determined by the quantity r_t^2/Λ and if z is expressed in units of this quantity, the calculated curves do not vary much with r_t or Λ provided the ratio r_t/Λ is large, as is almost always the case in practice, and the highly structured region close to the face of the transducer is excluded.

Experimental results (Reibold *et al* 1979) agree quite well with theory, especially for mechanically loaded transducers which may be expected to have a piston-like action, as is shown by the example in figure 2.4. The experimental values do not, however, follow the rapid variations close to the transducer where we should expect that they would be smoothed out by the finite extent of the laser beam.

Having determined v_0, the acoustic power P_A may be calculated (assuming

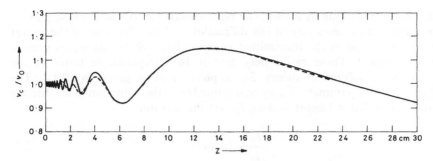

Figure 2.4 Comparison of the experimental and theoretical optical diffraction profiles in the near-field region of a 2 MHz circular mechanically loaded transducer 19.1 mm in diameter (Reibold *et al* 1979). The Raman–Nath parameter is plotted as a function of distance from the transducer. Full curve theoretical: broken curve observed, normalized to $z = 0$.

circular symmetry) from the formula

$$P_A = \frac{\lambda_0^2 v_0^2}{32 \pi \rho v (\partial \mu / \partial p)_S^2} \tag{2.8}$$

where ρ is the density of the medium and v the velocity of the acoustic wave. We note that the diameter of the beam does not enter into this equation and it is thus not necessary to know this quantity accurately as it would be if the actual acoustic pressure were required. For water at 20 °C we deduce

$$P_A = 0.130 v_0^2 \text{ (watts)}. \tag{2.9}$$

2.2.2 The use of optical near-field effects

For small values of v the fraction of light power diffracted out of the zeroth order is approximately $\frac{1}{2}v^2$. This is less than 4% for a power of 10 mW in water, for example. This would be quite difficult to measure accurately and more sensitive techniques have been suggested for the measurement of low powers, using the near-field diffraction pattern (Cook 1976, Riley 1980). In this region, the diffracted beams overlap and, since they have different frequency shifts, beating between them produces intensity fluctuations at the acoustic frequency f and harmonics.

It is well known that, close to a diffraction grating, interference fringes with a spacing equal to that of the grating are formed whose contrast varies periodically with distance from the grating. This is illustrated for a phase grating in figure 2.5. Fringes associated with a moving grating as formed by an ultrasonic wave evidently travel with the same velocity and points in the fringe regions experience intensity fluctuations at the acoustic frequency.

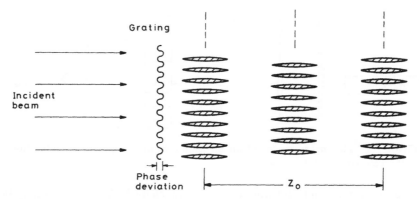

Figure 2.5 Fringe pattern in front of an optical phase grating illuminated by a parallel laser beam. z_0 is the repeat distance.

For small diffraction angles and small values of v where only the zeroth and ± 1 orders of diffraction are significant, it is easy to deduce the form of the pattern of interference between these orders. The maximum amplitude of intensity fluctuation in the fringes at frequency f is given by

$$I_F = 2I_1 v \tag{2.10}$$

occurring at a distance $\frac{1}{4}z_0$ from the grating and at $\frac{1}{2}z_0$ intervals thereafter, where z_0 is the full repeat distance:

$$z_0 = 2s^2/\lambda. \tag{2.11}$$

For an ultrasonic wave, the grating spacing s is equal to the acoustic wavelength Λ. Equation (2.10) is adequate for acoustic powers less than about 30 milliwatts, but at higher powers we need a more complete analysis (Colbert and Zorkel 1963, Cook 1976) from which we obtain the following expression for the variation of intensity with time at a distance z from the ultrasonic wave treated as a sinusoidal phase grating:

$$I(z, t) = I_1 \sum_{n=-\infty}^{n=\infty} J_n[2v \sin(2\pi nz/z_0)]\cos(2\pi nft - \tfrac{1}{2}n\pi). \tag{2.12}$$

A diagram of an arrangement for the determination of ultrasonic power from near-field optical diffraction is shown in figure 2.6. The aperture in front of the detector is placed at the correct position to obtain the maximum intensity fluctuation. The actual distance from the ultrasonic wave depends on the refractive index of the medium and hence on the fraction of the path in water and air. The diameter of the aperture must be small compared with the ultrasonic wavelength: it could typically be 25 μm. If too large an aperture is used, there is a reduction in the modulation depth of the signal due to

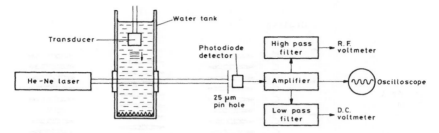

Figure 2.6 The measurement of an ultrasonic beam using optical diffraction in the near (optical) field region.

averaging over a significant fraction of the wavelength. A correction may be applied for this effect. The Raman–Nath parameter is determined from the ratio of the peak modulation to the mean level of the signal by equation (2.10) or (2.12).

2.2.3 The heterodyne method

Another method of measuring low-power ultrasonic waves employs the heterodyne principle. The arrangement used by Nagai and Iizuka (1982) and Nagai (1985) is shown in figure 2.7. A reference beam is split from the laser beam before its passage through the tank. This is frequency shifted by a Bragg cell (see §2.4.1) by 30 MHz for example, before being recombined with the main beam which has traversed the acoustic field. The accurate superposition of beams in this way produces intensity fluctuations or 'beats' at a frequency equal to the difference between the frequencies of the two beams.

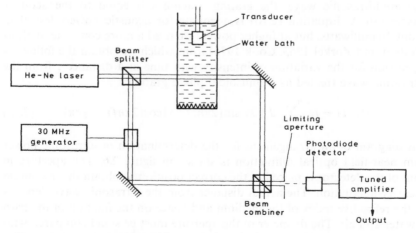

Figure 2.7 The measurement of the diffraction of a laser beam by an ultrasonic beam by optical heterodyning (Nagai and Iizuka 1982).

The use of a reference in optical interferometry is discussed in detail in §3.2.3 but we note here that, at a point in the region of superposition, the mixing of the reference beam with the nth diffraction order, which is frequency shifted by nf, results in the formation of a signal frequency f_s given by

$$f_s = f_B - nf \qquad (2.13)$$

where f_B is the frequency shift introduced by the Bragg cell. The amplitude of the intensity fluctuation and hence of the signal, V_{sn}, is proportional to the product of the electric-field amplitude of the reference E_{R0} and that of the diffracted beam E_{n0},

$$V_{sn} \propto E_{R0} E_{n0} \propto \sqrt{I_R I_n} \qquad (2.14)$$

where I_R and I_n are the corresponding intensities of the beams.

Each diffraction order is thus associated with a specific signal frequency that may be isolated by a suitable tuned filter. The width of the filter should be considerably less than f to avoid the overlapping of orders and f_B should be sufficiently high that the required beat signals are clear of any significant harmonics of f generated by direct beating between different diffraction orders.

Since each diffraction order is separated out in this way, the result is evidently independent of the exact position of the aperture in the beam (unlike the previous method, §2.2.2), but the aperture is restricted by the spatial coherence limitation discussed in §3.2.3, and beam alignment is crucial. It is not possible for the reference beam to be aligned with more than one diffracted beam at the same time. This leads to the same aperture limitation as in the previous method, i.e. that the diameter of the aperture in front of the detector must be small compared with the wavelength of the ultrasonic wave. It is also necessary that the difference between the optical path lengths of the signal and reference beams is less than the coherence length of the laser; see §3.5.1.

The Raman–Nath parameter may be calculated from the ratio of the signal at the frequency $f_B - nf$ to that at f_B without the acoustic wave present,

$$V_{sn}/V_{s1} = (I_n/I_1)^{1/2} = J_n(v) \qquad (2.15)$$

using equation (2.3). Normally the first-order diffraction, $n = \pm 1$, would be chosen to give the best sensitivity over most of the range of acoustic powers.

2.3 BRAGG DIFFRACTION

2.3.1 Limits of simple diffraction theory

So far in this chapter it has been assumed that the ultrasonic beam is sufficiently narrow and the angle of diffraction sufficiently small that light beams diffracted from all parts of the ultrasonic field illuminated by the laser are in phase. This permits the diffraction to be treated in the same way as that from a narrow phase grating. The region in which this approximation is valid is known as the Raman–Nath regime. This is generally valid in most practical situations at frequencies below about 5 MHz, but at higher ultrasonic frequencies and/or with wide ultrasonic beams, a more detailed treatment is necessary.

In figure 2.8, the generation of a first-order diffracted beam by the passage of a laser beam through an ultrasonic beam of width $\mathscr{L}$ is illustrated. In figure 2.8(a), it is assumed that the laser beam is incident normal to the direction of propagation of the ultrasound. The diffracted beams generated at the beginning and end of the passage are shown. Since the phase of the ultrasonic wave is the same at A and E, the phase difference $\Delta\phi$ between

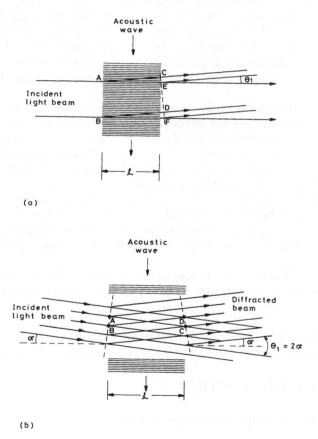

Figure 2.8 First-order diffraction of a laser beam by a wide acoustic beam. (a) Laser beam incident normal to the acoustic wave, (b) laser beam incident at the Bragg angle.

the two diffracted waves at the wavefront CEDF may be derived from the difference between the optical path lengths AC and AE,

$$\Delta\phi = \frac{\pi\mathscr{L}\theta_1^2}{\lambda} = \frac{\pi\lambda\mathscr{L}}{\Theta^2} = \frac{Q}{2} \tag{2.16}$$

where θ_1 is the first-order diffraction angle. If this is small we may put

$$\Delta\phi = \frac{\pi\mathscr{L}\theta_1^2}{\lambda} = \frac{\pi\lambda\mathscr{L}}{\Lambda^2} = \frac{Q}{2} \tag{2.17}$$

since $\theta_1 = \lambda/\Lambda$ where λ and Λ are the optical and acoustic wavelengths respectively. The parameter $Q = 2\pi\lambda\mathscr{L}/\Lambda^2$ is generally used as a measure of deviation from the Raman–Nath regime. A simple argument shows that for small values of v, the diffracted power is reduced by a factor $[(\sin(Q/4))/(Q/4)]^2$, an error of approximately 2% when $Q = 1$. For $Q = 4\pi$ diffraction effects disappear for small v and remain small even when v becomes quite large (see figure 2.10(b)).

2.3.2 Diffraction at the Bragg angle

For oblique incidence, the effect is similar except for the angle of incidence α given by

$$\sin\alpha = \lambda/2\Lambda \tag{2.18}$$

when the incident and diffracted beams are equally inclined to the acoustic wave. In this case, the phase of the diffracted beam is the same for diffraction in every part of the acoustic field. This is illustrated in figure 2.8(b). The optical path lengths between the incident and diffracted wavefronts shown are AC and BD for diffraction at opposite edges of the acoustic wave. These are evidently equal, both being $\mathscr{L}\cos\alpha$. As diffraction at B and C is from the same acoustic wavefront, the phase differences between the various elements of the diffracted beam are zero and a strong beam is formed. This effect is known as Bragg diffraction or reflection after the discoverer of the analogous phenomenon in the diffraction of x-rays by crystals (Bragg and Bragg 1915). As in the x-ray case, the effect may be regarded as the superposition of reflections from periodically spaced planes. This is illustrated in figure 2.9. In the opto-acoustic case, these are interfaces between regions of different refractive index. If the optical path difference between the reflections at successive interfaces is exactly one wavelength, reinforcement occurs and a strong reflection obtained. This is an alternative approach to Bragg diffraction giving exactly the same results as the conventional diffraction treatment. In figure 2.9, we see that reflections from successive planes spaced at the acoustic wavelength Λ add an optical path length BCD = $2\Lambda\sin\alpha$. This must be equal to λ for constructive interference and evidently leads to the Bragg condition, equation (2.18).

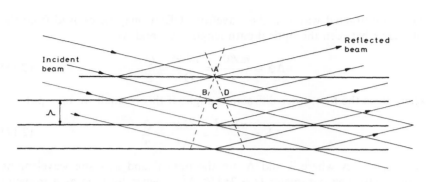

Figure 2.9 Bragg diffraction treated as the superposition of reflections from periodically spaced surfaces.

For small values of the parameter v, the fraction of the incident light intensity transferred to the favoured first-order diffracted beam (designated $+1$) is $\frac{1}{4}v^2$ just as in the Raman–Nath case, but when there is a significant intensity in the diffracted beam, diffraction back to the direction of the incident beam becomes important since this is also favoured by the Bragg condition. This leads, at high values of v, to the periodic transfer of light intensity between the 0 and $+1$ diffraction orders:

$$I_1 = I_1 \sin^2(\tfrac{1}{2}v) \qquad I_0 = I_1 \cos^2(\tfrac{1}{2}v). \tag{2.19}$$

Although derived for the limit $Q \to \infty$, these equations are good to within a few per cent down to about $Q = 10$. The value of Q does, however, affect the sharpness of tuning to the Bragg condition; i.e. the range of the angle α over which a strong diffraction peak is obtained. The breadth of this range is inversely proportional to Q.

For values of Q intermediate between the Bragg and Raman–Nath regions, it is not possible to give analytical formulae for the intensities of the diffracted beams, but various numerical computations have been made, particularly by Klein and Cook (1967). Examples of the variation of intensity in significant orders for the two angles of incidence usually of interest, normal incidence ($\alpha = 0$) and the Bragg angle, are given in figures 2.10 and 2.11 for two values of Q, 2π and 4π, the latter being often regarded as the lower limit of the Bragg regime.

2.3.3 Brillouin scattering

Bragg diffraction from thermally excited ultrasonic waves in a transparent medium gives rise to a weak optical scattering mechanism first predicted by Brillouin (1922). In thermal equilibrium, the vibrational excitation of a medium may be considered as a combination of acoustic waves travelling in all directions and covering a wide frequency band. Optical scattering shows

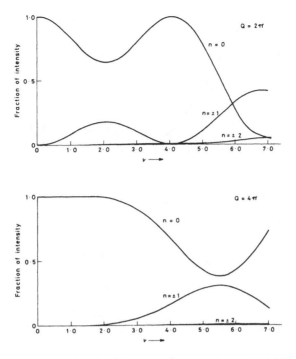

Figure 2.10 The intensities of beams diffracted from a wide ultrasonic beam as functions of ultrasonic excitation for normal incidence of the laser beam for two values of the parameter Q, intermediate between the Raman–Nath and Bragg regimes.

up only the very-low-frequency end of the band where excitation is very weak. Even so, the frequency of the waves contributing to the scattering is still high by normal ultrasonic standards and the Bragg condition is closely satisfied.

In an isotropic medium, scattering at an angle θ from the laser beam direction therefore only takes place from those waves whose direction of propagation makes an angle $\frac{1}{2}(\pi - \theta)$ with the laser beam and is in the plane of scattering. The wavelength Λ must satisfy the condition

$$\lambda/\Lambda = 2 \sin(\tfrac{1}{2}\theta). \tag{2.20}$$

The corresponding ultrasonic frequency is

$$f = \frac{2v}{\lambda} \sin(\tfrac{1}{2}\theta) \tag{2.21}$$

and thus light diffracted by this mechanism is frequency shifted by this amount (v is the acoustic velocity). Up- and down-shifted components are

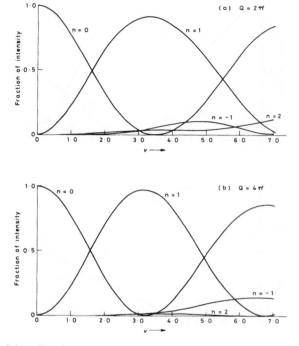

Figure 2.11 The intensities of beams diffracted by an ultrasonic beam as functions of ultrasonic excitation for incidence of the laser beam at the Bragg angle for two values of the parameter Q in the intermediate region.

present since ultrasonic waves travelling in opposite directions both satisfy the Bragg condition. This shift enables Brillouin scattering to be separated from other mechanisms that do not produce a shift, e.g. Rayleigh scattering and scattering from impurities and defects. Studies of Brillouin scattering in liquids using a helium–neon laser have been described by Benedek and Greytak (1965).

In the Brillouin scattering experiment shown in figure 2.12, the scattered light is analysed by a high-resolution Fabry–Perot spectrometer. (These spectrometers are discussed in more detail in §3.11.) For a scattering angle of 180° as used by Durrand and Pine (1968) in the arrangement illustrated, the frequency shift for $\lambda_0 = 514$ nm (argon laser) for silica for example would be 34 GHz. Frequencies of this order are beyond the range of generation by ultrasonic transducers and the study of Brillouin scattering enables this region of ultrasound to be investigated. The propagation velocity may be obtained from the frequency shift and information on attenuation from line-width measurements using very-high-resolution spectrometers.

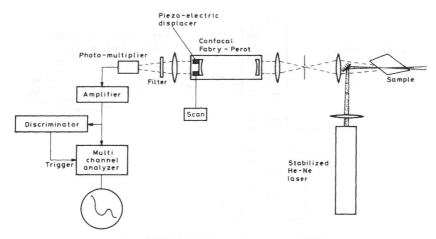

Figure 2.12 Experimental arrangement for observing Brillouin scattering from a solid sample (Durrand and Pine 1968).

2.4 ACOUSTO-OPTIC DEVICES

2.4.1 Bragg cells

The diffraction of light by ultrasonic waves finds many applications using diffraction cells of which many types are commercially available. Reviews of these devices and their uses have been given by Gordon (1966), Chang (1976) and Korpel (1981). Their primary functions are to modulate, deflect or frequency shift light beams. The cells do require well-collimated input beams and are consequently well suited to use with lasers.

The basic acousto-optic cell is illustrated in figure 2.13. It consists of a block of acousto-optic medium, glass or transparent crystal, in which ultrasonic waves are produced by a piezoelectric transducer bonded or deposited on the end. The light beam enters the cell through a good quality optical surface and travels transversely through the acoustic wave field. The resulting diffracted beams are available at the output on the other side. The cells normally use travelling waves and must be terminated by a deflector and/or absorber to prevent interference from acoustic reflections at the end of the cell.

To maximize bandwidths, deflection angles and efficiency, practical devices generally operate at as high a frequency of ultrasound as possible consistent with available generation techniques and the attenuation of the medium. High frequencies and thus short acoustic wavelengths enable high efficiencies to be obtained by concentrating the diffracted light energy into one of the first-order diffracted beams by Bragg reflection. The required value of the Q

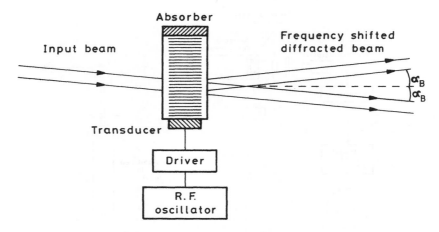

Figure 2.13 A basic acousto-optic cell.

parameter for this may then be obtained with reasonable cell dimensions. Since almost all commercial acousto-optic cells use this principle, they are often called simply 'Bragg cells'.

2.4.2 Acousto-optic materials

The choice of material for an acousto-optic device depends on its optical transmission in the wavelength range required, the availability of large crystals, acoustic absorption and particularly its acousto-optic efficiency. The merits of various materials have been discussed by Dixon (1967), Ushida and Niizeki (1973) and Young and Yao (1981). Some properties of commonly used materials are listed in table 2.3.

To compare diffraction efficiencies of different materials, we note that, for low excitation levels, the fraction of light transferred into the first-order diffracted beam is given by

$$\eta_0 = \tfrac{1}{4}v^2. \tag{2.22}$$

This is true both in the Raman–Nath regime (equation (2.3)) and for Bragg diffraction (equation (2.19)). Using equation (2.4) and the relation between acoustic power flux P_F and the peak pressure fluctuation Δp,

$$P_F = \frac{(\Delta p)^2}{2\rho v} \tag{2.23}$$

we obtain

$$\eta_0 = \frac{\pi^2}{2\lambda_0^2} \mathscr{L}^2 P_F M_2 \tag{2.24}$$

where $\mathscr{L}$ is the interaction length in the cell and M_2 the most commonly

Table 2.3 Acousto-optic materials. o = ordinary ray, e = extraordinary ray, l = longitudinal wave, s = shear wave. Refractive indices and figure of merit are for a wavelength of 633 nm unless otherwise stated.

Material	Crystal structure	Wavelength range (μm)	Density ($\times 10^3$ kg m^{-3})	Refractive index	Acoustic velocity (km s^{-1})	Acoustic attenuation (dB cm^{-1} GHz^{-2})	Figure of merit M_2 ($\times 10^{-15}$ s^3 kg^{-1})
SiO_2	Glass	0.2–2.5	2.20	1.457	5.97	12	1.52
$PbMoO_4$	Tetragonal, 4/m	0.42–5.5	6.95	2.36 (o) 2.25 (e)	3.75	15	36.3
TeO_2	Tetragonal, 442	0.35–5.0	6.00	2.26 (o) 2.45 (e)	4.20 (l) 0.617 (s)	15 290	34.5 (l) 793 (s)
As_2S_3	Glass	0.6–13	3.20	2.61	2.6	170	433
$LiNbO_3$	Trigonal, $\bar{3}$	0.4–4.5	4.64	2.46(1.15 μm) 2.20	6.57	0.15	347(1.15 μm) 7.0
Ge	Cubic, m3m	1.8–23	5.33	4.00(10.6 μm)	5.50	30	840(10.6 μm)
H_2O	Liquid	0.2–0.9	1.00	1.33	1.49	2400	129

used 'figure of merit' of the material,

$$M_2 = 4\rho v \left(\frac{\partial \mu}{\partial p} \right)_s^2 = \frac{4\rho}{v^3} \left(\frac{\partial \mu}{\partial \rho} \right)_s^2 = \frac{\mu^6 p^2}{\rho v^3}. \qquad (2.25)$$

The last expression is written in terms of the photoelastic coefficient p commonly used in solids, defined (for an isotropic material) as follows:

$$p = \rho \frac{\partial (1/\mu^2)}{\partial \rho}. \qquad (2.26)$$

Since p tends not to vary very much between different materials, a high acousto-optic efficiency is generally associated with high refractive index and low acoustic velocity.

For anisotropic materials, the photoelastic constant is a tensor of rank 4 and the analysis of the acousto-optic effect is quite complicated; the reader is referred to the appropriate textbooks (Sapriel 1979). The effective photo-elastic constant depends on the type and direction of propagation of the acoustic wave and the direction and polarization of the light wave. The correct choice of these parameters often enables advantage to be taken of special properties of some crystalline materials.

Most acousto-optic devices operate at frequencies in excess of 30 MHz. Thus thin crystalline piezoelectric transducers are required to generate the ultrasonic waves. Lithium niobate is probably the most efficient material below 1 GHz and is widely used. Deposited films of zinc oxide have applications at higher frequencies.

2.5   APPLICATIONS OF ACOUSTO-OPTIC DEVICES

2.5.1   Light modulators

Acousto-optic cells are extensively used to modulate light beams, particularly in applications like fast printers. They can be much faster than mechanical devices and are competitive with electro-optic modulators.

Modulation is normally accomplished by transferring light intensity from the direct ($n = 0$) to the first-order ($n = \pm 1$) diffracted beam by varying the acoustic excitation. The fraction of light transferred is given by equations (2.19). Theoretically 100% modulation depth is possible in the Bragg regime; 80% might be expected in practice. The maximum light in the diffracted beam occurs for $v = \pi$, for which the acoustic power density required is

$$P_F = \frac{\lambda_0^2}{2 \mathscr{L}^2 M_2}. \qquad (2.27)$$

An important consideration is the speed of response of the modulator. This is evidently governed by the time of travel of the acoustic wave through

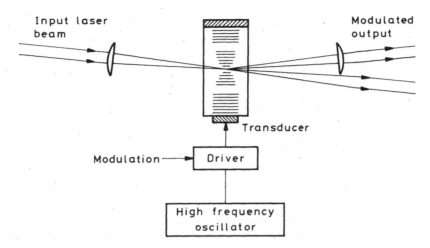

Figure 2.14 Diagram of an acousto-optic modulator.

the light beam. Fast response is favoured by a high acoustic velocity and a narrow beam, and for this reason the light beam is often focused into the modulator cell, as shown in figure 2.14. The fundamental limitation is the divergence of the laser beam which must be considerably less than the diffraction angle to ensure satisfactory separation of the diffraction orders. It was shown in §2.1.2 that to ensure this for a laser beam of Gaussian profile, there must be at least three cycles of acoustic wave within the diameter of the beam (i.e. the waist of the focused beam). This gives a minimum transit time of approximately $3/f$ where f is the acoustic excitation frequency, typically 15 ns for a drive frequency of 200 MHz, corresponding to a rise time of about 10 ns. By the use of higher frequencies, cells with response times down to 1 ns can be constructed.

2.5.2 Beam deflectors

The deflection of a light beam in an acousto-optic cell is a useful property enabling the scanning of a laser beam at speeds beyond those easily obtainable from mechanical devices. The angle of deflection is changed by frequency modulation of the acoustic drive. Although deflection angles are somewhat limited (a few degrees at the most), acousto-optic deflectors are extremely useful devices. A basic beam deflector is illustrated in figure 2.15.

The most important consideration is the number N of resolvable spots obtainable in a scan which may be defined as the ratio of the acceptable range of the deflection angle θ_R to the divergence of the laser beam α_d. Since $\alpha_d \simeq \lambda/d$ where d is the diameter of the beam in the acousto–optic cell, we have

$$N \simeq \frac{d\theta_R}{\Lambda\theta_1} \tag{2.28}$$

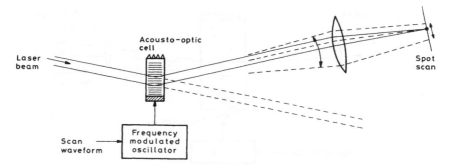

Figure 2.15 Deflection and scanning of a laser beam by an acousto-optic cell.

where θ_1 is the first-order or Bragg diffraction angle. Since the diffraction angle varies but the laser beam and acousto-optic cell are fixed, the Bragg condition cannot always be satisfied and we cannot expect to cover too wide a range of angular deviation without considerable variation in diffracted intensity unless a low value of Q is chosen with consequent loss in overall efficiency. Consequently a compromise has to be made between efficiency at the Bragg angle and tolerance to deviations from that angle. A Q value of around 8 enables the diffraction efficiency to be over 50% for the range $\theta_R = \frac{1}{2}\theta_1$.

Schemes for decreasing sensitivity to the Bragg condition have been devised, for example by varying the direction of the acoustic wave with frequency using phased arrays of transducers. Advantage can also be taken of special Bragg conditions in some anisotropic crystals (Young and Yao 1981).

A high value of N is achieved by using a short acoustic wavelength and a large diameter d of the laser beam. However, d is limited by practical crystal dimensions and acoustic absorption. Increasing d and/or decreasing the acoustic velocity also has the effect of decreasing the random access time $\tau_A = d/v$, often an important quantity. Since

$$\theta_R/\theta_1 = f_R/f \tag{2.29}$$

where f_R is the range of modulation of the excitation frequency f required to produce the full deflection range, we have (using equation (2.28))

$$N = \tau_A f_R. \tag{2.30}$$

A value of N of around 500 is available from a typical device with a frequency modulation range of about 500 MHz and a random access time of 1 μs.

Lithium niobate is the most commonly used material for fast deflectors in the visible range on account of its low acoustic absorption and availability. Although having a relatively high acoustic absorption, tellurium dioxide crystals are widely used in slower acousto-optic deflectors using frequencies below 100 MHz. These take advantage of the existence of shear waves with an exceptionally low velocity (617 m s^{-1}) and the high figure of merit of the material. The resulting short wavelength gives relatively high deflection angles, and enables a large number of waves to be accommodated within a reasonably sized crystal and thus a large value of N obtained. This is the only type of acousto-optic deflector from which values of N in excess of 2000 have been obtained.

Deflection in two orthogonal directions (X, Y) is normally accomplished by two simple cells in series. It is possible, however, to obtain simultaneous XY deflection from a single cell in which two ultrasonic waves are generated by crossing the laser beam in orthogonal directions (La Macchia and Coquin 1971).

2.5.3 Optical frequency shifting

Since beams diffracted from Bragg cells are frequency shifted (see §2.1.2), these cells have proved popular for supplying beams with off-set frequencies for use in optical heterodyning techniques. They are much used in laser Doppler velocimetry for directional discrimination and for matching signal frequencies to signal processor ranges. Their use in laser interferometry is described in §3.8.1. For these applications, Bragg cells are simply driven continuously at a constant ultrasonic frequency. With monochromatic laser radiation, the Bragg condition can always be satisfied and efficiencies over 80% obtained. Bragg cells can give much greater frequency shifts than mechanical devices and require less power to drive than electro-optic devices. It is also easier to obtain a pure shifted beam. Bragg cells do however give a beam deviation and, if inserted into an aligned optical system, a correcting prism will be required. The directly transmitted beam and other diffraction orders will always be present to some extent and may have to be removed by stops in the optical system. Since the separation of orders is usually quite small, a long beam path or extra optical components may be necessary to remove them. To obtain high efficiencies and good separation of orders, ultrasonic frequencies in excess of 30 MHz are necessary, giving frequency shifts too large for a number of applications. In these cases, it may be necessary to use an auxiliary electronic frequency down shift or more satisfactorily two Bragg cells in opposition, driven at frequencies whose difference is equal to the low frequency shift required.

2.5.4 Laser intra-cavity devices

Since acousto-optic cells may be operated as fast beam modulators, they may be used as 'Q' switches in laser cavities (see §1.2.4). However, since the response time is limited by the time taken for the acoustic wave to cross the beam, it is necessary to bring the beam to a focus in the acousto-optic cell, as described in §2.5.1, if a fast switching time is required.

An interesting variation is the use of an acousto-optic cell as an intra-cavity

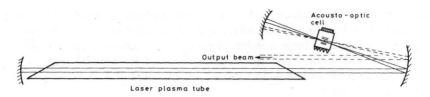

Figure 2.16   Cavity dumping of a gas laser using an intra-cavity acousto-optic deflector (Maydan 1970).

deflector (Maydan 1970). In this application, illustrated in figure 2.16, a gas laser is operated with high-efficiency fully reflecting mirrors. Unless the acousto-optic cell is activated, there is no output and optical oscillation builds up to a high level inside the cavity. Switching on the cell then deflects the energy out of the cavity for external use. Peak powers may be up to 50 times those available in continuous wave oscillation, with pulse repetition rates up to 1 MHz.

This mode of operation ('cavity dumping') differs from Q switching in that the energy is stored optically in the cavity rather than as excitation of the energy levels of the lasing medium. Note that, in the arrangement illustrated in figure 2.16, the output consists of two diffracted beams superimposed. These are frequency shifted in opposite directions and thus the output pulses are modulated at twice the ultrasonic frequency.

Acousto-optic materials for intra-cavity use must have very low optical absorption and be able to stand up to the high optical fields present. Silica is the favoured material in spite of its low figure of merit.

The most widely used application of acousto-optic cells in laser cavities is for mode locking (see §1.2.4). In this technique, the longitudinal modes of a laser are locked in phase so that energy is concentrated in pulses occurring at the mode-spacing frequency of the laser, c/L_c, where L_c is the cavity length; i.e. twice the separation of the mirrors. This is achieved by a Q switch pulsed at this frequency so that only light passing through the switch when it is open each time around is amplified.

The switching may be done by an acoustic-optic modulator cell placed close to one of the laser mirrors in the cavity (figure 2.17) and driven at a frequency $c/2L_c$. In this case, the acousto-optic modulator uses standing waves so that a stationary diffraction grating is generated whose amplitude varies sinusoidally with the pressure fluctuations in the medium, passing through zero twice per cycle. When this happens, no energy is diffracted out of the laser beam and the Q switch is 'on', otherwise diffraction losses quench the amplification of the laser.

The standing-wave pattern is produced by reflection of ultrasonic waves backwards and forwards across the block of acousto-optic medium which may thus be considered as oscillating in a high-order mode. The acousto-optic

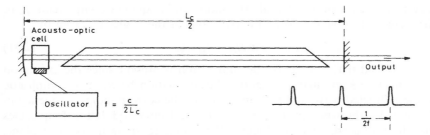

Figure 2.17   Acousto-optic mode locking of a gas laser.

cell must be designed to resonate at the correct frequency as a unit, including the exciting transducer. Exact matching to the mode spacing of the laser may be made by adjustment of the cavity length. Losses due to reflection at the surfaces of the cell may be minimized by anti-reflection coating or by incidence at the Brewster angle.

2.5.5   Spectrum analysis and signal processing

An important field of application of acousto-optic devices is to the analysis and processing of electronic signals. The use of an acousto-optic intermediate stage in signal processing can often give advantages in performance and enable methods of analysis to be used which are difficult to implement by purely electronic means, particularly at high frequencies. This field has been reviewed by a number of authors (Korpel 1976, Flores and Hecht 1977).

Since sinusoidal variations of the refractive index of the medium produce diffracted beams at specific angles, the acousto-optic effect can be used to separate out different wavelength components corresponding to frequency components in the exciting waveform, thus performing a spectral analysis or Fourier transformation. The arrangement is illustrated in figure 2.18. The amplified signal voltage is applied to a wide-band ultrasonic transducer which produces a travelling wave in the acousto-optic medium. At a given time, the pressure fluctuations in the wave are a spatial representation of the signal

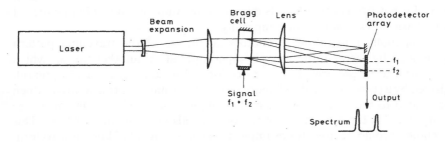

Figure 2.18   Spectrum analysis of a signal using a Bragg cell.

waveform over a short interval. The individual frequency components f_m in the signal thus give rise to diffracted beams at the angles θ_m given by

$$\sin \theta_m = f_m \lambda / v. \tag{2.31}$$

It is assumed that only first-order diffraction is present since the excitation level must be kept small to ensure a linear relation between the amplitude of the diffracted wave and the amplitude of the corresponding frequency in the signal; otherwise spurious harmonics and intermodulation frequencies may be produced. The lens focuses the beams to spots in its focal plane, producing a pattern representing the power spectrum of the signal. This may be conveniently read out by an array of photodetectors, as shown in figure 2.18.

The resolution of the spectral analysis is inversely related to the time window of the optical Fourier transformation. This is directly related to the illumination profile of the acoustic beam by the laser beam through the acoustic velocity.

It should be noted that the diffracted beams are shifted by the frequencies of the components that give rise to them and also contain phase information on these components that may be useful in further signal processing. If a beam is optically heterodyned with the original light frequency, the corresponding original component may be recovered in filtered form from the remainder of the signal.

Another field of application is to the correlation of signals (Sprague 1977); i.e. the evaluation of the function

$$\mathrm{Cor}(t) = \int S_1(t')S_2(t' + t)\,\mathrm{d}t' \tag{2.32}$$

where S_1 and S_2 are two time-dependent signals. These applications do not specifically rely on diffraction or the coherence properties of the light source. though diffraction is used as a way of converting the phase modulation produced in the acousto-optic cell to modulation of the light intensity so that light transmission may be used as a basis for integration. The acousto-optic cell thus provides a convenient spatial representation of the signal, a moving 'mask' whose transmission varies in accordance with the signal waveform. Correlation integrals may then be evaluated optically in either of two basic ways, spatially or temporally.

In spatially integrating correlators, an image of the travelling pattern generated by the acoustic wave is superimposed on another 'mask' which may be a static preformed one, or one formed by another acousto-optic cell driven by a (time-reversed) reference signal. The total light transmitted, which is related to the correlation integral, is observed by a photodetector whose output yields the correlation of the two signals as a function of time. This technique could be used, for example, to resolve a signal of known waveform from background noise.

In time-integrating correlators, the light source itself is modulated by one of the signals and an image of the acousto-optic cell (driven by the other signal) formed on an array of photodetectors, with the zeroth-order diffraction stopped. The total light received by each detector thus depends on the product of the two signals with a different delay time. The correlation function is thus formed as a pattern in the integrated response of the photodetector array.

The main advantage of acousto-optic cells in signal processing of this type is that frequencies up to 1 GHz or more can be accommodated giving a very large bandwidth capability with a rapid processing time that is not easy to achieve by other methods.

2.6 THE INTERACTION OF LIGHT WITH SURFACE WAVES

In view of the importance of electronic devices based on surface acoustic waves (SAW), a number of studies of surface-wave phenomena have been made by optical techniques. This field has been reviewed by Lean (1973) and Stegeman (1976). Here we consider techniques based on the diffraction or deflection of laser beams. These require reasonably good optically reflecting surfaces, which are usually possible to arrange when investigating SAW devices. The application of interferometric techniques to surface waves is discussed in §4.4. These do not necessarily require polished surfaces though good reflectivity is an advantage.

2.6.1 The diffraction of light by surface waves

The corrugations of a surface of a solid or liquid caused by the passage of a surface acoustic wave evidently form a phase grating which will diffract light reflected from the surface in the same way that a transmitted beam is diffracted by bulk acoustic waves travelling through a transparent medium (see §2.1.2). This is illustrated in figure 2.19. The situation here is, however, simpler since the effective length $\mathscr{L}$ of the diffracting region is virtually zero and hence the Raman–Nath approximation is almost always valid.

The angle of reflection of the nth diffraction order θ_n is given by

$$\sin \theta_n = \sin \theta_1 + n\lambda/\Lambda \qquad (2.33)$$

where θ_1 is the angle of incidence. λ and Λ are as usual the wavelengths of the light and acoustic waves, respectively. Assuming a surface wave of small amplitude a, the amplitude of the phase modulation of the grating (equal to the Raman–Nath parameter; see §2.1.2) is given by

$$v = \frac{2\pi a}{\lambda} (\cos \theta_1 + \cos \theta_n) \qquad (2.34)$$

or $v \simeq 4\pi a/\lambda$ for near normal incidence and reflection.

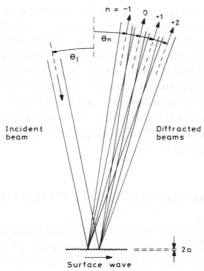

Figure 2.19   The diffraction of a laser beam by a surface wave.

Assuming a sinusoidal waveform and perfect reflectivity and an incident intensity I_1, the intensity I_n in the nth diffraction order is given by equation (2.3). A more general analysis has been given by Lean (1973). Usually surface-wave amplitudes are so small compared with the optical wavelength that only first-order diffraction is significant,

$$I_{\pm 1} = \tfrac{1}{4} I_1 v^2 \simeq I_1 \left(\frac{2\pi a}{\lambda} \right)^2 . \tag{2.35}$$

Thus the intensity diffracted into the first order is proportional to the square of the displacement of the surface wave.

When the amplitude is large enough for higher orders to be significant non-linear acoustic effects are likely to occur. The harmonics thus generated also contribute to the higher diffraction orders and modify their intensity distribution, and asymmetric distributions are possible (Neighbors and Mayer 1971). These effects may be used to study non-linear phenomena in surface acoustic wave propagation.

A simple application of diffraction is the measurement of the velocity of surface waves. An experimental arrangement for this (Auth and Mayer 1967) is shown in figure 2.20. The light beam from a helium–neon laser is expanded and focused to a small spot on an image plane in front of a photodetector after reflection from the surface of the sample on which waves are generated. This must of course be a good mirror surface. The diffraction pattern is scanned by a slit in the image plane in front of the detector. The surface-wave velocity v_s is given in terms of the angular separation of the first diffraction

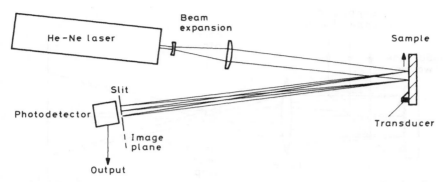

Figure 2.20 The measurement of surface wavelength and hence velocity by the diffraction of a laser beam (Auth and Mayer 1967).

orders from equation (2.33). Thus if f is the excitation frequency,

$$\frac{2\lambda f}{v_s} = \sin \theta_{+1} - \sin \theta_{-1}. \tag{2.36}$$

If in addition the diffraction angles are small

$$\frac{2\lambda f}{v_s} = (\theta_{+1} - \theta_{-1})\cos \theta_1. \tag{2.37}$$

An accuracy of 0.5% or better is possible (Stegeman 1976). By the measurement of the relative intensities of the first and zeroth diffraction orders for different parts of the surface, studies of the propagation and attenuation of surface waves may be made.

Full-field visualization techniques based on diffraction have been used by a number of workers (Saffer *et al* 1969, Zuliani *et al* 1973). The principle, illustrated in figure 2.21, is based on the illumination of the whole area of the surface on which it is desired to detect waves, by an expanded laser beam. The surface is imaged by a lens or lens system onto a viewing screen or photographic plate. To prevent light reaching the screen in the absence of surface waves, a stop is placed at the focus of the lens to intercept the zeroth-order diffraction, i.e. the direct beam. Only diffracted light is transmitted. The regions of the surface giving rise to diffraction show up as bright areas on the image plane giving a 'map' of the distribution of surface waves. The intensity in the image is proportional to the power in the waves. This arrangement is clearly akin to the Schlieren technique for visualizing refractive index gradients due to shock waves in fluids.

Note that the ± 1 diffraction orders are Doppler shifted by the acoustic frequency f in opposite directions and if they are both passed as shown in figure 2.21, interference in the image plane has a beat frequency $2f$. This results in interference fringes moving with the velocity of the surface waves

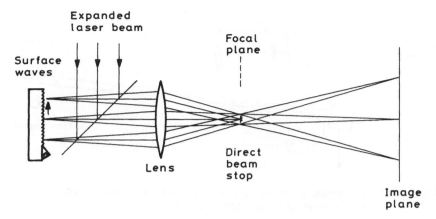

Figure 2.21 Technique for the visualization of surface waves based on the diffraction of a laser beam.

(adjusted for the optical magnification of the system) but with a spacing corresponding to half an acoustic wavelength.

2.6.2 The knife-edge technique

The techniques described in the last section used the diffraction of laser beams of width covering many acoustic wavelengths where the diffraction orders are consequently clearly separated. In another class of techniques for investigating surface waves, the laser beam is focused to a small spot on the surface of diameter comparable with or less than the acoustic wavelength. In this situation, the divergence of the diffracted beams is of the order of or greater than the diffraction angle. The diffraction orders then overlap. This is illustrated in figure 2.22. Because the diffracted beams are frequency shifted, interference in the overlap regions results in intensity fluctuations by which the acoustic waves may be detected.

For simplicity we consider here the case of near-normal incidence and diffraction from a perfectly reflecting surface. A detailed analysis of the general case has been given by Lean (1973). The displacement amplitudes of the ultrasonic waves are usually small compared with their wavelength and this will be assumed. Only the first-order diffraction is then significant and this is small compared with the directly reflected (zeroth-order) beam; i.e. $v \ll 1$. We consider the diverging beams overlapping in the far-field region at a large distance z from the surface as shown in figure 2.22. Intensity fluctuations at the acoustic frequency arise from the overlap of the $+1$ and -1 orders with the zeroth order (direct reflection) in the regions O_+ and O_- shown. Because the total reflected light must be constant, the contributions from these two regions are equal and opposite. Thus to obtain the best possible

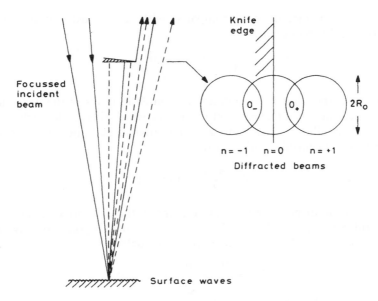

Figure 2.22 Analysis of the knife-edge technique for the observation of surface acoustic waves in terms of the overlapping of diffraction orders.

signal, one half of the field should be cut off by a central stop or 'knife edge' as shown.

To calculate the beat signal, we use the result (derived in §3.2.1) that the amplitude of the intensity fluctuation I_F of interference beats between beams of intensity I_1 and I_2 is given by

$$I_F = 2(I_1 I_2)^{1/2}. \tag{2.38}$$

Integrating over the overlapping beams 0 and $+1$ over the unobstructed area and subtracting the contribution from the 0 and -1 overlap in the same area, we obtain the amplitude of the fluctuation in power reaching the detector:

$$\Delta W_F = \frac{8\sqrt{\pi}aW_0}{\lambda}\, e^{-\beta^2} \int_0^\beta e^{-x^2} dx \tag{2.39}$$

where W_0 is the laser power. The diffracted beams are assumed to be Gaussian (see §1.3.4) having a cross section of radius R_0 at distance z and radius r_0 at the waist which is assumed to be at the surface (both to $1/e^2$ intensity points).

The parameter β is here given by

$$\beta = \frac{z\lambda}{\Lambda R_0 \sqrt{2}} = \frac{\pi r_0}{\Lambda \sqrt{2}} \tag{2.40}$$

using equation (1.14).

For waist diameters small compared with the acoustic wavelength Λ, equation (2.39) reduces to

$$\Delta W_{\rm F} = \frac{4\pi\sqrt{2\pi}aW_0 r_0}{\lambda\Lambda}. \tag{2.41}$$

In this case, the effect may be described simply in terms of the deflection of a beam reflected at a plane reflecting surface which tilts as the surface wave passes. This is illustrated in figure 2.23. The angular deflection of the reflected beam from its central position at time t is

$$\zeta = \frac{4\pi a}{\Lambda}\cos(ft + \phi) \tag{2.42}$$

where a is the displacement amplitude and f the frequency of the surface wave. Integrating over the light beam not intercepted by the centrally placed knife edge, we see that the light reaching the detector is

$$W_{\rm D} = \frac{W_0}{2} + \frac{W_0}{\sqrt{\pi}}\int_0^{\beta'} e^{-x^2}dx \tag{2.43}$$

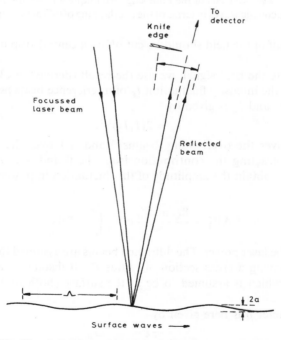

Figure 2.23 The deflection of a finely focused laser beam by a surface wave. A knife edge placed centrally produces a light output modulated at the ultrasonic frequency.

where

$$\beta' = \frac{\zeta z \sqrt{2}}{R_0} = \frac{\pi \zeta r_0 \sqrt{2}}{\lambda} \qquad (2.44)$$

using equation (1.14) with $\lambda z \gg r_0^2$. For small displacement amplitudes, this reduces, as we expect, to a sinusoidal variation having an amplitude given by equation (2.41).

Both of the above methods of analysis are based on the treatment of the far optical field, i.e.

$$z \gg \pi r_0^2 / \lambda \qquad (2.45)$$

and do not predict any variation of the phase of the observed signal with the position of the knife edge in the reflected beam. However, if the above condition (2.45) is not obeyed, this effect has been observed and can be explained by the Fresnel treatment of diffraction in the near optical field (Aharoni et al 1985).

2.6.3 The sensitivity of the knife-edge technique

To determine the theoretical minimum SAW amplitude that can be detected, we return to equation (2.39) which is valid for small displacement amplitudes, for any spot size. The maximum signal is found to be obtained for $\beta = 0.620$ which, using equation (2.40), corresponds to

$$r_0 / \Lambda = 0.279. \qquad (2.46)$$

The maximum signal is then

$$W_F = 0.422 \left(\frac{4\pi a}{\lambda} \right) W_0. \qquad (2.47)$$

We note, however, that the sensitivity can be increased by using a split photodiode as is done by some workers (Engan 1978). Subtraction of the outputs of two sections of the diode then gives a signal double that from a simple centrally placed knife edge. For this calculation of the ultimate sensitivity, we will assume that advantage is taken of this improvement.

The limit of sensitivity is ultimately determined by the random photon noise of the light beam (see §1.2.5). From equation (1.8), we see that the photon noise associated with the beam power W_0 is equivalent to a mean square power fluctuation of

$$\overline{W_N^2} = \frac{2 h v \Delta f}{\eta} W_0 \qquad (2.48)$$

when h is Planck's constant, v the light frequency, η the quantum efficiency and Δf the bandwidth. Hence, using equation (2.47) (and allowing the extra factor of 2), the maximum mean square signal to noise ratio becomes

$$\left(\frac{S}{N}\right)^2 = 0.356\left(\frac{4\pi a}{\lambda}\right)^2 \frac{W_0\eta}{2h\nu\Delta f}. \tag{2.49}$$

The RMS noise equivalent displacement is thus given by

$$\delta x_N = \frac{\delta a_N}{\sqrt{2}} = 1.18 \frac{\lambda}{4\pi}\left(\frac{2h\nu\Delta f}{\eta W_0}\right)^{1/2} \tag{2.50}$$

where δa_N is the amplitude of sinusoidal displacement of the surface for unity RMS signal to noise ratio. For example, for a 2 mW helium–neon laser with a silicon diode detector ($\eta = 0.75$) and our usual 10 MHz bandwidth, the theoretical minimum detectable displacement amplitude $\delta a_N = 5.4$ pm. However, this bandwidth is much larger than would normally be required for surface-wave measurements.

The above results have been calculated for a beam of Gaussian cross section but are substantially the same as those deduced by Whitman and Korpel (1969) for a beam of uniform intensity. Comparison of equation (2.50) with equation (3.13) shows that the knife-edge and interferometric techniques have virtually the same maximum theoretical sensitivity under the conditions assumed here. For interferometry, however, there is no dependence on the beam diameter at the surface.

2.6.4 Scanned laser probes

The knife-edge technique forms the basis of methods that have been used by a number of workers for the visualization of surface waves by optical scanning. The principle, due to Adler *et al* (1968), is illustrated in figure 2.24. The laser beam is scanned across the surface by means of a deflecting mirror (or mirrors), but to ensure that the beam is always incident at the same

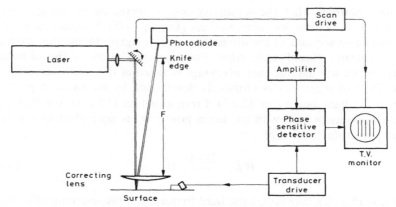

Figure 2.24 A scanned laser probe for visualizing surface waves.

angle, the deflecting mirror is placed in the focal plane of a correcting lens as shown. The angle of reflection is then constant and the reflected beam always intercepts the knife edge when this is also in the focal plane of the lens. The fluctuating signal from the detector may be amplified and displayed on the television screen. By mixing the signal with a reference from the transducer drive, in a phase-sensitive detector, an output dependent on the phase of the oscillation relative to the reference is obtained at each point on the surface. Displaying this on the TV screen then gives a stationary pattern of fringes representing the surface waves.

It is interesting to note that, if the scan speed is high as is possible with a normal TV scan, there may be significant 'Doppler' shift of the signal due to the velocity of the probe spot over the surface. This is up or down depending on whether the scan direction is with or against the direction of propagation of the surface waves, and may be used to differentiate between directions of wave travel (Adler *et al* 1968).

The direction of wave travel may also be determined by the use of a frequency-shifted reference signal, when the fringes will move with or against the direction of wave travel depending on whether the frequency shift is down or up. For standing waves, the fringes will simply modulate in amplitude with alternations of sign, whereas for partially standing waves a combination of travel and modulation will be observed. To obtain quantitative measurements in this situation, it is desirable to have a two-phase synchronous detection system as used by Engan (1978) who also employed a system of electro-optic modulation of the laser power at the acoustic excitation frequency due to Parker (1974). This provides a stroboscopic effect which enables the electronic measurement circuits to operate at the lower shift frequency rather than at the actual ultrasonic frequency.

Alers *et al* (1973) have demonstrated the use of a scanned laser visualization technique for showing up surface and subsurface defects by the scattering of Rayleigh and Lamb waves on glass and various metal surfaces. Using Lamb waves on a thin steel sheet it was possible to detect defects on the surface opposite to the transducer and optical probe. This could be useful if this surface was otherwise inaccessible. The technique has the ability to inspect an area 100 mm or so across in less than a minute but the scanning area is limited by the diameter of the correcting lens available and the need for a flat surface with an optically reflecting finish.

The technique can also be used for studying surface waves on solid–liquid interfaces. Figure 2.25 shows the arrangement used by Davids (1980) for studying the attenuation of waves on a quartz surface due to leakage into a water medium. This illustrates the use of a water tank, the return of the beam along the same path as the incident beam and the use of separate mirrors for X and Y scanning. Parallel scanning (see also §4.2.2) is achieved by placing the Y scan mirror at the focus of the final lens L_3. The previous lens L_2 images the X scan mirror to the same point and thus ensures a

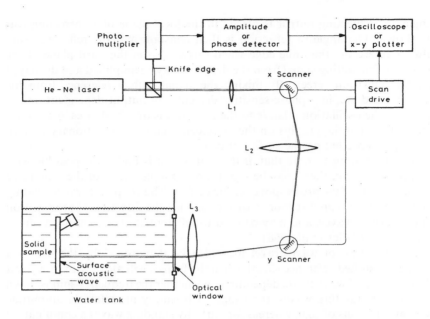

Figure 2.25   An arrangement for studying waves on a solid–liquid interface by a scanned laser beam using the knife-edge technique (Davids 1980).

parallel X scan also. Very similar optical and scanning arrangements are used with interferometric systems (§§4.1.1 and 4.2.2). The knife-edge technique is convenient for the observation of waves on polished surfaces as it is often found to be less sensitive to alignment than interferometric systems and if the beam width is close to its optimum value of about $\frac{1}{2}\Lambda$, it has similar sensitivity (see §2.6.3).

Sensitivity to bulk waves may be achieved by using inclined incidence to a surface. A dynamic ripple is then produced on the surface of a wavelength comparable with that of the ultrasound and suitable for effective use of the knife-edge technique. A commercial scanning laser acoustic microscope (SLAM) based on this principle has been developed by Kessler and Yuhas (1979). To achieve the horizontal scan rates required for the standard television output, an acousto-optic deflector (§2.5.2) was used, the vertical deflection being obtained by the linear movement of a mirror. In common with other methods of acoustic microscopy, SLAM produces images giving different information from that obtained by optical microscopy. It has many applications in biomedical research and in non-destructive testing, where subsurface defects and delaminations may be detected. For example, Kupperman *et al* (1980) have demonstrated the use of the technique for the detection of cracks and natural flaws in specimens of SiC and Si_3N_4. The ultrasonic frequency used was 100 MHz, which was varied a little to minimize

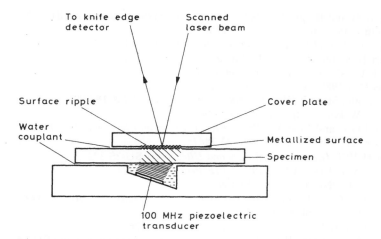

Figure 2.26 Arrangement for visualizing bulk ultrasonic waves by the laser scanning of a surface using a knife-edge technique (Roth *et al* 1986).

(acoustic) speckle effects. The high ultrasonic velocity and low absorption in these ceramic materials makes them very suitable for study by this method.

The scheme as used by Roth *et al* (1986) for the detection of voids in ceramic specimens is illustrated in figure 2.26. 100 MHz ultrasonic waves are incident on a specimen plate at an angle of about 10°. Shear waves produced on refraction into the plate are inclined at around 45° to the surface and give rise to a short-wavelength ripple on reaching the far surface. As the surface finish of the specimens is not usually smooth enough for good quality optical reflection, a plastic cover slip is placed on top. This is gold plated on the underside to provide a good reflecting surface for the knife-edge technique. High resolution can be achieved with good focusing of the laser spot and voids down to 100 μm in diameter can be detected.

The knife-edge technique is also an option in the SLAM technique of Rudd *et al* (1987) but in this, sensitivity to displacement rather than tilt is usually required, and interferometric detection is generally more satisfactory. This will be discussed in §4.2.3.

REFERENCES

Adler R A, Korpel A and Desmares P 1968 *IEEE Trans. Sonics and Ultrasonics* **SU-15** 157
Aharoni A, Gover A and Jassby K M 1985 *Appl. Optics* **24** 3018
Alers G, Tennison M A, Thompson R B and Tittman B R 1973 *Ultrasonics* **11** 174
Auth D C and Mayer W G 1967 *J. Appl. Phys.* **38** 5138
Benedek G and Greytak T 1965 *Proc. IEEE* **53** 1623

Bragg W H and Bragg W L 1915 *X-rays and Crystal Structure* (Bell) ch 2
Brillouin L 1922 *Ann. Phys., Paris* **17** 88
Chang I C 1976 *IEEE Trans. Sonics and Ultrasonics* **SU-23** 2
Colbert H M and Zorkel K L 1963 *J. Acoust. Soc. Am.* **35** 359
Cook B D 1976 *J. Acoust. Soc. Am.* **60** 95
Davids D A 1980 *Rev. Sci. Instrum.* **51** 1059
Debye P and Sears F W 1932 *Proc. Natl Acad. Sci. USA* **18** 409
Dixon R W 1967 *J. Appl. Phys.* **38** 5149
Durrand G C and Pine A S 1968 *IEEE J. Quantum Electronics* **QE-4** 523
Eisenberg H 1965 *J. Chem. Phys.* **43** 3887
Engan H 1978 *IEEE Trans. Sonics and Ultrasonics* **SU-25** 372
Flores L N and Hecht D L 1977 *SPIE Proc.* **118** 182
Gordon E I 1966 *Proc. IEEE* **54** 1391
Haron H E, Cook B D and Stewart H F 1975 *J. Acoust. Soc. Am.* **57** 1436
Ingenito F and Cook B D 1969 *J. Acoust. Soc. Am.* **45** 572
Kessler L W and Yuhas D E 1979 *Proc. IEEE* **67** 526
Klein W R and Cook B D 1967 *IEEE Trans. Sonics and Ultrasonics* **SU-14** 123
Korpel A 1976 *Optical Information Processing* ed Yu E Nesterikhin and G W Stokes
    (New York: Plenum) p 171
—— 1981 *Proc. IEEE* **69** 48
—— 1988 *Acousto-optics* (New York: Marcel Dekker)
Korpel A, Adler R, Desmares P and Watson W 1966 *Appl. Optics* **5** 1667
Kupperman D S, Pahis L, Yuhas D and McGraw T E 1980 *Am. Ceramic Soc. Bull.*
    59 814
La Macchia J T and Coquin G A 1971 *Proc. IEEE* **59** 304
Lean E G H 1973 *Progress in Optics XI* ed E Wolf (Amsterdam: North-Holland) p 123
Lorentz L 1880 *Am. Phys., Lpz* **11** 70
Lucas R and Biquard 1932 *J. Phys. Rad.* **3** 464
Maydan D 1970 *J. Appl. Phys.* **41** 1552
Nagai S 1985 *Ultrasonics* **23** 77
Nagai S and Iizuka K 1982 *Japan J. Appl. Phys.* **21** L505
Neighbors III, T H and Mayer W G 1971 *J. Appl. Phys.* **42** 3670
Parker T E 1974 *Proc. IEEE Ultrasonic Symp.* 365
Quate C F, Wilkinson C D W and Winslow D K 1965 *Proc. IEEE* **53** 1604
Raman C V and Nath N S N 1935 *Proc. Ind. Acad. Sci.* A **2** 406
Raman V and Venkateraman K S 1939 *Proc. R. Soc.* **171** 137
Reibold R 1976/77 *Acustica* **36** 214
Reibold R, Molkenstruck W and Swamy K M 1979 *Acustica* **43** 253
Reisler E and Eisenberg H 1965 *J. Chem. Phys.* **43** 3875
Riley W A 1980 *J. Acoust. Soc. Am.* **67** 1386
Roth D J, Klima S J, Kiser J D and Baaklini G Y 1986 *Mater. Eval.* **44** 762
Rudd E P, Mueller R K, Robbins W P, Skaar T, Soumekh B and Zhou Z Q 1987
    Rev. Sci. Instrum. **58** 45
Saffer B H, Close D H and Pedinoff M E 1969 *Appl. Phys. Lett.* **15** 339
Sapriel J 1979 *Acousto-optics* (New York: Wiley)
Sprague R A 1977 *Opt. Eng.* **16** 467
Stegeman G I 1976 *IEEE Trans. Sonics and Ultrasonics* **SU-23** 33
Ushida N and Niizeki N 1973 *Proc. IEEE* **61** 1073

Waxler R M and Weir C E 1963 *J. Res. NBS (Phys. & Chem.)* **67A** 163
Waxler R M, Weir C E and Schamp H W Jr 1964 *J. Res. NBS (Phys. & Chem.)* **68A** 489
Whitman R L and Korpel A 1969 *Appl. Optics* **8** 1567
Young E H and Yao S 1981 *Proc. IEEE* **69** 54
Zuliani M, Ristic V M, Vella P and Stegeman G 1973 *J. Appl. Phys.* **44** 2964

3 Laser Interferometry

3.1 PRINCIPLES OF LASER INTERFEROMETRY

3.1.1 Introduction

The techniques to be considered in this chapter employ the principle of optical interferometry using light reflected from or scattered by a surface subject to ultrasonic displacement. They depend on the wave nature of light, as do the diffraction techniques described in Chapter 2. Optical interferometry is a very sensitive way of measuring displacement, but to be practicable for general use, it requires a highly monochromatic light source and thus the use of lasers is virtually essential.

Interferometers for the detection of ultrasonic movements of surfaces may be divided into two distinct types. In the first type, light scattered or reflected from a surface is made to interfere with a reference beam, thus giving a measure of optical phase and hence instantaneous surface displacement. The second type of interferometer is designed as a high-resolution optical spectrometer to detect changes in the frequency of the scattered or reflected light. It thus gives an output dependent on the velocity of the surface. The first type is the more widely used and the most practical at lower frequencies and with reflecting surfaces. The second type offers a potentially higher sensitivity with rough surfaces, particularly at high frequencies.

For the detection of ultrasonic waves at a surface, the techniques are admittedly rather insensitive compared with piezoelectric devices. They do, however, offer a number of advantages.

(a) They are non-contacting and thus do not disturb the ultrasonic field. The point of measurement may be quickly moved and there are no fundamental restrictions on surface temperature.

(b) High spatial resolution may be obtained without reducing sensitivity. The measurements may be localized over a few micrometres if necessary.

(c) As the measurements may be directly related to the wavelength of the light, no other calibration is required.

(d) They can have a flat broadband frequency response, something difficult to achieve with piezoelectric transducers, particularly at high frequencies.

3.1.2 The Michelson interferometer

Interferometers of the first type have various designs but may usually be described as variations of the classic Michelson interferometer illustrated in

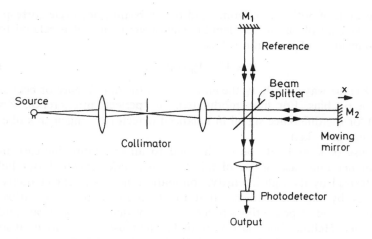

Figure 3.1 The basic Michelson interferometer.

figure 3.1. Light from a monochromatic source is collimated to form a parallel beam which is divided by a beam splitter, part going to a static reference mirror M_1 and part to M_2 whose displacement is required. The beams are reflected back to the beam splitter where they are recombined, a fraction of each going to the observer or detector unit. The relative phase of the signal and reference beams at the detector depends on the difference between the optical paths. If the difference is an integral number of wavelengths, the beams are in phase and constructive interference occurs. If the difference is an integer $+\frac{1}{2}$, the interference is destructive and the light intensity is a minimum. Thus as the mirror M_2 moves, the detector output varies sinusoidally, as shown in figure 3.2. Since the optical path difference must

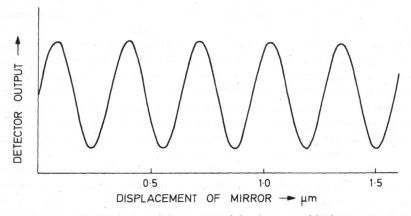

Figure 3.2 The variation of the output of the detector with the movement of a mirror in a Michelson interferometer (light wavelength = 633 nm).

take account of both the outgoing and return beams, one cycle corresponds to a movement of half a wavelength. The detector output is related to the displacement, x, of the surface thus:

$$V = \bar{V} + V_0 \sin(4\pi x/\lambda) \qquad (3.1)$$

where λ is the wavelength, V_0 the amplitude of the interference or beat signal and $\bar{V}$ a DC bias. For fully modulated, i.e. 'perfect' interference, $\bar{V} = V_0$. ($\bar{V} = 0$ for the difference signal in balanced detector arrangements like that shown in figure 3.6.)

The use of laser light sources has led to an enormous increase in the performance and ease of use of this type of interferometer. Laser light is characterized by its high intensity, collimation and monochromaticity and it is only by the use of lasers that the high signal to noise ratios and submicrosecond response times necessary for the detection of ultrasound can be achieved. Helium–neon lasers are by far the most commonly used source for these experiments, being very practical, convenient and inexpensive and having excellent coherence properties and a very well defined wavelength.

3.1.3   Coherence considerations

The concepts of spatial and temporal coherence of the light source have been introduced in §1.3.2. They are very important in interferometry. Non-laser light is invariably derived from an extended source and must be severely collimated to obtain a sufficiently parallel beam. Lasers, however, can provide perfectly parallel beams which are equivalent to those derived from idealized point sources. No further collimation is required and thus laser light may be used very efficiently.

Most non-laser sources (and some lasers) have an appreciable spread of optical frequency or wavelength. When such a source is used in interferometry, the contributions to the interference from the various wavelengths present will evidently get out of step after a number of cycles, resulting in a decrease in the interference signal. The optical path difference that may be accepted without appreciable loss of interference quality, the 'coherence length', may possibly be only 1 mm or so even with a nominally monochromatic source. The path lengths of the signal and reference beams must therefore be equalized to this accuracy. Lasers used in interferometry have a coherence length of several centimetres or more and thus the matching of optical paths is not at all critical, though some attention must be paid to this unless a single-mode laser is used. This question will be discussed in more detail in §3.5.

Spatial coherence between the signal and reference beams is an important consideration in the satisfactory observation of light interference. To obtain large variations of output from the detector and thus good sensitivity, the light beams interfering must be coincident and parallel to a fairly high accuracy. Consider the superposition of two plane parallel beams of exactly

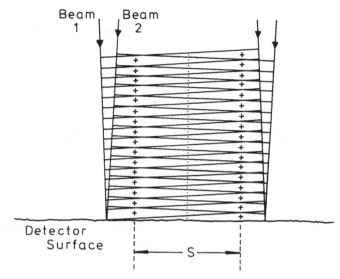

Figure 3.3 The effect of angular misalignment on the interference between two light beams incident on a detector.

the same wavelength but inclined at a small angle α. We suppose that they are incident on a detector as shown in figure 3.3. By consideration of the superposition of the wavefronts, it may be easily shown that there is constructive interference in the planes marked $+ + +$ which are separated by a spacing given by

$$S = \frac{\lambda}{2\sin(\frac{1}{2}\alpha)} \approx \frac{\lambda}{\alpha}. \tag{3.2}$$

Similarly, destructive interference occurs in the planes . . . . As the relative phase of the beams changes, the planes of interference or 'fringes' move transversely and a point on the detector will experience fluctuations of intensity whose phase clearly varies across the beam. If the fringe spacing is smaller than the width of the beams, contributions from different parts will differ greatly in phase and tend to cancel. To obtain full efficiency, it is essential that the width of the beam is no more than half a fringe spacing. Thus beams of diameter d must be parallel to an accuracy of $\lambda/2d$. Typically, for laser beams of diameter 1 mm and wavelength 0.6 μm, an alignment accuracy of 0.3 milliradian or 1 minute of arc is required.

When using an interferometer to measure the displacement of a mirror surface, it is possible by the use of the correct optics and accurate alignment to get the returned signal beam into exact coincidence with the reference beam. This gives perfect interference conditions where the whole of the light available is utilized (except for some unavoidable losses in the optics) and

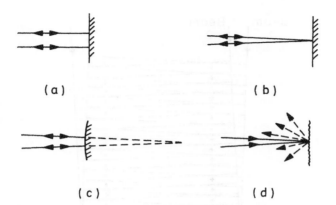

Figure 3.4 The return of a light beam to an interferometer from various surfaces. Fully coherent return may be obtained from the mirror surfaces (*a*), (*b*) and (*c*) but not from the rough surface (*d*).

maximum sensitivity obtained. The incident beam may or may not be focused onto the plane surface shown in figures 3.4(*a*) and (*b*).

Essentially perfect coherence may also be obtained if the reflecting surface is spherical or cylindrical provided it has a mirror finish. In figure 3.4(*c*) we see that if the incident beam is focused towards the centre of curvature of the surface, the light beam is returned along the same path without degradation. From a rough surface, however, the returned light may be spread out over a wide angle. Even if the incident light is focused onto the surface as shown in figure 3.4(*d*), the illuminated region has finite dimensions (see §1.3.4). Consequently, the returned light cannot be considered as coming from a point source. It is then possible to get only a very small fraction of this light into coherence with the reference beam. The potential signal to noise ratio is thus very much reduced.

3.1.4 Practical interferometer arrangements

Many variations of the basic interferometer are possible to maximize the use of the available light, simplify the alignment and reduce sensitivity to noise. Some examples are shown in figures 3.5 – 3.8.

The signal beam is almost invariably focused onto the surface whose displacement is required as shown in figure 3.5. This enables good performance to be achieved with surfaces not having a good optical finish and in any case makes the signal much less sensitive to the tilting of the surface. If the surface has a diffusing or matt finish, focusing is essential for obtaining reasonable sensitivity.

The use of a polarizing beam splitter in conjunction with a polarized laser and quarter-wave plates, as shown in figure 3.6, enables best use to be made

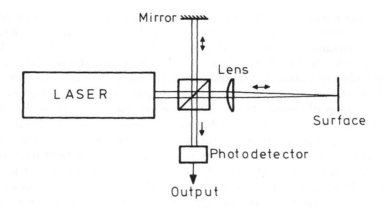

Figure 3.5 Michelson interferometer with beam focusing onto the surface.

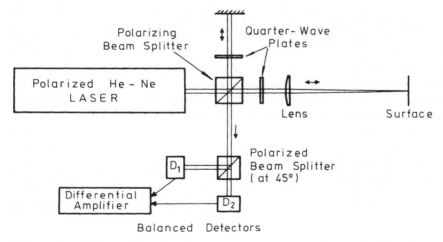

Figure 3.6 A light-efficient interferometer arrangement using a polarizing beam splitter and balanced detectors.

of the available light power. If there are no polarization changes on reflection, the reference and signal beams are returned with their polarizations changed, since they have both passed twice through a quarter-wave plate whose axis is inclined at 45° to the polarization directions. They thus emerge with no appreciable loss in intensity at the side of the beam splitter, no light being returned to the laser. This incidentally avoids feedback to the laser, which can be a source of instability. The ratio of reference beam to signal beam can be readily controlled by changing the orientation of the laser and hence the angle its plane of polarization makes with the axes of the beam splitter.

It will be noted that the signal and reference beams are orthogonally polarized after recombining in the beam splitter, and thus no intensity changes due to interference will be observed at the detector unless a polarizer is inserted to select components of the same polarization. This fact may be turned to advantage by the use of a second polarizing beam splitter whose axes are at 45° to the polarization axes of the beams. Interference signals are then obtained from both outputs of the beam splitter and they are 180° out of phase. Taking the difference of the outputs of the two detectors then gives a 'balanced' output. This is a considerable advantage as the balanced output is insensitive to stray incoherent light and to intensity modulation on the reference beam. In applications where there is a poor light return from the surface, a very high ratio of reference to returned light intensity is often used to improve sensitivity and quite small modulations on the reference beam can be a troublesome source of unwanted signals if the balanced scheme is not used. Modulation levels are typically 1% for the helium–neon lasers due to power supply ripple and intermode beating effects (see §3.5.2) and can be much larger with high-power lasers.

In the above arrangements, the alignment of the mirror returning the reference beam is quite critical and a convenient alternative is the use of a corner cube retroreflector. This is a device containing three reflecting surfaces mutually at right angles; it has the property that any beam incident on it and reflected by the three surfaces in turn is returned along a parallel path. The accuracy of parallelism is dependent on the construction of the device and not on its orientation. This makes alignment of the interferometer easy since the need for accurate setting of the reference beam mirror is avoided. An interferometer arrangement using a corner cube reflector is shown in figure 3.7. Since the outgoing and return beams are not coincident, it is essential to use the focusing technique with this arrangement. For poor light returns, only a small fraction of the light would be required for the reference beam and the beam splitter could transmit most of the light to the surface. The beam splitter could then be simply a plane glass surface. If a 50% beam

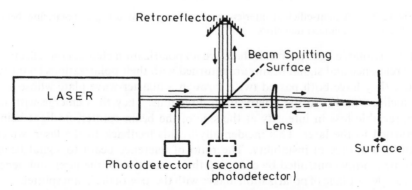

Figure 3.7 An interferometer arrangement using a retroreflector to return the reference beam.

splitter were used a second detector placed as shown would enable a balanced output to be obtained.

It is not normally possible for a retroreflector to return a small laser beam exactly along its incident path because of the difficulty of preserving the plane reflecting surfaces right into the apex of the cube. If the apex is avoided, a displacement is necessarily introduced which means that the light interfering with the reference beam is not the light returned on the same path as the illuminating beam. One effect is to reduce the range of distances over which an interference signal may be obtained, i.e. to reduce the 'depth of focus'. In some circumstances, this may be an advantage as it minimizes interference from light returned from intervening or more distant surfaces whose movement it is not intended to measure. However, with this system the optics has to be large enough to accommodate separate outgoing and return beams. This may be inconvenient especially when scanning over a surface is required.

To enable a coincident outgoing and return beam geometry to be obtained using a retroreflector, an additional beam displacement device is required to compensate for the displacement introduced by the retroreflector. Figure 3.8 shows one method using a parallel-ended calcite prism in conjunction with a polarizing beam splitter arrangement. This prism separates the laser beam into two parallel beams of orthogonal polarization separated by a few millimetres. These are then directed into the reference and signal beam paths by the polarizing beam splitter. The equal and opposite displacement produced by the retroreflector then positions the reference beam correctly for interference with light returned along the path of the outgoing beam. The orientation of neither the prism nor the retroreflector is critical. A balanced detector system may again be used as in the arrangement of figure 3.6.

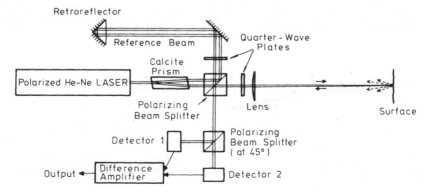

Figure 3.8 An interferometer using a retroreflector and a calcite beam displacement prism to restore the coincidence of incident and reflected beams to and from the surface.

3.1.5   Relationship to the laser Doppler technique

It should be noted that the measurement of the movement of surfaces by laser interferometry may be described as a laser Doppler technique in the reference beam mode. Light back-scattered from the surface is Doppler shifted by the movement and this small shift in frequency is detected by mixing the light with a reference beam. Referring to figure 3.9, we see that the angle of scattering is $(\pi - \zeta)$ where ζ is the angle between the illuminating beam and the direction in which the scattered light is received. Using the well known formula for Doppler shift (Drain 1980) we find that it is given by

$$\delta v = \frac{2u}{\lambda} \cos \beta \sin \tfrac{1}{2}(\pi - \zeta) = \frac{2u}{\lambda} \cos \beta \cos(\tfrac{1}{2}\zeta) \qquad (3.3)$$

where β is the angle the velocity vector **u** makes with the bisector of the illuminating and receiving directions. As ζ in practice is often zero and certainly not more than one or two degrees, $\cos(\tfrac{1}{2}\zeta) = 1$ for all practical purposes and the Doppler shift is

$$\delta v = \frac{2u}{\lambda} \cos \beta. \qquad (3.4)$$

This then is the heterodyne beat frequency obtained from the detector. It shows that there is one cycle of the beat frequency to a half wavelength of movement in the direction of the incident beam, exactly as we found in the earlier derivation in §3.1.2.

There is, however, one important respect in which laser interferometry differs from laser Doppler anemometry of fluids or continuously moving surfaces. It is that the random phase noise is usually negligible. Even if the surface is rough, the same part of it remains constantly illuminated. There is normally not sufficient sideways movement or tilt of the surface to upset the phase and amplitude of the interference signal. This makes signal processing very much simpler since the random phase changes and the

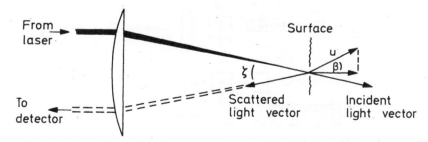

Figure 3.9   Diagram illustrating the Doppler shifting of light returned from a surface moving with a velocity represented by the vector **u**.

possibility of signal 'drop-outs' do not have to be allowed for. It also allows small fractions of an interference cycle to be measured. In these circumstances, we are dealing with very small phase shifts only and the concept of a Doppler frequency shift, though formally valid, is not a natural way of describing the operation of an interferometer.

3.2 THEORY OF INTERFERENCE BETWEEN LIGHT BEAMS

3.2.1 Superposition of electromagnetic fields

To calculate the effect of the superposition of light waves, we must add the individual contributions to the electromagnetic field at any given point. In accordance with normal practice, we consider only the electric field component as it is this that is almost always concerned in absorption or detection processes.

Consider two light beams having the same polarization and frequency, v, whose electric fields at a given point are given by

$$E_1 = E_{10} \cos(2\pi vt + \phi_1) \qquad E_2 = E_{20} \cos(2\pi vt + \phi_2) \qquad (3.5)$$

where E_{10} and E_{20} are the amplitudes of the electric fields and ϕ_1 and ϕ_2 their phases. The resultant is

$$E_T = E_1 + E_2 = (E_{10} \cos \phi_1 + E_{20} \cos \phi_2)\cos 2\pi vt$$
$$- (E_{10} \sin \phi_1 + E_{20} \sin \phi_2)\sin 2\pi vt. \qquad (3.6)$$

The amplitude of the resultant, E_{T0}, is given by summing the squares of the sine and cosine terms. Thus

$$E_{T0}^2 = E_{10}^2 + E_{20}^2 + 2E_{10}E_{20} \cos(\phi_1 - \phi_2). \qquad (3.7)$$

Since the light intensity is proportional to the square of the electric field amplitude (the coefficient of proportionality need not concern us here) we obtain

$$I_T = I_1 + I_2 + 2(I_1 I_2)^{1/2} \cos(\phi_1 - \phi_2) \qquad (3.8)$$

where I_1 and I_2 are the intensities of the individual beams and I_T that of the resultant. The interference term which varies with the phase difference between the beams is thus proportional to the square root of the individual intensities.

Assuming that the signal and reference beams in an interferometer have the same intensity profile and are parallel and coincident, we may deduce the amplitude of the interference term in the output current of the detector:

$$i_1 = 2i_0(W_R W_s)^{1/2}/(W_R + W_s) \qquad (3.9)$$

where W_R and W_S are the powers of the reference and signal beams respectively and i_0 is the mean detector current. For a full light return from a mirror surface, all the laser power is utilized and the sum $(W_R + W_S)$ is equal to the total laser power available. Then the maximum interference signal is obtained when the light is split equally between the signal and reference beams, i.e. $W_R = W_S$. This corresponds to 100% modulation of the detector current.

When there is only a very small light return however, very much less power is required in the reference beam and the highest efficiency is obtained when most of the light is directed towards surface illumination.

3.2.2 Photon noise limitations

Although the interference signal for a given light return approaches a square root dependence on the reference beam intensity, the detectability of the signal does not increase indefinitely because of noise considerations. The fundamental limit is set by photon noise and, because this increases as the square root of the light power on the detector, the signal to noise ratio cannot exceed a certain limiting value.

From equation (1.7) we note that the mean square noise current is given by

$$\langle i_N^2 \rangle = i_0^2 \, \frac{2\,h\nu\Delta f}{\eta(W_R + W_S)} \tag{3.10}$$

where i_0 is the mean current from the detector, η its quantum efficiency, h Planck's constant and Δf the bandwidth of the detection system. From equations (3.9) and (3.10) we obtain

$$\frac{i_1^2}{\langle i_N^2 \rangle} = \frac{2\eta W_R W_S}{h\nu\Delta f(W_R + W_s)}. \tag{3.11}$$

As the reference beam intensity is increased, the ratio of the interference signal amplitude to the RMS photon noise approaches $(2\eta W_S/h\nu\Delta f)^{1/2}$. As $W_R/W_S = 4$, for example, gives a value within 89% of the limit, there is clearly no fundamental advantage in increasing the reference beam to more than a few times the intensity of the signal beam. Practical considerations when optimizing the performance of the detectors and amplifiers may however modify this conclusion. This question is discussed in §3.4.

Since the reference beam level and the amplification used are arbitrary, the output signal level is also arbitrary and it is convenient to use the ultimately more significant quantity, signal level to photon noise ratio, in general discussions of sensitivity and in judging the performance of laser interferometers.

The maximum possible RMS signal to noise ratio for a given laser power W_0, assuming full light return, perfect coherence and no losses, is obtained by putting $W_R = W_S = \frac{1}{2} W_0$ in equation (3.11) and converting to RMS values:

$$\left[\frac{S}{N} \right]_{max} = \left(\frac{\eta W_0}{4h\nu\Delta f} \right)^{1/2}. \tag{3.12}$$

In fact it is not physically possible to divide a laser beam into two equal parts and recombine them on one detector as implied above. However, an equivalent result may be obtained by using two detectors in a balanced arrangement, as illustrated in figures 3.6 and 3.8, in which half of each beam goes to each detector.

Using a 1 mW helium–neon laser with silicon diode detectors, for example, the maximum signal to noise ratio in a bandwidth of 10 MHz is approximately 7700. Assuming that the interferometer is operating at its most sensitive point (which corresponds to zero output for the balanced detector arrangements) the RMS photon noise equivalent displacement δx_N may be deduced with the aid of equation (3.1),

$$\delta x_N = \frac{\lambda}{4\pi} \left(\frac{2h\nu\Delta f}{\eta W_0} \right)^{1/2}. \tag{3.13}$$

For a 2 mW helium–neon laser and a bandwidth of 10 MHz we find $\delta x_N = 3.2$ pm.

3.2.3 Spatial coherence of interfering light beams

The above discussion concerned interference at a point or if the beams were exactly coincident in position, direction and convergence, over an arbitrary detector aperture. If, however, the beams are not fully spatially coherent, the interference signal over a finite aperture will be reduced and the resulting sensitivity will be less than might have been expected. The net result may be evaluated by calculating the interference signal at each element across the aperture and then integrating, taking account of phase.

The result for the case of two parallel beams making a small angle α with each other incident on a circular aperture will be required and will now be derived. We suppose that the beams make an angle $\pm\frac{1}{2}\alpha$ with the normal to the aperture, as illustrated in figure 3.10, though this does not make any fundamental difference to the argument. x and y are cartesian coordinates across the aperture referred to the centre O. By consideration of the wavefront ABCO, we see that the phase of beam 1 at P is ahead of its phase at O by $2\pi x \sin(\frac{1}{2}\alpha)/\lambda$ (radians). Similarly, the phase of beam 2 at P is behind its value at O by the same amount. Thus the phase difference between the beams at P is

$$\phi_1 - \phi_2 = \frac{4\pi x \sin(\frac{1}{2}\alpha)}{\lambda} + \phi_0 \tag{3.14}$$

where ϕ_0 is the phase difference at the centre O. Substituting equation (3.14) in equation (3.8) and integrating over the circular aperture of radius ρ, we find that the total light power received by the detector is

$$W_D = (I_R + I_s)\pi\rho^2 + 2(I_R I_s)^{1/2} \left(\iint_0 \cos\frac{4\pi x \sin(\frac{1}{2}\alpha)}{\lambda} \, dx \, dy \right) \cos\phi_0 \tag{3.15}$$

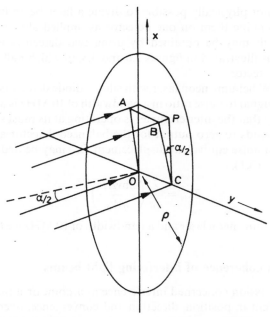

Figure 3.10 Diagram illustrating the calculation of the interference between two parallel beams slightly inclined to each other incident on circular aperture. (For clarity the second beam inclined at $-\alpha/2$ is not shown.)

where I_R and I_s are the intensities of the two beams. Substituting $x = \rho \sin \psi$, the integral may be reduced to

$$\frac{2\rho^2}{z} \int_{-\pi/2}^{+\pi/2} \sin(z \sin \psi) \sin \psi \, \mathrm{d}\psi = \pi\rho^2 \frac{2J_1(z)}{z} \qquad (3.16)$$

where $J_n(z)$ is a Bessel function of the first kind of order n and $z = 4\pi\rho \sin(\tfrac{1}{2}\alpha)/\lambda$. Thus the interference term is reduced by a factor $2J_1(z)/z$ from its value if the beams were parallel. In figure 3.11, this factor is plotted as a function of ρ/ρ^* where $\rho^* = \lambda/[4 \sin(\tfrac{1}{2}\alpha)]$ is a measure of the aperture radius for which coherence is substantially retained. It is assumed that the widths of the beams are great enough to fill the aperture. Since the interference signal is proportional to the expression (3.16) we deduce that higher values may be obtained by continuing to increase the diameter of the aperture. This is not, however, true for the basic signal to noise ratio. We note from equation (1.8) that photon noise varies as the square root of the optical power reaching the detector, and hence (in this calculation) as the square root of the area of the receiving aperture. We therefore find that the fundamental signal to photon noise ratio varies as $|J_1(\pi\rho/\rho^*)|$ and thus has its highest value for $\rho/\rho^* = 0.586$.

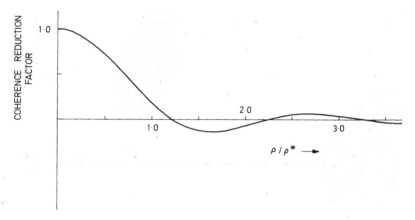

Figure 3.11 The reduction in interference amplitude due to the non-parallelism of beams incident on a circular aperture of radius ρ. ρ^* is defined in the text.

3.3 REFERENCE-BEAM INTERFEROMETRY WITH ROUGH SURFACES

3.3.1 Laser speckle

In applications of laser interferometry, we are frequently concerned with surfaces that are not optically finished and are irregular on the scale of the wavelength of light. Light scattered from different parts of the surface is not phase related and is consequently diffused over a wide angle. In a given direction of scattering, the resultant may be considered as made up of contributions from a large number of independent sources having the same frequency but a random phase relation. In a slightly different direction, the relative phases change and the resultant is different. This gives rise to an irregular angular distribution of the intensity of scattered light, a 'speckle pattern' characteristic of monochromatic illumination. This is illustrated in figure 3.12 where light scattered from a rough surface illuminated by a small spot of laser light is shown projected on a screen. The size of the speckle is inversely proportional to the dimensions of the area illuminated. This effect is analogous to that commonly observed when viewing an object illuminated by laser light through a limited aperture, e.g. the pupil of the eye. The object has a speckled appearance, the angular size of the speckle being inversely proportional to the aperture of observation. At a specific point on the screen, the optical electric field is the resultant of many contributions from independent sources on the illuminated surface. We may represent the addition of these contributions of different phase and amplitude on a

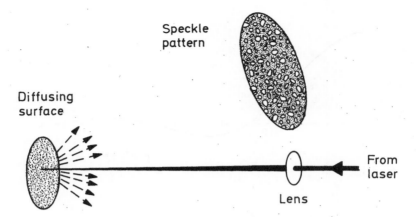

Figure 3.12 The production of a speckle pattern by scattering of laser light from a diffusing surface.

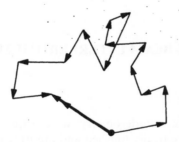

Figure 3.13 Vector diagram illustrating the combination of light waves of random phase and amplitude.

two-dimensional vector diagram as shown in figure 3.13. This is an example of the well known 'random walk' problem and leads to a two-dimensional Gaussian probability distribution of the resultant amplitude vectors. All phases are equally probable and the distribution of intensity has the form

$$P(I) = \frac{1}{I_{M}} \, e^{-I/I_{M}} \tag{3.17}$$

where $P(I)\mathrm{d}I$ is the probability of the intensity having a value between I and $I + \mathrm{d}I$, and I_M is the mean intensity.

3.3.2 Optical properties of scattering surfaces

For an ideal uniform scatterer of diffusing surface, light is multiply scattered or reflected on a fine scale in the surface layer producing a basic intensity distribution of scattered light:

$$W(\theta) = \frac{\Gamma W_i}{\pi} \cos\theta \qquad (3.18)$$

using Lambert's cosine law of diffusion. $W(\theta)$ is the scattered light power per unit solid angle in a direction making an angle θ with the normal to the surface and W_i is the incident power on the surface. Light is also completely depolarized by this type of scattering. The parameter Γ is unity for a perfectly white surface, and less for absorbing grey or black surfaces. This intensity distribution is illustrated in figure 3.14(a). It is independent of the direction of the incident beam. For illumination with laser light, the small-scale random speckle pattern modulation is of course superimposed.

Perhaps more commonly encountered are surfaces giving a basically specular reflection, but which are not completely smooth on the scale of the wavelength of the light. We may consider such a surface as composed of a large number of tiny reflecting facets whose inclinations to the mean plane

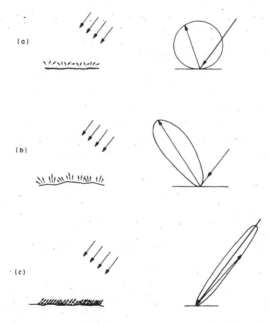

Figure 3.14 Polar diagrams of the variation of light intensity with direction for scattering from the various types of surfaces illustrated on the left. (a) A perfectly diffusing surface, (b) a rough surface with some degree of specular reflection, (c) retroreflecting paint. (The angular distribution is usually much narrower than can be illustrated in the diagram.)

of the surface are small but randomly distributed (figure 3.14(b)). Because the facets are small, the beams reflected from them spread out by diffraction and overlap each other in a random manner producing the usual speckle pattern effect superimposed on the basic scattered intensity distribution, which may be as illustrated in figure 3.14(b). If the facets are large and/or few in number, the speckle pattern may be modified. Any regular surface structure can have a considerable effect and may give rise to well defined diffraction peaks.

A useful surface finish for obtaining good signals in laser interferometry is a retroreflecting surface. With this, light is returned to the source independently of the orientation of the surface. This may be realized by the use of small glass spheres of high refractive index, as illustrated in figure 3.14(c). The returned light is usually spread over a few degrees and the reflections from the randomly distributed spheres overlap, giving the usual speckle pattern effect. Retroreflecting surface finishes are available in adhesive tape or paint form but their usefulness is probably limited to low ultrasonic frequencies as the reflective layer is necessarily rather thick (about 0.1 mm), and at high frequencies the response of the surface to ultrasonic waves could be affected.

3.3.3 Angular correlation of speckle

The angular scale of a speckle pattern from an illuminated surface may be estimated by consideration of the effect of progressively advancing the angles between the vectors in figure 3.13. It will be seen that the phase and intensity become uncorrelated approximately when the phase difference between contributions from opposite sides of the surface changes by one cycle. This occurs for an angular separation of scattering directions of λ/d where d is the diameter of the illuminated spot.

Quantitatively, it may be shown (Goodman 1984) that the correlation function of intensity in the speckle pattern is related to the Fourier transform of the intensity distribution across the illuminated spot. For a Gaussian distribution of intensity, as is usual with laser illumination, the intensity correlation (in directions not too far from back scatter) becomes

$$\overline{I_A I_B} = I_M^2 \{ 1 + \exp[-(\pi r_0 \theta/\lambda)^2] \} \tag{3.19}$$

where I_A and I_B are the intensities in directions A and B separated by an angle θ. I_M is the mean intensity (i.e. averaged over the speckles) and r_0 is the radius of the illuminated spot to $1/e^2$ intensity points. We assume that $\lambda \ll r_0$. The correlation of the fluctuating part of the intensity thus falls to $1/e$ of its maximum value when $\theta = \lambda/\pi r_0$. Thus the 'speckle size' on this definition may be taken to be $2\lambda/\pi r_0$. Referring to equation (1.15), we find that this is also equal to the convergence angle (to $1/e^2$ intensity points) of a Gaussian beam focused to a minimum radius r_0. Thus with the above definitions, a rough surface placed at the waist of a focused Gaussian laser beam gives rise to an angular speckle pattern whose size is equal to the convergence angle of the incident beam.

3.3.4 Calculation of the average interference signal

Since the phase of the scattered light varies randomly from speckle to speckle, the interference signals produced by the speckles also have random phase variation. Thus if N speckles are included in the reception aperture, the average signal will be proportional to $N^{1/2}$. However, the light on the detector varies as N and thus the basic photon noise is also proportional to $N^{1/2}$ (see §1.2.5). There is therefore no basic advantage to be gained by the use of more than one speckle. This puts a fairly severe limitation on the signal to noise ratio available with a given laser power, beam convergence and surface finish. The limiting signal to noise ratio for an idealized interferometer using scattering from a rough surface will now be calculated.

We consider a rough surface at the focus of a laser beam, scattered light from the illuminated spot being returned to the detector through the lens, of focal length F, as shown in figure 3.15. (The illuminating beam is not shown.) A reference beam is also directed to the detector by means of a beam combiner. For the purpose of this calculation, we consider an idealized system in which the loss of light at optical surfaces is negligible. We also assume that there is enough intensity available in the reference beam greatly to exceed that of the scattered light even if the reflection coefficient of the beam combiner is small. (In practice a polarizing beam splitter and/or balanced detectors may be used to enable the reference beam to be used more efficiently but this does not affect the formal result.)

The reference beam is assumed to be aligned in the best possible way so that light returned from the centre, O, of the illuminated spot through the lens is accurately parallel to the reference beam. We suppose that the detector

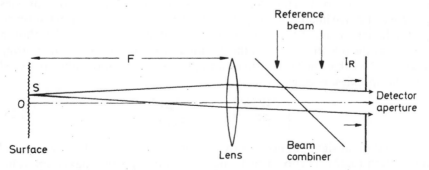

Figure 3.15 Idealized geometry assumed in computing the limiting signal to noise ratio of a laser interferometer using scattered light from a diffusing surface.

area is restricted by a circular aperture of adjustable radius ρ and that a parallel reference beam of uniform intensity I_R is available with sufficient width to cover the maximum aperture of the detector.

Suppose that the illuminating beam has a Gaussian profile as is usual when using a laser source. Let the radius to $1/e^2$ intensity points be r_0. Consider the beam formed from light scattered from an element of the surface dA at a distance r from the centre of the illuminated spot. Due to the Gaussian profile, the intensity at the detector aperture is

$$I_{dA} = \frac{W_{so}}{F^2} \frac{2}{\pi r_0^2} \exp(-2r^2/r_0^2) \, dA \qquad (3.20)$$

where W_{so} is the mean light power per unit solid angle scattered from the whole illuminated spot in the direction of the detector (i.e. averaged over the speckles), and having the same polarization as the reference beam.

The angle between this beam and the reference beam is $\tan^{-1}(r/F)$. Assuming this angle to be small and using equations (3.15) and (3.16), we find that the amplitude of the interference beat signal expressed as a fluctuation of light power falling on the detector is

$$W_{dA} = 2\pi\rho^2 (I_R I_{dA})^{1/2} \frac{2J_1(z)}{z} \qquad (3.21)$$

where $z = 2\pi r\rho/\lambda F$.

The phases of the contributions are random and we may integrate to obtain the average of the mean square signal from the surface as it moves to give all phases of interference,

$$\langle W_S^2 \rangle = \frac{4\rho^2 I_R W_{so} \lambda^2}{r_0^2} \int_0^\infty \{\exp[-2(\rho^*/\pi\rho)^2 z^2]\} \frac{2[J_1(z)]^2}{z} \, dz \quad (3.22)$$

where $\rho^* = \lambda F/2r_0$ is a measure of the radius for maintaining coherence. The integral is plotted as a function of ρ/ρ^* in figure 3.16. The limiting value as $\rho/\rho^* \to \infty$ is unity. Thus, for large ρ, the mean square signal continues to increase approximately in proportion to the area of the aperture. However, the fundamental signal to noise ratio is limited. Assuming that the photon noise from the reference beam is the dominant source of noise, we find, using equation (1.8), that the maximum mean square signal to noise ratio, averaged over the speckles, is given by

$$\left[\frac{\langle W_S^2 \rangle}{\langle W_N^2 \rangle} \right]_{max} = \frac{2\eta W_{so} \lambda^2}{\pi r_0^2 h\nu\Delta f}. \qquad (3.23)$$

The variation in signal to noise with aperture is proportional to the integral in equation (3.22), as shown in figure 3.16. Although the maximum value continues to increase with aperture, it is not normally advantageous to use an aperture radius much in excess of ρ^*, as problems of handling poorly

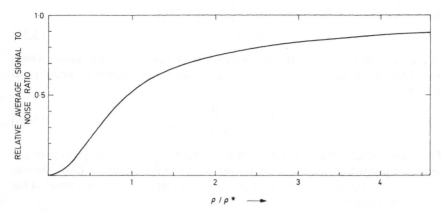

Figure 3.16 Variation of the mean square signal to noise ratio with the radius ρ of a circular detector aperture for interfering light scattered from a diffusing surface with a wide uniform reference beam. $\rho^* = \lambda F/2r_0$ where r_0 is the radius of the focused spot on the surface and F the distance to the surface (see figure 3.15).

modulated signals outweigh any marginal advantage in theoretical signal to noise ratio.

It should be remarked that as the calculated signal to noise ratio is an average over the speckles, a useful increase in signal to noise ratio is possible by the choice of a favourable speckle. Conversely, for an arbitrary aperture, the signal can of course be less than expected and, occasionally, vanishingly small.

We also consider the case where the reference beam has a Gaussian profile and the receiving aperture is effectively controlled by the extent of the beam. The calculation is straightforward and we find that the average signal to noise power ratio is given by

$$\left[\frac{\langle W_S^2 \rangle}{\langle W_N^2 \rangle} \right]_A = \left(\frac{2 W_{so} \lambda^2 \eta}{\pi r_0^2 h \nu \Delta f} \right) \left(\frac{R_0^2}{R_0^2 + (\lambda F/\pi r_0)^2} \right) \qquad (3.24)$$

where R_0 is the radius of the beam to $1/e^2$ intensity points. In particular, if the reference beam has the same diameter as the illuminating beam before the focusing lens (or beam expansion system), as is normal with most beam splitting arrangements, $R_0 = \lambda F/\pi r_0$ and hence the average signal to noise power is just half the maximum value given by equation (3.23).

3.3.5 Limits of sensitivity

This signal to noise ratio clearly sets a limit to the smallest displacement that can be measured. Assuming the normal interferometer arrangement, we take the example of a standard rough surface, a perfectly diffusing, non-absorbing, 'white' surface (see figure 3.14(a)). Assuming complete depolarization, we then have from equation (3.18)

$$W_{so} = W_0/2\pi \qquad (3.25)$$

where W_0 is the power of the beam (assumed to be that of the laser) which is incident normally on the surface. The mean square signal to noise ratio is then

$$\frac{\langle W_s^2 \rangle}{\langle W_N^2 \rangle} = \frac{W_0 \lambda^2 \eta}{2\pi^2 r_0^2 h v \Delta f}. \qquad (3.26)$$

The relation betweeen the surface displacement, x, and the amplitude of the beat component of the detector output is given by equation (3.1). At the position of maximum slope where the interferometer should be operated for maximum sensitivity,

$$\delta x = \frac{\lambda}{4\pi} \frac{\delta V}{V_0}. \qquad (3.27)$$

Hence the displacement corresponding to the RMS noise level, i.e. the 'noise equivalent displacement', is given by

$$\delta x_N = \frac{\lambda}{4\pi} \left(\frac{\langle V_N^2 \rangle}{\langle 2V_s^2 \rangle} \right)^{1/2} = \frac{\lambda}{4\pi} \left(\frac{\langle W_N^2 \rangle}{\langle 2W_s^2 \rangle} \right)^{1/2} \qquad (3.28)$$

where $\langle V_N^2 \rangle$ is the mean square noise at the output of the detector and $\langle V_s^2 \rangle$ the mean square interference signal. Thus using equation (3.26) we find for the case of the diffusing surface

$$\delta x_N = \tfrac{1}{4} r_0 \left(\frac{h v \Delta f}{\eta W_0} \right)^{1/2}. \qquad (3.29)$$

For example, with a 2 mW illuminating beam from a He–Ne laser focused to a spot of diameter 0.05 mm, the theoretical minimum RMS noise level (averaged over the speckles) in a 10 MHz bandwidth corresponds to a displacement of 0.3 nm. A quantum efficiency $\eta = 0.75$ has been assumed for silicon diode detectors at this wavelength. This result may be compared with that when the full laser power can be utilized with a reflecting surface, quoted in §3.2.2 (equation (3.13)). There is about a hundred-fold difference in sensitivity. The advantage of having a good surface reflection is thus considerable. Fortunately, it is possible to arrange to have reasonably good reflecting surfaces in many ultrasonic experiments. In some circumstances, the use of retroreflecting paint may be allowed to increase back scattering.

The relatively poor performance with a rough surface is basically due to the fact that only a small fraction of the scattered light (approximately one speckle) can be used because of coherence considerations. This problem can be minimized by the use of a highly convergent illuminating beam, thus reducing the spot size (radius r_0) and consequently increasing the speckle size. This is probably the most practical approach, although it reduces the depth of focus and increases the sensitivity to sideways displacement of the surface. This can lead to modulation of the signal strength and hence to variable sensitivity to normal surface displacements.

In principle it is possible to bring the signals from all the speckles into phase (or select only those with the correct phase) by the use of a correcting plate, i.e. a hologram formed by exposing a suitable sensitive material, e.g. holographic film or thermoplastic, to light scattered from the surface and combined with the reference beam. This will probably be a useful technique for some specific applications, but is clearly not suitable for rapid scanning over a surface. A technique based on this principle has been described by Aharoni *et al* (1987). Another approach based on optical phase conjugation using the non-linear optical material $BaTiO_3$ has been described by Paul *et al* (1987).

For rough surfaces, use may also be made of the other types of interferometer in which there is a large optical path difference between the interfering beams. These measure the velocity of the surface rather than the displacement and will be discussed in §§3.10 and 3.11. They do not have the intrinsic sensitivity limitation of the simple interferometric method for rough surfaces, but can be difficult to implement in practice.

3.4 LIGHT DETECTION AND SIGNAL AMPLIFICATION

3.4.1 Choice of photodetectors

The theoretical sensitivity limit set by quantum noise is the ultimate goal of interferometric equipment and this can often be approached in practice, at least in the visible region, provided attention is paid to good optical and electronic design. It is particularly important to avoid adding extra noise in the detection and amplification circuits. Practical detectors in the visible and near infrared region are photomultipliers and silicon photodiodes, either simple or avalanche types.

The great advantage of photomultipliers is the very high, virtually noiseless amplification possible by electron multiplication. Thus amplifier noise is no problem and the quantum noise limit can be easily attained. However, the quantum efficiency is rather poor compared with photodiodes, particularly in the red part of the spectrum (see figure 1.8) which is of special interest because of the availability and suitability of helium–neon lasers. Whilst photomultipliers are by far the most sensitive devices at very low light levels, the maximum amount of light that can be accepted is very limited, typically about $1 \mu W$. This can often be exceeded even by the signal beam with a moderately good light return. Even if the signal beam is very weak, a much larger reference beam is usually available. This gives an effective signal

amplification and in most interferometers as good or better performance can
be obtained by the use of photodiodes. There can however be difficulties in
using a very high reference to signal beam ratio due to modulation noise on
the reference beam, and the option of using photomultipliers should not be
overlooked when the scattered light return is low, particularly in the green,
blue or ultraviolet parts of the spectrum where photomultiplier efficiency is
much better.

3.4.2 Balanced photodetectors

It is generally possible to approach the quantum noise limit with simple
silicon diode PIN detectors, but high light levels on the detector may be
necessary if response to high frequencies is required. To minimize the influence
of modulation on the laser beam under these conditions (and for other
reasons), a balanced detector arrangement, as shown in figure 3.6 for example,
is often used. The difference between the two detectors may be taken by the
use of a differential amplifier as shown in figure 3.17(*a*) or by connection to
a common load resistor as shown in figure 3.17(*b*).

The scheme employing a differential amplifier is simpler and enables the
inputs to be balanced easily electrically. One voltage supply is sufficient for
reverse biasing both diodes. At high diode outputs, however, the dynamic
range available may be a problem and the common-mode rejection ratio

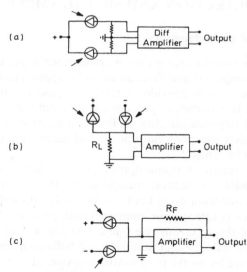

Figure 3.17 Circuit arrangements for balanced optical detection. (*a*)
The use of a differential amplifier, (*b*) subtraction of the detector currents
before amplification, (*c*) use of feedback to reduce the input impedance
of the amplifier to improve high-frequency response.

may not be adequate. Also this type of amplifier may not be available with a very low noise figure. In the second arrangement (figure 3.17(b)), the subtraction of the diode currents is made before amplification and the demands made of the amplifier are less severe and a very low noise FET circuit can be chosen. This is thus the most satisfactory scheme when very small displacements are to be measured. However, voltage supplies of opposite sign are required and the unsymmetrical arrangement makes it more difficult to achieve a balance independent of frequency. In addition, electrical balancing is not easy to arrange (unless using avalanche diodes), but this may be done optically, e.g. in the arrangements shown in figures 3.6 and 3.8, by rotating the polarizing beam splitter with the detector units or by adjusting a rotatable half-wave plate in front of the beam splitter.

3.4.3 Sources of noise

In addition to the basic photon or quantum noise, other sources can be of importance in practical circuits. These are as follows.

(1) Dark current noise from the photodiode,
(2) Thermal or Johnson noise from the load resistor,
(3) Noise generated in the amplifier.

The influence of these depends on the frequency and the characteristics of the devices and circuits used but a simplified general treatment will be given.

Dark current noise from photodiodes, though much larger than from photomultipliers, is not usually a significant contribution. For a typical small PIN silicon photodiode of 1 mm^2 sensitive area, the current noise is approximately 10^{-12} A Hz$^{-1/2}$, equivalent to about 2×10^{-13} W Hz$^{-1/2}$ of incident light power fluctuation at 633 nm. From equation (1.8), we find that photon noise of this magnitude is obtained with an incident light level of 0.04 μW. It is thus usually easy to arrange for the photon noise to dominate over photocurrent noise. This is all that is required at low frequencies where the load resistor R_L may be large. There are, however, other considerations when a high-frequency response is required for ultrasonic applications.

3.4.4 Frequency response considerations

The basic response of a small silicon photodiode has a rise time of typically 3 ns, corresponding to a frequency response 3 dB down at 120 MHz, but a practical limitation arises from the capacitance, C, of the diode which it is difficult to reduce below about 5 pF in any practical circuit. The photocurrent has to feed into this capacitance and this considerably reduces the voltage output available at high frequencies, thus increasing the relative importance of circuit noise. Considering the circuit shown in figure 3.17(b), we see that the load resistor R_L corresponding to a response 3 dB down at frequency f

is given by

$$R_L = 1/4\pi f C. \tag{3.30}$$

Assuming this value of load resistor, the noise equivalent power for a typical detector–amplifier combination is plotted as a function of the light power incident on the detector in figure 3.18 for different frequency responses. In this example, the spectral density of the noise generated by the amplifier has been assumed to be equivalent to $3 \text{ nV Hz}^{1/2}$ at its input, a value expected for a good low-noise amplifier though not the best that can be achieved. Also shown in figure 3.18 is the limiting photon noise, assuming a quantum efficiency of 0.75. Photomultiplier performance up to a power level that may reasonably be employed is also shown. Over the range shown, photon noise is everywhere the limiting factor but the quantum efficiency is comparatively low. A value of 0.05 has been assumed, typical of the response of an S20 cathode at the normal helium–neon wavelength of 633 nm.

The frequency response of a detector system may be considerably improved by the use of a preamplifier with negative feedback as shown in figure 3.17(c). This offers a very low effective impedance to the diode thus minimizing the effect of its capacity. A good frequency response is obtained whilst retaining a good signal to noise ratio at low frequencies. However, the basic circuit noise considerations for broadband use are substantially the same.

Also available are silicon diode–amplifier modules in which the matching of the diode and amplifier is optimized and the effect of circuit capacity

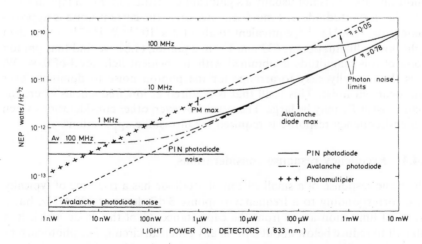

Figure 3.18 Dependence of noise equivalent light power (NEP) on the incident light level for various detectors, assuming an amplifier input noise level of $3 \text{ nV Hz}^{-1/2}$ and a load resistor chosen to give the frequency response indicated.

reduced to the absolute minimum. These devices are useful when the light power on the detector is not too large but may not be convenient in balanced detector arrangements.

Exceptionally good frequency response and reduction in the effect of circuit noise may be obtained from the avalanche type of silicon photodiode, in which a considerable internal amplification is achieved by avalanche multiplication of carriers in regions of the semiconductor where conditions approach those for electrical breakdown. The amplification may be controlled by the reverse bias voltage; values of around 100 are usual. Avalanche diodes are used in essentially the same way as simple diodes but the photon noise limit may be reached at up to 100 times lower power level on the detector for the same frequency response. The estimated noise equivalent power for a typical avalanche photodiode is shown in figure 3.18, making the same assumptions as for the simple diode example. The maximum light power that may be used is, however, more limited and avalanche diodes are most useful with poor or moderate light returns.

3.5 EFFECT OF LASER MODE STRUCTURE

3.5.1 Coherence length

Ideally, interferometric measurements should be made using single-mode lasers giving an extremely long coherence length with no problems from fluctuations due to mode sweeping and intermode beating. These lasers are essential for maintaining interference over very long path lengths but are not really necessary where small displacements are to be measured and the reference beam path is adjustable. In general, the coherence length (path difference over which interference coherence is retained) is given by

$$L_{coh} \approx c/\delta v \qquad (3.31)$$

where δv is the frequency spread of the light source and c the velocity of light (see also §1.3.2). It is not practicable to use a source with a large frequency spread because the coherence length would be extremely short and it would be an unrealistic task to match the signal and reference paths sufficiently closely. Most gas lasers and many solid state and semiconductor lasers have coherence lengths of several centimetres; this is perfectly adequate and convenient for most acoustic and ultrasonic applications.

In particular, helium–neon lasers are readily available and very widely used. The ordinary type is quite suitable for general interference work with path differences up to about 5 cm or so. The use of a single-mode version usually entails extra expense and a sacrifice in available power. An ordinary unstabilized small helium–neon laser normally operates with between two and five longitudinal modes. As an example of coherence behaviour, we

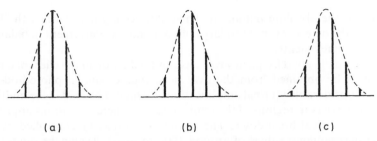

(a)   (b)   (c)

Figure 3.19 Diagram of the high-resolution optical spectrum of a typical helium–neon laser showing the mode structure within the overall gain profile (dotted curve). Unless the cavity length is controlled, the structure drifts between these forms.

consider a laser operating with basically three modes, with a spectrum as shown in figure 3.19. The separation between the modes would typically be around 500 MHz. (These patterns may be seen by a scanned Fabry–Perot optical spectrum analyser.) The pattern changes as the modes 'sweep' through as the cavity length changes with temperature. One complete cycle, e.g. from (a) in figure 3.19 to (b) to (c) to the mirror image of (b) and back to (a), occurs for a change in the separation of the mirrors of half a wavelength. The patterns shown in the figure are separated by changes of one eighth of a wavelength in the separation of the mirrors. The modes generally sweep through rapidly as the laser warms up after switching on, and then drift about slowly when thermal equilibrium is attained.

The dependence of the interference signal on the path length difference between the signal and reference beams is shown in figure 3.20. This curve

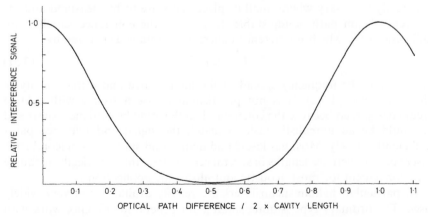

Figure 3.20 Reduction in interference signal due to the multiple longitudinal-mode structure of a laser. The curve shown corresponds to the structure shown in figure 3.19. (Cavity length here refers to the separation of the mirrors $= \frac{1}{2}L_c$ where L_c is the round trip optical path length.)

was calculated for the mode structure shown in figure 3.19. The differences between (a), (b) and (c) are too small to show but it should be noted that for case (c) the signal must fall to exactly zero. It will be seen that, after initially falling to very low values, the interference signal increases again, coherence being completely regained when the path length difference becomes equal to twice the laser cavity length. This is also true for all even multiples of this length under the assumptions made. This property is a useful one, as it means that an interferometer using a multimode laser can be used to measure small movements or vibrations of surfaces at long distances without the need for inconveniently long reference paths. The maximum distance over which this can be done probably depends on the uniformity of the mode spacing and the stability of the cavity length. It is certainly 100 m or more.

3.5.2 Intermode beating

The use of multimode lasers does, however, lead to one undesirable effect: the occurrence of spurious signals generated by beating between the modes. Direct beating between adjacent modes leads to frequencies equal to the mode spacing, which is generally several hundred megahertz; for example, 375 MHz for a cavity length (mirror separation) of 400 mm. These basic beat frequencies are above the range normally of interest in ultrasonics and are not usually transmitted by the detector and electronic circuits. They may in any case be filtered out. Beat signals are, however, also observed below 1 MHz. Spurious signals appear cyclically, sweep through a range of frequencies and disappear with the mode sweeping of the laser. These arise from the fact that the modes are not always exactly equally spaced in frequency. The basic beat frequencies between adjacent modes are then not exactly the same. Signals at the difference between these basic frequencies can then appear (Watrasiewicz and Rudd 1976). The uneven mode spacing is due to dispersion in the laser medium. There is a variation of refractive index associated with any absorption (and hence stimulated emission) line. This is decreased below its normal value just below the line and increased just above, resulting in the mode frequencies at these positions being shifted slightly away from the line centre. This has most effect in the mode pattern shown in figure 3.19(c), where the spacing of the two centre modes is increased relative to the other spacings. In the symmetrical pattern (a), however, the three central modes are dominant and remain equally spaced and low-frequency intermode beats are not observed in this case. Clearly, at least three modes are necessary to produce this effect. In many small ($\leqslant 2$ mW) helium–neon lasers, the effective number of modes may be less than this and the low-frequency beats may not be present.

In an interferometer, the intermode beats may generate spurious signals via the following mechanisms.

(i) A direct modulation of the laser intensity,

(ii)   modulation of the coherence,
(iii) frequency modulation of the laser.

The first mechanism is not usually important because the modulation depth usually only amounts to about 1%, and with a balanced detector arrangement is cancelled out anyway. It then only appears as a small modulation of the displacement signal. This is the only source of intermode beats when the signal and reference paths are equal. When there is a path difference, a further modulation of the displacement signal may arise from a fluctuation of the coherence properties of the laser but this is in practice only significant when the optical path difference deviates considerably from zero or an even multiple of the cavity length. This situation would be avoided in practice.

The most common source of trouble from intermode beating appears through an effective frequency modulation of the laser. The spurious signal from this varies linearly with the optical path difference and may be significant for quite small differences. It may sometimes be necessary to match the paths to within a millimetre or so to reduce the effect sufficiently. The effective frequency modulation depth may typically be around 5 parts in 10^8 which is equivalent to a movement of the surface whose displacement is being measured of 0.25 nm for every 10 mm difference in optical path.

3.6   SIGNAL PROCESSING

The basic problem in processing the output of a laser interferometer is the fact that it is effectively linear over only a very small range. It is thus possible to use the direct output as a measure of displacement only if it is small enough and the interferometer is close to a point of maximum sensitivity or balance point for balanced detectors. To ensure this in spite of extraneous low-frequency fluctuations normally present involves the use of either a vibration-free table or, more usually, one of the active compensation methods which will be described in §3.7. Since the relation between output and displacement is sinusoidal (equation (3.1)), the maximum displacement that can unambiguously be measured in this direct way is $\pm\frac{1}{8}\lambda$ (± 80 nm for the He–Ne laser wavelength). This does in fact cover the majority of ultrasonic applications but larger displacements may be encountered, especially with powerful transducers at low frequencies. Then some other method of processing must be used to obtain an output unambiguously related to displacement or velocity.

There are two techniques commonly used for processing signals arising from these larger amplitude displacements, as follows:

(i)   optical frequency shifting,
(ii)   two-phase or quadrature interferometry.

Either of these overcomes the intrinsic ambiguity in phase or direction of motion associated with a simple single-phase interferometer. These methods are most suited to displacement amplitudes of many wavelengths but they can allow the interpolation of phase to a fraction of a cycle with suitable processing. The measurement of very small displacements is difficult, however, and the direct method is preferred for these.

The region 50 nm to 1 μm or so is an unsatisfactory one for interferometry. These displacements are too high for the direct method but are at the low end of the range covered by conventional counting or frequency-tracking techniques. There is probably only one technique that satisfactorily spans the range from nanometres to micrometres, that described in §3.9.4. No one technique can cover the whole range of displacements it is possible to measure.

3.7 STABILIZED INTERFEROMETERS

3.7.1 The problem of low-frequency vibration

A troublesome problem in the use of laser interferometry to measure very small ultrasonic displacements is the presence of low-frequency background vibration, often with an amplitude of a few micrometres or more in many laboratory and industrial environments. At these levels, the response of the interferometer is by no means linear and the low-frequency signal cannot be simply filtered out leaving the ultrasonic signal undistorted. The example shown in figure 3.21 shows signals due to the same ultrasonic pulse in the

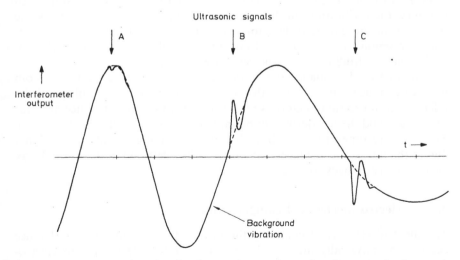

Figure 3.21 Ultrasonic pulse signals superimposed on a low-frequency background from an unstabilized laser interferometer.

presence of a low-frequency background. It will be seen that there are great variations in sensitivity to ultrasonic displacements including changes of sign depending on the timing of the pulse with respect to the low-frequency displacement. This is perhaps the most fundamental problem in the practical application of laser interferometry to the measurement of very small amplitude high-frequency displacements.

In some circumstances, it is possible to allow for these variations of sensitivity, particularly when the ultrasonic vibration is continuous and only the amplitude is required. Mezrich *et al* (1976) deliberately added phase fluctuations to the reference beam by 'wiggling' the return mirror with an amplitude greater than $\frac{1}{4}\lambda$ to ensure that the sensitivity passes through the full range of values during a measurement period. By measuring the maximum value of the signal at the ultrasonic frequency, the periods of full sensitivity are selected and we can thus get a good measurement of the ultrasonic vibration amplitude. The 'wiggler' frequency, typically 20–100 kHz, must be low compared with the ultrasonic frequency but high enough to ensure that the full range of sensitivity is sampled during the measurement time, i.e. the time during which the amplitude of the ultrasonic vibration is assumed to be constant.

The above method does not permit the determination of phase and is not suitable for short or intermittent pulses. It is beneficial to stabilize an interferometer actively so that full sensitivity is always maintained and the response is effectively linear for displacement amplitudes below about 30 nm. This is achieved by changing the effective optical path length of the reference beam to compensate for the low-frequency components of the motion of the surface, having of course taken all reasonable precautions to minimize these vibrations in the first place. The principle is to feed back the output from the balanced detectors (and the use of this type of system is almost essential) to control an electro-mechanical, piezoelectric or electro-optic phase shifting device in the reference beam path.

Near one of the balance points of the interferometer, a deviation from balance produces an output of the correct sign to change the phase of the reference beam so that balance is restored. Deviations from balance are thus suppressed and the interferometer maintained in its most sensitive condition. The response time of the feedback is not short enough to affect the signal due to the ultrasonic displacement which is thus detected in the normal way against a steady background.

3.7.2 Electro-mechanical compensation

The most direct way of changing the optical path length of the reference beam is by physically moving the mirror or retroreflector by an electro-magnetic or piezoelectric displacement device. In the scheme shown in figure 3.22, the mirror is attached to a moving coil loudspeaker movement. Problems

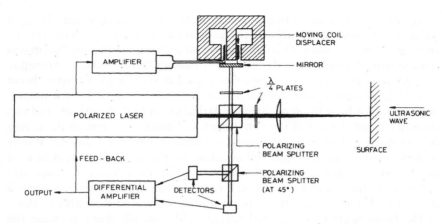

Figure 3.22 Mechanically compensated interferometer for detecting small ultrasonic displacements.

of instability may be experienced with mechanical devices due to phase changes connected with mechanical resonances but this type of stabilization technique has been successfully implemented (Cooper *et al* 1986, Moss 1987, White and Emmony 1985), the last authors using a piezoelectric bimorph displacement device. Quite large low-frequency displacements may be compensated for in this way, though there is the possibility of misalignment occurring due to the tilting of the mirror. A simple piezoelectric device such as a PZT tube is perhaps better in this respect but the range is limited to a few micrometres. Compensation up to higher frequencies can also be obtained from piezoelectric systems and this possibly contributes to better stabilization. Stabilized interferometers based on piezoelectric compensation have been used in many laboratories (Bondarenko *et al* 1976, Palmer and Green 1977a,b, Jette *et al* 1977, Kline *et al* 1978, Kim and Park 1984).

3.7.3 Electro-optic compensation

Stabilization may be accomplished without mechanical movement by means of electro-optic cells. These are devices containing a material in which birefringence may be induced by the application of an electric field. In most materials, the dependence of birefringence on electric field is quadratic. This is known as the Kerr effect. It is only large enough to be useful in some liquids, notably nitrobenzene. It is more usual now, especially in laser applications, to make use of the linear Pockels electro-optic effect which can occur in crystals with a non-centrosymmetric structure. Commonly used materials for the visible and near infrared region are lithium niobate, and ammonium and potassium hydrogen phosphates (ADP and KDP). Potassium dideuterium phosphate (KD*P) is often used instead of KDP as it has a larger electro-optic coefficient.

The electrically induced birefringence means an increased velocity for light polarized parallel to an axis of the cell and a decreased velocity for the orthogonal polarization. This fact may be used directly to change the optical path length of a suitably polarized beam passing through the electro-optic cell. We note, however, that if the cell is placed directly in the reference beam of an interferometer with the beam returned through the cell by a mirror, the beam must have the same linear polarization in both directions, and this does not allow the use of the convenient and efficient system of polarizing beam splitter and quarter-wave plates as shown in figures 3.6 and 3.8. It may be noted that a magneto-optic cell based on the Faraday effect could be used in this way as an applied magnetic field can modify the velocity of a circularly polarized beam. An electro-optic cell can be used with the above system if it is placed before the polarizing beam splitter with its axes parallel to those of the beam splitter. An applied voltage will then effectively increase the optical path of the reference beam and decrease that of the signal beam or vice versa. Because of the limitations of the electro-optic effect, the range of control available in this way is one or two cycles, only barely sufficient to compensate for vibration levels usually encountered.

3.7.4 Electro-optic phase locking

An alternative way of using the electro-optic effect does not have this limitation. In this two or more cells are combined in a modulator arrangement forming a phase- or frequency-shifting device. An analysis of these devices for generating frequency-shifted beams has been given (Drain and Moss 1972). The principle is similar to that of single sideband modulators used in radio circuits. In the arrangement shown in figure 3.23 the plane-polarized laser beam is converted to circular polarization by a quarter-wave plate and passed through two electro-optic Pockels cell units whose axes are at $45°$ to each other and driven by voltages $90°$ out of phase, $V_E \cos \phi$ and $V_E \sin \phi$, where ϕ is a function of time. It may be shown that when V_E is small the effect is to produce a component of the reverse circular polarization with a phase shifted by ϕ relative to the component with the original polarization which is unshifted. Reconversion to plane polarization by the second quarter-wave plate enables the shifted and unshifted components to be separated by the polarizing beam splitter, the phase-shifted beam forming the reference. A direct calculation of the polarization changes at the various birefringent components shows that the phase shift, ϕ_R, in the reference beam produced by applied voltages of amplitude V_E is given by

$$\tan \phi_R = \frac{\tan(\tfrac{1}{2}\Delta \sin \phi)}{\tan(\tfrac{1}{2}\Delta \cos \phi)} \tag{3.32}$$

where Δ is the peak retardation on each cell. The result for $\Delta = 45°$ ($\tfrac{1}{8}$ wave) is shown in figure 3.24. Also shown is the intensity of the beam relative to

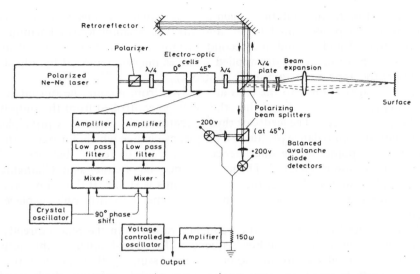

Figure 3.23 Laser interferometer for the detection of ultrasonic displacements with electro-optic compensation for low-frequency vibration.

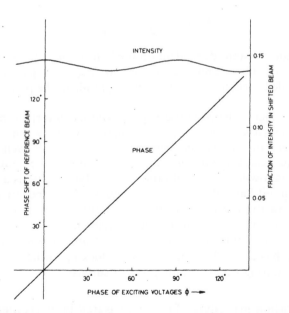

Figure 3.24 The variation of phase and intensity of the reference beam with the phase of the exciting voltages in the electro-optic arrangement shown in figure 3.23. In this example, the amplitude of the exciting voltages correspond to $\frac{1}{8}$ wave retardation in each cell.

the input. For small excitations such as that shown, the mean intensity is approximately proportional to V_E^2 and there is a small fourth harmonic modulation. (Fundamental and second-harmonic modulation produced by misalignment or errors in the applied voltages may be removed by adjustments of the appropriate amplifier controls.)

The voltages applied to the cells are derived from mixing the outputs of two oscillators, one fixed and the other voltage controlled from the output of the interferometer. The feedback thus operates on the phase-lock principle. A deviation from balance produces a frequency change which, for one of the two balance points in the interference cycle, tends to restore balance and thus stabilization is achieved. The feedback may be of either sign. Changing this simply changes which of the balance points is stable. It does, however, reverse the sign of the response of the interferometer to ultrasonic displacement.

Since the feedback changes the frequency rather than the phase directly, it is naturally more effective at low frequencies. Thus low-frequency vibrations are attenuated relative to ultrasonic signals. Assuming that the deviation from balance is small, the response to a sinusoidal variation of phase of amplitude ϕ_0 and frequency f produced by movement of the surface has an amplitude given by

$$\phi_s = \frac{2\pi f \tau \phi_0}{(1 + 4\pi^2 f^2 \tau^2)^{1/2}} \tag{3.33}$$

where τ is a time constant dependent on the light return, the response of the detectors and amplifiers and the voltage to frequency parameter of the voltage-controlled oscillator. ϕ_s is the net amplitude of the phase fluctuation after feedback.

At high frequencies, the response is proportional to displacement, whilst at low frequencies the output is proportional to velocity. The feedback is arranged so that the transition frequency, $f_T = 1/2\pi\tau$, is well below any ultrasonic frequencies that are being detected but above the frequency range of the interfering low-frequency vibrations. The above analysis assumes that the feedback circuit has a flat frequency response. A sharper transition may be obtained with the use of filters but the result is then more difficult to calculate and there is a risk of instabilities occurring.

The advantages of this stabilization method are its unlimited range of compensation, basic stability and speed of locking, which are useful in applications requiring rapid scanning over a surface.

3.7.5 Calibration procedures for compensated interferometers

Compensated interferometers are well suited to the observation and measurement of very small displacements and there is no fundamental limit to the frequency response set by the use of a carrier frequency. Signal processing is simple and gives displacement directly. The proportionality constant between output and displacement does, however, depend on the light return and, for absolute measurement of displacement, calibration under operating conditions is necessary. This may be easily accomplished by switching off the feedback, driving the compensating device through its range or allowing background vibration to take the interference through a number of cycles. (This may be done automatically if the interferometer is under computer control.) The peak signal voltage is then the parameter V_0 discussed in §3.1. If the feedback is restored without otherwise changing the conditions, the ultrasonic displacement, x, is related to the output V as follows:

$$x = \frac{\lambda}{4\pi} \sin^{-1}(V/V_0) \approx \frac{\lambda}{4\pi}\frac{V}{V_0} \qquad \text{if } V \ll V_0. \qquad (3.34)$$

It is evident that displacement amplitudes of more than $\frac{1}{8}\lambda$ give rise to difficulties. Although it is possible to stabilize an interferometer with displacement (peak to peak) swings of somewhat more than $\frac{1}{4}\lambda$, the response becomes very non-linear and with higher amplitudes stabilization soon becomes impossible. Other techniques must then be used.

3.8 OPTICAL FREQUENCY SHIFTING

3.8.1 Directional discrimination

To facilitate signal processing, especially with larger amplitude displacements, the technique of frequency shifting the reference beam is often used. This may be regarded as a method of rapidly sampling the phase of the interference signal using the shift as a high-frequency 'carrier' or as a method of directional discrimination when signal processing, by counting cycles or tracking the signal frequency. In simple interferometry by these methods, a sign ambiguity exists in the relation of frequency to velocity or, if counting cycles, whether counts should be added or subtracted. This problem may be overcome by shifting the frequency of the reference beam. The interference signal frequency (or 'Doppler' beat frequency; see §3.1.5) due to the velocity of the surface is then either added to or subtracted from the shift frequency, depending on the direction of motion. If the shift frequency is greater than the maximum beat frequency produced by the moving surface, the frequency at the detector never falls to zero and thus sign ambiguity does not arise. Frequency shifts up to approximately 10 MHz are required for vibration amplitudes commonly encountered.

The effect of frequency shifting on an interferometric signal from a sinusoidally vibrating surface is illustrated in figure 3.25. The unshifted signal is shown in (a) with the output obtained from a signal processing system giving a frequency analogue output. The result is a rectified waveform

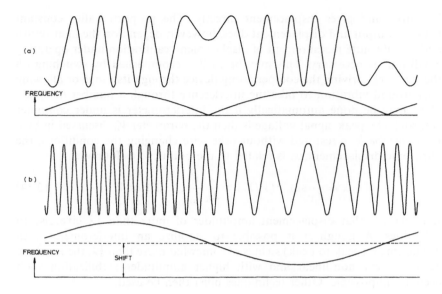

Figure 3.25 The use of frequency shifting for directional discrimination. An interferometer signal from a sinusoidally vibrating surface (*a*) without frequency shift, and (*b*) with frequency shift.

proportional to the modulus of the velocity. The frequency is the same whether the surface is moving towards or away from the interferometer. For sinusoidal displacements, we know that the sign alternates regularly but, with more complex waveforms, changes of sign may not be easily discerned. Interpretation is then difficult and the output is of limited value. With an adequate frequency shift, the signal shown in figure 3.25(*b*) is obtained. Now opposite velocities are distinguished by high and low frequencies. The frequency analogue output is then a faithful representation of the velocity variation except that it is offset by the shift frequency which may be easily subtracted.

Frequency shifting is the preferred method of directional discrimination in laser Doppler anemometry. Practical shifting techniques use acousto-optic or 'Bragg' cells, rotating diffraction gratings or electro-optic cells. Electro-optic (Drain *et al* 1977) and acousto-optic (De la Rue *et al* 1972, Buchave 1975, Monchalin 1985a) cells have been applied in laser interferometry. Acousto-optic cells use diffraction by ultrasonic waves and their principle of operation has been described in Chapter 2. The most efficiently produced shift frequencies are of the order of 40 MHz and these are very suitable for ultrasonic interferometers.

3.8.2 Heterodyne interferometers

Interfering or 'heterodyning' a light beam reflected from a vibrating surface with a frequency-shifted reference beam evidently produces an output signal centred on the shift frequency f_S but phase modulated in proportion of the displacement of the surface. Thus for a sinusoidal displacement of the surface, $x = a\cos(2\pi ft)$, and normal incidence and reflection, the output of the interferometer has the form

$$V = \bar{V} + V_0 \cos\left(2\pi f_S t + \frac{4\pi a}{\lambda} \cos 2\pi ft \right). \qquad (3.35)$$

This is equivalent to a frequency modulation by the velocity and produces the well known system of sideband frequencies $f_S \pm nf$ with amplitudes $V_0 J_n(4\pi a/\lambda)$ where J_n is a Bessel function of order n.

For small ultrasonic displacements, $a < \tfrac{1}{2}\lambda$, only the sidebands $f_S \pm f$ are significant and their amplitude reduces to

$$V_{\pm 1} = V_0 \frac{2\pi a}{\lambda}. \qquad (3.36)$$

One of these sidebands may be selected by a narrow-band amplifier or superheterodyne receiver yielding an output proportional to the ultrasonic displacement.

The heterodyne technique was used in early interferometric ultrasonic visualization experiments (Massey 1968, Whitman *et al* 1968) using water-filled Bragg cells for frequency shifting. The technique was found to be easier to implement and interpret than 'homodyning', i.e. interfering or 'beating' with an unshifted reference beam. In the simple arrangement employed by Whitman *et al* (1968) and illustrated in figure 3.26, the Bragg cell is used as a beam splitter and combiner as well as for frequency shifting. In the

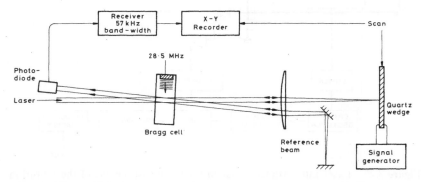

Figure 3.26 Heterodyne interferometer using a water-filled Bragg cell (Whitman *et al* 1968).

configuration shown, the reference beam is shifted by $-f_S$ and the signal beam by $+f_S$, producing a frequency-modulated output centred on $2f_S$. To facilitate the detection of very small signals, they could be amplitude modulated at 1 kHz by modulating the excitation of the Bragg cell.

Since these single sideband signals are phase modulated by background vibration, direction and phase information cannot be obtained simply and these signals usually only yield basic displacement amplitude information. However, we note (De la Rue *et al* 1972, Monchalin 1985a) that the 'carrier' component of the signal (normally at frequency f_S) is also phase modulated by the background vibration and thus if this is mixed with the sideband signal in a lock-in amplifier, an ultrasonic displacement signal may be recovered free of background vibration. Very narrow bandwidths can be used and this enables continuous ultrasonic displacements to be observed down to picometre levels even from rough metal surfaces (Monchalin 1985a).

For displacements not small compared with $\frac{1}{2}\lambda$, the phase-modulated signal from a heterodyne interferometer would be suitable for processing by an FM receiver or a frequency tracker (see next section). An output dependent on the velocity of the vibrating surface (rather than the displacement) would then be obtained and direction, phase and waveform information preserved. A diagram of a general purpose heterodyne interferometer or 'vibrometer' using an acousto-optic device for frequency shifting is shown in figure 3.27. The beam from a small polarized helium–neon laser is split into a reference beam and signal beam by the first beam splitter. The acousto-optic Bragg cell placed in the reference beam produces a frequency shift of around 40 MHz

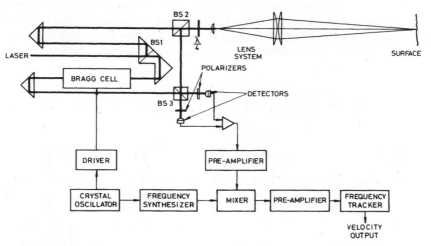

Figure 3.27 Laser interferometer system for the measurement of the vibration of a surface using an acousto-optic Bragg cell for frequency shifting the reference beam (Buchave 1975).

with high efficiency. The signal beam is passed through the polarizing beam splitter (bs) 2. The quarter-wave plane ensures that if there is no depolarization on scattering, all the returned light has the orthogonal polarization on reaching beam splitter 2 and is deflected towards the detectors. The reference beam is combined with the scattered light in beam splitter 3. This, and the polarizers and two photodiode detectors, form a balanced detector arrangement with the advantages discussed in §3.1.4. The resulting signal frequency is the shift plus/minus the Doppler beat frequency due to the motion of the surface. In most applications, the frequency swing around 40 MHz is small and this range of frequencies is in any case to high for commonly available signal processors. The signal is therefore converted down to the range 1–10 MHz or less by the mixer. This also enables the effective frequency shift to be varied without changing the optical shift produced by the Bragg cell. The range of the signal processor may then be matched to the frequency spread of the signal to give the best performance.

3.8.3 Frequency tracking

To process interferometer (frequency-shifted) signals which have a definite but variable frequency but possibly a poor signal to noise ratio, frequency trackers are most appropriate. These are in effect frequency modulation receivers giving an output proportional to the instantaneous frequency of the input but independent of the input level. They generally operate by combining the signal with the output of a voltage controlled oscillator (vco) in a balanced mixer. The output of the mixer which depends on the phase difference between the signal and the vco is fed back to control the frequency, thus phase locking the vco to the signal. This is illustrated in figure 3.28. When the circuit is 'in lock' the feedback voltage gives a measure of the instantaneous frequency. In addition the vco output is virtually noise free and cycles may be readily counted or otherwise processed. The signal to noise ratio that can be tolerated by a tracker depends on the speed of response required. The tracking of signals of poor quality may be improved by the

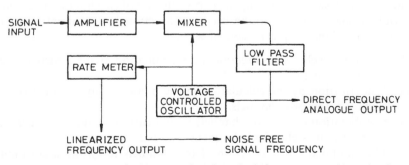

Figure 3.28 Schematic diagram of a phase-lock frequency tracking circuit.

use of a low-pass filter in the feedback loop, but at the expense of frequency response.

The feedback may alternatively be controlled by frequency deviation rather than phase, and this may have some advantages with intermittent signals in laser Doppler anemometry, but the phase lock principle is better for the constant amplitude signals normally obtained in interferometry. Frequency trackers designed for laser anemometry need to give protection against errors due to frequent losses of signal. This is not a problem in interferometry and simpler circuits may be employed, possibly using integrated tracking circuits.

Normally phase deviations of many cycles are expected when using frequency trackers, but with a reasonably good signal strength, frequency deviations associated with phase changes of only a fraction of a cycle can also be measured, and this can be a useful way of overlapping into the range of compensated interferometry. However, it should be ensured that fast fluctuations in frequency are being followed. At high displacement amplitudes, loss of lock will indicate an inadequate signal strength, but with displacements corresponding to phase shifts of less than a quarter of a cycle, lock may be retained without fast fluctuations necessarily being tracked faithfully. This may be tested by attenuating the signal by a factor of two, say, and checking that the output is not appreciably affected. High-frequency displacements down to about 10 nm amplitude may be detected in this way but the frequency response is clearly limited to a small fraction of the shift (or effective shift) frequency. With a frequency shift of 5 MHz, the limit would certainly be less than 1 MHz, and possibly a good deal lower in practice. The frequency shifting and tracking technique is thus of most value in the lower part of the ultrasonic range (and of course at audio frequencies).

3.9 QUADRATURE INTERFEROMETERS

3.9.1 Processing techniques using two-phase interferometry

A number of processing techniques involve the use of two interference signals 90° out of phase. These may be used for directional discrimination or the linearization of the output in several ways.

(i) Counting cycles with directional discrimination giving a displacement resolution to $\frac{1}{8}\lambda$.

(ii) Linearization by the calculation of the phase at any instant from the signals in the two channels.

(iii) Combination of the two channels with a carrier frequency to generate a signal that is effectively frequency shifted.

(iv) The combination or selection of the channels to ensure a continuing maximum sensitivity to small ultrasonic displacements.

3.9.2 Optical arrangements for quadrature interferometry

To ensure an accurate phase relation between the two channels, it is important that they are derived from the same reflection or speckle. This is not difficult with a mirror reflector, but with diffuse scattering it is necessary to arrange that the beams interfering in the two channels follow exactly the same paths and that the same polarization component is used. Practical schemes make use of polarizing optics to separate the channels and introduce the required $\frac{1}{4}\lambda$ optical path difference. Two arrangements that have been used are shown in figure 3.29. The scheme in figure (a) uses a non-polarizing beam splitter and a suitably oriented eighth-wave plate in the reference beam. Because of the double passage through the plate, it is effectively quarter wave and converts the reference beam to circular polarization. This is combined with reflected light of the original polarization in the beam splitter. (If polarization

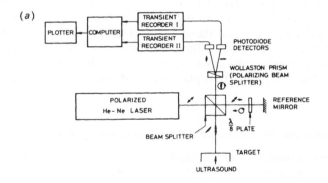

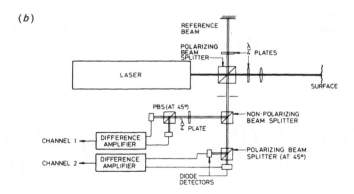

Figure 3.29 (a) Quadrature interferometer detection system (after Reibold and Molkenstruck 1981). (b) Interferometer system with balanced quadrature detection.

changes can occur on reflection, the use of a polarizer in the signal path would be an advantage.) The combined beams are then split into two channels by the polarizing beam splitter whose axis is at 45° to the original polarization direction. Both the reference and signal beams are split evenly between the channels but the phases of the reference beam components differ by 90°. Thus the interference signals in the two channels also have this phase difference.

In the above arrangement, the channels are not balanced. As previously discussed (§3.1.4), this is a desirable feature particularly with poor light returns and may be obtained with the optical arrangement shown in figure 3.29(b). This employs the usual polarizing beam splitter for separating the signal and reference beams and quarter-wave plates in the beam paths to convert their polarizations so that they emerge at the side of the beam splitter. The recombined beam, which contains the reference and signal components coded by polarization, is split into the two channels by a non-polarizing beam splitter. Balanced detector units with polarizing beam splitters at 45° to the polarization directions are used in each channel but in one a phase shift between the reference and signal beams is introduced by a quarter-wave plate whose axes are parallel to the polarization directions. (In practice, a small additional phase shift may be introduced by the non-polarizing beam splitter and thus the retardation plate required may not be exactly quarter wave.)

3.9.3 Displacement measurement by counting cycles

The direction of motion may be readily inferred from the signals in the two quadrature channels by noting which lags or leads the other. This is illustrated in figure 3.30 where the outputs of the channels are shown when a change in direction of motion occurs. This enables a decision to be made concerning whether to add or subtract when registering displacement by counting zero crossings of the signals. The decision depends on whether the crossing is

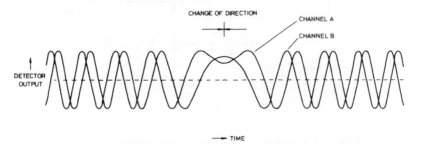

Figure 3.30 Signals from two channels of a quadrature interferometer showing how the direction of movement may be discriminated.

from positive to negative or vice versa and also depends on the sign of the other phase at the time. Counting both types of zero crossing on both phases enables a resolution of a quarter of a cycle to be obtained. This is equivalent to $\frac{1}{8}\lambda$ in displacement. One scheme for generating a displacement output is shown in figure 3.31. Signals from channels 1 and 2 are squared off and differentiated to obtain pulses of alternate signs. The signs may or may not be reversed in the mixers, depending on the current sign of the other channel. Subtraction of the mixer outputs then gives a series of pulses whose sign depends on the direction of movement. These may be fed to the up or down inputs of the counter in accordance with their sign. A hysteresis or inhibit feature should be incorporated in the squaring circuits to avoid the confusion of a rapid succession of pulses if one of the channels happens to be jittering about a zero-crossing point.

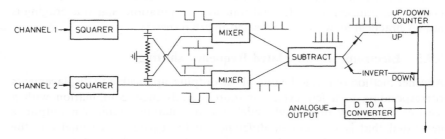

Figure 3.31 Directional counting scheme for interferometer signals using quadrature detection.

3.9.4 Computer evaluation of displacement

It is possible to obtain very much higher resolution of displacement by using the actual values of the interference signals in each channel at any instant. The exact phase and hence fractional wavelength movement may be calculated. For ultrasonic pulse excitation, this may be done conveniently by using signals stored in a transient recorder. The computer does not then have to be fast enough to analyse the signals in real time. Reibold and Molkenstruck (1981) employed the optical arrangement shown in figure 3.29(*a*) with a small polarized helium–neon laser. The waveforms recorded are the quadrature interference signals superimposed on a mean level. They are normalized by observing the maximum and minimum values when the interference swings through more than one cycle. If necessary, the reference mirror may be oscillated to achieve this. The phase between the channels is trimmed by rotating the $\frac{1}{8}\lambda$ plate so that $(V_1^2 + V_2^2)$ remains constant, where V_1 and V_2 are the normalized interference signals. The displacement at a given time is then given by

$$x = \frac{\lambda}{4\pi}[\tan^{-1}(V_1/V_2) + m\pi] \tag{3.37}$$

where m is an integral number of half cycles which may be tracked from the continuity of the motion.

Displacement as a function of time is thus obtained with directional discrimination, without distortion and without practical limitation. Any undesirable low-frequency components may be filtered out if necessary. The eight-bit digitization of the transient recorder enables a displacement of 0.5 nm to be resolved. A sampling frequency of 100 MHz allows ultrasonic frequencies up to 20 MHz to be followed. This technique is valuable as it seems to be the only one that can satisfactorily cover displacements from a very small fraction of a wavelength to many wavelengths. It does, however, need a fast transient recorder and time for computation. Signal averaging is not so practical as with compensated interferometers.

3.9.5   Electronically simulated frequency shifting

Another method of using the two phase outputs, applicable to displacements corresponding to a few cycles or more, entails their combination with a 'carrier' frequency in a single sideband modulation circuit. An output is generated that is the sum or difference of the carrier and signal (i.e. the Doppler) frequency. The carrier thus acts as a frequency shift and the resulting shifted signal may be frequency tracked or otherwise processed in the same way as an optically shifted signal (see §3.8.3).

A block diagram of the arrangement is shown in figure 3.32. One phase, $V_0 \cos(2\pi f_D t)$, is mixed with the carrier and the other, $V_0 \sin(2\pi f_D t)$, with the carrier shifted by 90°. The outputs of the balanced mixers are therefore proportional to $V_0 \cos(2\pi f_D t)\cos(2\pi f_c t)$ and $V_0 \sin(2\pi f_D t)\sin(2\pi f_c t)$ where f_D and f_c are the signal and carrier frequencies respectively. Subtraction

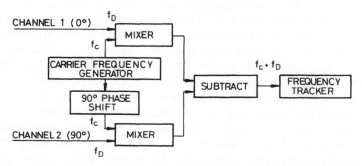

Figure 3.32   Frequency tracking with directional discrimination using quadrature signals.

yields $V_0 \cos[2\pi(f_c + f_D)t]$, i.e. the sum frequency. It may be verified that this is true also if f_D is negative which would result from a reversal of the direction of motion.

This scheme is not a perfect replacement for optical frequency shifting but can give similar performance and is often acceptable in interferometry (but not usually in laser Doppler anemometry). To avoid leakage of the carrier, which would seriously upset frequency tracking, the modulators must be carefully balanced. This is sensitive to the DC level of the signal inputs and may make the use of DC coupling in the signal preamplifiers unpractical. Signals from poorly reflecting surfaces may need considerable amplification, and ideally DC coupling should be used to permit response down to zero velocity. It may, however, be necessary to accept a very-low-frequency cut-off at a fraction of a hertz, which could lead to malfunctioning in exceptionally vibration-free environments.

3.9.6 Selection and combination of quadrature signals

When dealing with ultrasonic pulses of amplitude small compared with an eighth of a wavelength, superimposed on a low-frequency background, simple methods for selection or combination of the quadrature signals are useful. We suppose that there is a considerable frequency difference between the ultrasonic signals and the background so that they may be clearly separated by filters. Small signals occurring not too close to a turning point (such as B and C in figure 3.21 but not A) are not severely distorted, and the background level may be removed by a high-pass filter to give a fair representation of the signal with a sensitivity factor proportional to $\sin \phi$, where $V_0 \cos \phi$ is the current value of the low-frequency component. ϕ is the current mean phase value which is varying due to the background vibration. Sensitivity in the quadrature phase is proportional to $\cos \phi$. We thus see that there is always one channel in which the sensitivity is at least $\sqrt{2}$ of its maximum value. Thus, by choosing the most favourable channel, ultrasonic signals may always be received with reasonable sensitivity and not too much distortion. The selection may be based on the actual magnitudes of the signals or on the low-frequency components. Assuming a balanced output, the best choice is the phase nearest balance at the time of the signal. The sign of the ultrasonic response may also be registered if required by noting the sign of the other phase.

A simple way of combining the two phases directly is by squaring and adding the high-frequency components. The ultrasonic sensitivity is then evidently independent of the low-frequency vibrations but directional information is lost. This may not be important in many practical circumstances when we are only interested in the time of arrival and the peak-to-peak amplitudes of pulses, which probably contain many alternations of sign anyway.

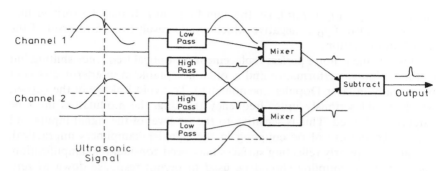

Figure 3.33 The use of quadrature detection for separating low-amplitude ultrasonic signals from low-frequency background vibration.

A more complicated combination technique allows the directional information to be retained. The scheme is shown in figure 3.33. The high- and low-frequency components are separated by filters and the high-frequency component of one phase is multiplied by the low-frequency component of the other in balanced mixers. The high-frequency signals in each channel from a small ultrasonic displacement Δx are proportional to the displacement derivative of the signal in that channel. Supposing the current values of the low-frequency components in the two channels are $V_0 \cos \phi$ and $V_0 \sin \phi$, then the high-frequency inputs to the mixers are $-(4\pi\Delta x/\lambda)V_0 \sin \phi$ and $(4\pi\Delta x/\lambda)V_0 \cos \phi$. Thus the outputs are proportional to $-\Delta x V_0^2 \sin^2 \phi$ and $\Delta x V_0^2 \cos^2 \phi$. Subtraction thus yields a sensitivity independent of ϕ. At any instant the calibration constant V_0^2 may be determined from the sum of the squares of the unfiltered signals in the two channels, but because the calibration and signal processing circuits cannot be exactly the same, high accuracy in the determination of displacement should not be expected.

In the above scheme, it is necessary that the filters do not introduce significant phase shifts and this is only possible if the ultrasonic and background signals are in well separated frequency ranges. An alternative approach, more generally applicable but based on the same principle, is to multiply the derivative of the signal in one phase with the signal in the other. Subtraction then yields an output proportional to the product of V_0^2 and the instantaneous frequency of the signal and hence the velocity of the surface.

This type of signal processing technique is most useful for processing signals from a moving surface where the background may be varying randomly in phase and amplitude due to the changing speckle pattern. Compensated interferometers may not be able to adjust rapidly enough to the changing conditions. Keeping track of the continuously varying calibration factor could be a problem.

3.10 LONG-PATH-DIFFERENCE INTERFEROMETRY

3.10.1 High-resolution spectroscopy

An alternative interferometric technique for detecting ultrasound does not use a reference beam but instead is based on the direct determination of the small frequency changes in the light scattered from the moving surface, i.e. the Doppler shift (see §3.1.5). The interferometer thus produces an output basically proportional to the velocity of the surface rather than the displacement. For highly reflecting surfaces, velocity interferometers have no particular advantage over those using a reference beam previously described. Because of the long optical path required, they may be much less practical, particularly at low frequencies. Velocity interferometers are, however, not affected by low-frequency vibrational motion of the surface, and for rough surfaces the sensitivity is not fundamentally limited by the speckle effect (see §3.3.5). They thus offer a potentially very useful advantage in sensitivity when dealing with poorly scattering or diffusing surfaces, particularly when the illuminated spot area cannot be made small. The technique is most practical for high ultrasonic frequencies where the sensitivity of the reference-beam technique tends to become inadequate.

The changes in frequency to be measured are extremely small, e.g. a fractional change of 10^{-8} for a velocity of $1.5 \, \mathrm{m \, s^{-1}}$, which is the maximum velocity associated with an amplitude of 120 nm at 2 MHz. This degree of resolution can only be achieved by interferometric spectroscopy with a large path length difference. This may be accomplished by employing either a simple interferometer with a very long path, or an interferometer using a large number of multiple reflections, i.e. a Fabry–Perot interferometer (§3.11). Both techniques have been used in laser Doppler studies of supersonic flow (Jackson and Paul 1971, Smeets and George 1978). Various interferometer arrangements for the detection of ultrasonic surface displacements have been described by Kaule (1977, 1983).

3.10.2 Time delay interferometry

In the scheme shown in figure 3.34, the surface is illuminated by a laser beam and some of the scattered light collected and made into a parallel beam which is then divided into equal parts by the first beam splitter. The beams are recombined in the second beam splitter, one of the beams having traversed a considerable distance. The phase difference between the two beams recombined at the detector D is given by

$$\phi_L = \frac{2\pi L}{\lambda} = \frac{2\pi L \nu}{c} \tag{3.38}$$

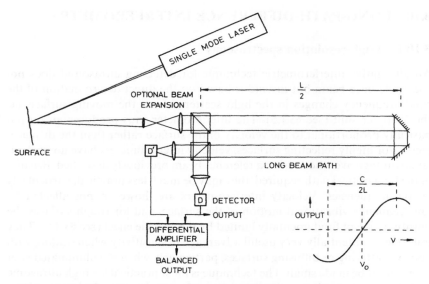

Figure 3.34 The direct measurement of the frequency shift of scattered light from a moving surface by an interferometer with a long optical-path-length difference.

where L is the optical path difference between the two beams and c is the velocity of light. Hence the output of the detector has a sinusoidal dependence on the light frequency v, as shown in the inset of figure 3.34. The maximum sensitivity occurs when the output is halfway between its maximum and minimum values. This occurs when $v = (n \pm \frac{1}{4})c/L$, or

$$L = (n \pm \tfrac{1}{4})\lambda \tag{3.39}$$

where n is an integer. Assuming that L is chosen to satisfy this condition for the unshifted laser frequency (and any adjustment needed would be very small) the output of the detector when the frequency is shifted by Δv is

$$V = \bar{V} \pm V_0 \sin \frac{2\pi \Delta v L}{c} = \bar{V} \pm V_0 \sin \frac{4\pi L u}{c\lambda} \tag{3.40}$$

where u is the velocity of the surface and V_0 the peak interference signal. For small frequency deviations as is usually the case, the output voltage approximates to a linear dependence on u:

$$V = \bar{V} \pm \frac{4\pi L u V_0}{c\lambda}. \tag{3.41}$$

It will be seen that the interferometer arrangement shown in figure 3.34 readily allows the addition of a second detector D' whose output is antiphase with that of D. This enables a balanced detector scheme to be used with the advantages discussed in §3.1.4. The interferometer may then be balanced to give zero output for the unshifted laser light.

An alternative way of describing the operation of the interferometer is to note that interference is produced between the scattered light input and is itself delayed by a time τ_D, the difference in time taken by light to traverse the two optical paths in the interferometer. The phase difference ϕ_D produced by the delay is clearly proportional to the distance moved by the surface in time τ_D:

$$\phi_D(t) = \frac{4\pi}{\lambda} [x(t) - x(t - \tau_D)] \qquad (3.42)$$

where $x(t)$ is the displacement of the surface at time t. Normally τ_D is very much less than the period of the ultrasonic movement and then the velocity $u (= dx/dt)$ is effectively constant and, since $\tau_D = L/c$, equation (3.42) evidently leads to (3.40). τ_D may typically be 10 ns corresponding to an optical path difference of 3 m. If τ_D is not small compared with the ultrasonic period, analysis in terms of the delay time is the more satisfactory method of describing the operation of the interferometer.

Clearly, the main problem with interferometers of this type is the stabilization of the path to a fraction of a wavelength in possibly several metres. Although reasonable stability can be obtained by rigid construction and temperature control, an active stabilization technique as used in the compensated interferometers discussed in §3.7 is desirable. The methods available are, however, more limited because of the width and/or angular spread of the beam in the interferometer.

The stability of the laser is of course equally important and a laser whose cavity length is insensitive to or stabilized against environment factors should be used. It is usual to select one longitudinal mode by means of an étalon (see §§1.2.3 and 3.11.1). The coherence length is then very great and large path length differences may be tolerated without loss of coherent interference. However, it is probable that the use of a single-mode laser is not absolutely necessary if L is chosen to be close to an integral multiple of twice the cavity length as discussed in §3.5.1.

3.10.3 Sensitivity considerations

For a sinusoidal ultrasonic displacement $x = a \cos(2\pi f t)$, we find, using equations (3.8) and (3.42), that the change in the light power reaching the detector (or each detector in a balanced scheme) is

$$\Delta W_D = \tfrac{1}{2} W_I \sin\left(\frac{8\pi a}{\lambda} \sin(\pi f \tau_D) \sin 2\pi f(t - \tfrac{1}{2}\tau_D)\right) \qquad (3.43)$$

where W_I is the light power entering the interferometer. As in the last section, we have assumed that there is an optimum 50% beam split in the interferometer and that the optical path length L satisfies equation (3.39). For small displacement amplitudes the variation of ΔW_D is sinusoidal with an amplitude

$$\Delta W_{D0} = \frac{4\pi a}{\lambda} W_I \sin(\pi f \tau_D). \qquad (3.44)$$

The response is thus a maximum for $\tau_D = 1/2f$, i.e. a time delay equal to half the period of the ultrasonic wave.

If there is perfect reflection from the ultrasonically vibrating surface, all the light from the laser is available for the interferometer and we may put $W_I = W_0$, the power of the laser, in equation (3.44). It is interesting to compare the response under these conditions with that of a reference-beam-type interferometer which may be deduced from equations (3.1) and (3.9), noting that half the laser power goes to each detector:

$$\Delta E_{D0}(\text{RBI}) = \frac{2\pi a}{\lambda} W_0. \qquad (3.45)$$

It is thus seen that in these circumstances the maximum sensitivity of the velocity interferometer is twice that of the reference-beam type, as pointed out by Mueller and Rylander (1979). This may be understood by noting that the light power in the interferometer is the same in both cases, but the reference-beam type essentially uses interference between the peak displacement and the mean, whereas in the long-path-difference type the signal results from interference between the peak positive and peak negative displacements. However, if only a fraction of the laser light is returned to the interferometer, this result is no longer valid and, for small light returns, the theoretical maximum sensitivity limited by photon noise tends to the same value for both types of interferometer.

There is thus no real advantage in the use of a velocity interferometer if a reflection from a good optical surface is available. The requirement for a long optical path in the interferometer is a considerable inconvenience particularly below about 100 MHz. Although sensitivity to low-frequency vibrations of the surface is avoided, stabilization of the interferometer is still necessary because of the long optical path involved.

The use of a fibre-optic delay loop is a way of providing a large path difference in a compact space. The optimum delay could then be arranged to suit most circumstances. A single-mode fibre is, however, necessary to retain coherence in the interferometer. The coupling requirements for these fibres (core diameters of only a few micrometres) place similar restrictions on the useful reception aperture for a given focused spot size as in reference-beam interferometry. There is thus the same limitation on sensitivity with rough surfaces, and signal fluctuations due to speckle (§3.3.1) will be experienced. This is confirmed by measurements of the sensitivity of a fibre-optic time-delay interferometer with various types of surface made by Bruinsma and Vogel (1988). As optical path lengths in fibres are very sensitive to external conditions, stabilization would be required in most interferometers. However, a special type of interferometer using an optical fibre delay loop which does not require stabilization is described in §4.4.3.

3.10.4 Long-path-difference interferometry with rough surfaces

For the detection of ultrasound on an optically rough surface, there is potentially a considerable advantage in the use of a velocity type of interferometer. This is because they are not restricted by the spatial coherence condition to which interferometers of the reference-beam type are subject, and are thus able to use light effectively even though it does not come from a single finely focused spot. In the interferometer shown in figure 3.34 for example, beams in the interferometer which are produced from light coming from a point on the surface remain parallel on recombination. There is thus no loss of interference signal due to lack of parallelism.

There is still, however, a restriction on the angular spread of the beams in the interferometer and hence on the diameter of the illuminated spot that can be accommodated. This is due to the fact that the path length difference in the interferometer is slightly dependent on the inclination of the beams to the axis of the interferometer, and this can result in a smearing of the phase of the interference signal and hence loss of sensitivity.

For interferometers using plane surfaces for beam splitters and mirrors, the path difference for an inclination ζ is $L \cos \zeta$. This may be seen most easily for interference between reflections from two parallel surfaces, one partially and one fully reflecting, as shown in figure 3.35, but it is also true for the interferometer arrangement shown in figure 3.34. A deviation of $\frac{1}{2}\lambda$ and hence a change in phase by 180° occurs for $\zeta = (\lambda/L)^{1/2}$ (since ζ is very small).

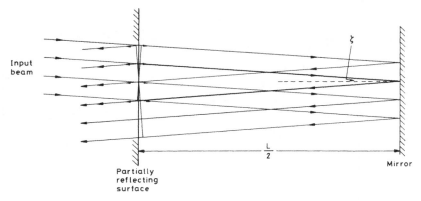

Figure 3.35 The effect of beam inclination on interference in a long-optical-path-difference interferometer.

For example, if $\lambda = 0.5\,\mu$m and $L = 500$ mm, $\zeta = 1.0$ mrad or $3.4'$. This is the maximum beam inclination allowable if a satisfactory interference signal is to be retained.

This restriction is not a fundamental one though, since the angular spread of the beam may be reduced by beam expansion using a lens system such as that shown in figure 3.34. This is, however, at the expense of a greatly increased beam diameter which is likely to make the optical components of the interferometer rather impracticable or expensive. A detailed calculation shows that to equal the sensitivity of a reference beam for the same spot size, laser power, etc, an interferometer of this type must have an optical delay arm of volume at least V_{int} where

$$V_{int} = \frac{\lambda c^2}{2\pi^2 f^2} \tag{3.46}$$

where f is the ultrasonic frequency and c the velocity of light. Thus at 10 MHz with $\lambda = 0.5\,\mu$m, mirrors of diameter approximately 8 mm would be required for a separation of 500 mm. The signal to noise ratio increases in proportion to the mirror diameter. Thus at this frequency, a modest improvement over the reference-beam technique should be obtainable without too much difficulty if an adequate reception aperture is available.

There is, however, a simple way of considerably improving the tolerance of a long-path-difference interferometer to beam inclination. This may be done by the use of two long path arms in the interferometer, each arm being in media of different refractive index (Gillard 1970). The arrangement is shown in figure 3.36. The two arms of the interferometer consist of lengths

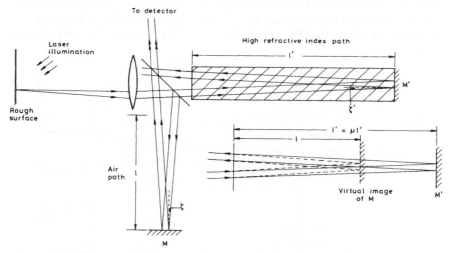

Figure 3.36 Velocity interferometer with long paths in media of different refractive index to ensure insensitivity to beam inclination.

l in air and l' in a medium of high refractive index (μ). This could be a glass or liquid. The optical path difference between the arms of the interferometer is then

$$L = 2\mu l' \cos \zeta' - 2l \cos \zeta \qquad (3.47)$$

where ζ and ζ' are the angles of inclination of the beam paths in air and the medium respectively. Evidently by Snell's law $\sin \zeta = \mu \sin \zeta'$. Expanding equation (3.47) in even powers of $\sin \zeta$, we find that the term in $\sin^2 \zeta$ vanishes for $l' = \mu l$. The path difference is then given by

$$L = L_0 \left(1 + \frac{\sin^4 \zeta}{8\mu^2} + \cdots \right) \qquad (3.48)$$

where

$$L_0 = 2l' \left(\mu - \frac{1}{\mu} \right) = 2l(\mu^2 - 1). \qquad (3.49)$$

Assuming that an optical path difference of $\frac{1}{2}\lambda$ is allowable, the maximum beam inclination angle ζ_{max} for the efficient operation of the interferometer is then given by

$$\sin^4 \zeta_{max} = 4\mu^2 \lambda / L_0. \qquad (3.50)$$

Thus if $L_0 = 500$ mm, $\lambda = 0.5$ μm and $\mu = 1.6$, then $\zeta_{max} = 3.24°$. This is a very much less severe restriction which is unlikely to be a limitation in practice. It corresponds in the example given to an acceptable spot diameter of about 11 mm viewed at a distance of 100 mm. Resolution requirements would normally restrict the spot size to a value much less than this. In any case, the accommodation of the inclined beams within the width of the interferometer arms is likely to be more of a problem.

Note that, in this arrangement, the mirrors are in equivalent image positions as shown in the inset in figure 3.36 where the path in air has been superimposed on the other arm of the interferometer. Thus the inclined beams in the two arms have the same lateral displacement and superimpose efficiently in this respect as well as suffering a minimum deviation of optical path difference.

3.10.5 Sensitivity calculation for optically rough surfaces

Assuming that the spot size is not a restricting factor, the sensitivity of an interferometer of this type is governed by the length of the optical path difference and the light power W_I used by the interferometer, which in turn depends on the distance of the collecting lens from the surface, aperture and interferometer optics. If a balanced detector system is used, the signal from the interferometer produced by an ultrasonic displacement of frequency f

and amplitude a corresponds, by equation (3.44), to a light fluctuation

$$2\Delta W_{D0} = \frac{8\pi a}{\lambda} W_1 \sin\left(\frac{\pi Lf}{c}\right). \tag{3.51}$$

The RMS fluctuation produced by photon noise in a bandwidth Δf is, by equation (1.8),

$$\delta W_N = (2W_1 h\nu\Delta f/\eta)^{1/2}. \tag{3.52}$$

Thus displacement amplitude δa_N for a unity signal to noise ratio (the noise equivalent amplitude) is

$$\delta a_N = \frac{\lambda}{4\pi \sin(\pi Lf/c)} \left(\frac{h\nu\Delta f}{\eta W_1}\right)^{1/2}. \tag{3.53}$$

For a perfectly diffusing 'white' surface, the light scattered per unit solid angle in a normal direction is W_0/π (see equation (3.18)) where W_0 is the power of the illuminating laser beam. Hence the noise equivalent ultrasonic vibration amplitude is

$$\delta a_N = \frac{c}{2\pi^2 fL}\left(\frac{F}{D}\right)\left(\frac{hc\lambda\Delta f}{\eta W_0}\right)^{1/2} \tag{3.54}$$

where we have now assumed that the practical value of L is small compared with its optimum value $c/2f$ as will almost certainly be the case below about 100 MHz.

The ratio of the diameter D of the receiving optics to the focal length F (the inverse of the 'f no') is a measure of the light collection efficiency. In view of the need for a long path in glass or liquid and to provide for beam inclinations up to a degree or so, there are likely to be practical restrictions on the value of the optical path length difference L. A value of 500 mm is suggested as a reasonable figure, corresponding for $\mu = 1.6$ to arms of length $l = 160$ mm (air) and $l' = 256$ mm (glass or liquid) (equation (3.49)). Assuming this and the use of a 1 W argon laser, $\lambda = 0.5 \mu$m and $\eta = 0.5$, we deduce

$$\delta a_N = 1.36 \times 10^{-9}\left(\frac{F}{D}\right)\frac{(\Delta f)^{1/2}}{f} \tag{3.55}$$

where δa_N is in metres and the frequencies are in hertz ($f < 100$ MHz). For $D/F = 0.1$ and an ultrasonic frequency of 10 MHz the estimated minimum detectable displacement amplitude in a 5 MHz bandwidth making the above assumptions would be about 30 pm.

3.10.6 Comparison with reference-beam interferometry

The limits of sensitivity of reference-beam interferometry under the same conditions were discussed in §3.3.5. The RMS noise equivalent displacement δx_N is given by equation (3.29). This corresponds to an amplitude of sinusoidal motion of $\delta x_N \sqrt{2}$. Comparison with the two types of interferometer for the same laser power, detector efficiency, wavelength, surface reflectivity and bandwidth for a sinusoidal movement of frequency f yields

$$\frac{\text{S/N(LPDI)}}{\text{S/N(RBI)}} = \frac{\pi}{\sqrt{2}} \left[\sin\left(\frac{\pi L f}{c}\right)\left(\frac{D}{F}\right) \right]\left[\frac{r_0}{\lambda} \right] \tag{3.56}$$

where the variables in the first and second sets of square brackets are the significant parameters for the long-path-difference (LPDI) and reference-beam interferometers (RBI) respectively. Note that the signal to noise value used for reference-beam interferometry is an average value of a quantity that varies widely owing to the speckle effect (§3.3.1).

The comparative sensitivity of the two types of interferometry in a particular application thus depends on the ultrasonic frequency f, the diameter of the RBI illuminated spot ($2r_0$), and practical values of the aperture D/F and the optical path length difference L for the LPDI. Taking a typical RBI illuminated spot diameter of 0.1 mm, a reasonable value of $D/F = 0.1$ and assuming maximum sensitivity (i.e. $\sin(\pi L f/c) = 1$) for the LPDI, we find that the sensitivity of the LPDI is about 20 times greater than that of the RBI. Assuming a reasonable value of $L = 500$ mm, the maximum sensitivity occurs at 300 MHz, which is well above the normal ultrasonic frequency range. We note that, for $L = 500$ mm, the sensitivities become equal at an acoustic frequency around 10 MHz. Equality occurs at 1 MHz for a spot diameter of 1 mm, which is about the maximum that would normally be used. Hence below 1 MHz, reference-beam interferometers would almost always be preferred.

Above 100 MHz, however, the situation is reversed and LPD velocity interferometers may be expected to have a greater sensitivity even if the spot size is as small as 10 μm. In microscopic applications, though, the necessity for having a very small spot may mean that the reference-beam type of interferometer remains competitive. The short wavelengths of high-frequency ultrasonic waves may also enforce the use of very small spot sizes, especially if surface waves are being observed. On the other hand, large spot sizes may be unavoidable due to imperfections in optical components or refractive index variations in the transmission medium. Then the long-path-difference type of interferometer would be superior down to much lower frequencies.

An important advantage of the velocity type of interferometer is its freedom from the wide fluctuations in sensitivity from the speckle effect, although signals may be less easy to interpret and calibrate. The advantage of insensitivity to low-frequency vibration of the surface is offset by a sensitivity to vibration in the long optical path, for which active stabilization is almost certainly required. A favourable factor when using the reference-beam system is that a strong reference beam provides a form of optical amplification, enabling the detector to operate at higher light levels. This can make it easier to approach the theoretical photon-limited sensitivity and reduce the susceptibility of the interferometer to background light.

3.11 FABRY–PEROT INTERFEROMETERS

3.11.1 Principle of operation

A way of increasing the effective path length difference in an interferometer is multiply to reflect the beam across the interference volume or 'cavity' by the use of highly reflecting surfaces as illustrated in figure 3.37. This is the principle of the Fabry–Perot interferometer or 'étalon', a device well known in high-resolution spectroscopy. For detailed information the reader is referred to recent books on this subject by Hernandez (1988) and Vaughan (1989). This type of interferometer makes use of interference between a large number of reflected beams to produce a response curve consisting of a sequence of narrow peaks, as illustrated in figure 3.37. Fully constructive interference is obtained when all the transmitted beams are in phase which occurs when

$$2h = n\lambda = nc/v \qquad (3.57)$$

where h is the thickness of the cavity, i.e. the separation of the reflecting surfaces. This occurs for any integral value of the order of interference n.

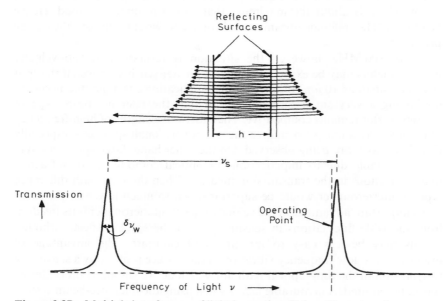

Figure 3.37 Multiple interference of light beams in a Fabry–Perot interferometer and an example of a spectral response curve.

Thus the response curve is periodic with peaks corresponding to adjacent orders separated by a frequency spacing

$$v_s = c/2h. \tag{3.58}$$

The refractive index of the medium in the interferometer cavity is assumed to be unity. The amplitudes of the transmitted beams decrease exponentially with the number of reflections and the phase difference advances by

$$\phi_D = 4\pi h v/c - 2\pi n. \tag{3.59}$$

For convenience, an integral number of cycles has been subtracted so that the ϕ_D lies between $\pm\pi$ for frequencies near the nth-order peak. By summing the infinite series of vector amplitudes, we may derive the following expression for the fraction of the incident intensity transmitted:

$$T_E = \frac{T^2}{1 + R^2 - 2R\cos\phi_D} \tag{3.60}$$

where T and R are the transmission and reflection coefficients for each reflecting surface. ($T = (1 - R)$ for zero absorption.) The transmission falls to a half of its maximum value on either side of the nth-order peak when

$$\sin(\tfrac{1}{2}\phi_D) = \pm\frac{1 - R}{2\sqrt{R}}. \tag{3.61}$$

Thus when R is close to unity, the width to half height is given (using equation (3.59)) by

$$\delta v_w = \frac{c(1 - R)}{2\pi h \sqrt{R}} = \frac{c}{2h\mathscr{F}} \tag{3.62}$$

where $\mathscr{F}$ is the 'finesse', the ratio of order spacing to line width. For a plane Fabry–Perot interferometer

$$\mathscr{F} = \frac{v_s}{\delta v_w} = \frac{\pi\sqrt{R}}{1 - R}. \tag{3.63}$$

To obtain good resolving power, it is necessary to make R as close to unity as possible. A low-absorption multiple dielectric coating would normally be used. For example, a reflectivity of 98.5% gives a 'finesse' of about 200 and for a thickness $h = 50$ mm the bandwidth would be 14 MHz.

When measuring small changes in the frequency of light reflected or scattered from a moving surface, the spacing of the interferometer should be adjusted so that the laser frequency is near a point of maximum slope of the response curve as shown in figure 3.37. For convenience and maximum efficiency, a single-mode laser should be used with a Fabry–Perot interferometer, although in principle the resolving power of the device is sufficient to separate the individual modes of the laser, so that one mode may be selected. An unstabilized laser, however, usually has too much frequency drift to maintain constant operation on the correct part of the response curve. The Fabry–Perot interferometer itself also drifts unless its temperature is stabilized. Thus an active compensation technique to maintain the correct operating point on the response curve is very desirable.

Analysis of the response of the interferometer is much simplified if we assume that it is operated about a half-height point of the response curve, e.g. $\frac{1}{2}\delta v_w$ below the central peak, as illustrated in figure 3.37. The slope of the response curve is then

$$\frac{\mathrm{d}T_E}{\mathrm{d}v} = \frac{2\pi h\sqrt{R}}{c(1-R)} = \frac{2h}{c}\,\mathscr{F} \tag{3.64}$$

where h is the separation of the mirrors and $\mathscr{F}$ the finesse. The change in frequency of light reflected or scattered normally from a moving surface is given by the usual Doppler shift formula, i.e. $\delta v_D = 2u_n/\lambda$ (see equation (3.4)), where u_n is the normal component of the surface velocity. Thus the change in the output of a Fabry–Perot interferometer receiving this light is proportional to surface velocity (provided the timescale of the variation of velocity is not too high: see §3.11.2). Assuming no light losses, the change in output is given by

$$\Delta W_D = \frac{4\pi h u_n\sqrt{R}}{c\lambda(1-R)}\,W_I \tag{3.65}$$

where W_I is the light input to the interferometer.

Because of the multiple use of the cavity in the Fabry–Perot interferometer, this arrangement is considerably less cumbersome than the long-path interferometers described in §3.10. Comparison of equations (3.40) and (3.65), assuming full use of the light in both cases (i.e. $V = V_0$ in equation (3.40)), shows that the ratio of the slope of the response curve at the operating point to the maximum output is the same for the two types of interferometer when the optical path difference in the long-path interferometer is given by

$$L^* = \frac{2h\sqrt{R}}{1-R} = \frac{2h}{\pi}\,\mathscr{F}. \tag{3.66}$$

L^* may be regarded as the effective path difference for multiple reflections in the Fabry–Perot interferometer. The corresponding effective time delay is thus

$$\tau_D^* = \frac{2h\sqrt{R}}{c(1-R)} = \frac{2h}{\pi c}\,\mathscr{F}. \tag{3.67}$$

3.11.2 The frequency response of Fabry–Perot interferometers

As with the long-path-difference interferometer, analysis simply in terms of frequency shift is not valid if the timescale of the fluctuations of surface displacement is greater than or comparable with the effective time delay in the interferometer. The response must then be calculated by noting that each of the interfering multiple beams in the output suffers an extra phase shift associated with the displacement of the surface when it was reflected or scattered there. The effects of these must be added for each contribution to the amplitude of the output beam. Assuming a large number of interfering beams, i.e. $(1 - R)$ is small, operation about the half-height point below the peak of the response curve, and no light losses, we may calculate the response to a small sinusoidal displacement of the surface of amplitude a and frequency f. The change in the output of the interferometer is, in this approximation,

$$\Delta W_D =$$

$$\frac{4\pi a \Delta [\sin \varepsilon \cos \varepsilon (\Delta^2 - 2 \sin^2 \varepsilon)\cos(2\pi f t) + 2 \sin^2 \varepsilon(\Delta + \sin^2 \varepsilon)\sin(2\pi f t)]}{\lambda(\Delta^4 \cos^2 \varepsilon + 4 \sin^4 \varepsilon)} W_I$$

(3.68)

where $\Delta = \pi/\mathscr{F}$, $\varepsilon = \pi f/v_s$ and W_I is the light input to the interferometer. The displacement of the surface is assumed to be $a \sin(2\pi f t)$ as observed at the input to the interferometer (i.e. allowing for the delay time from the surface if necessary).

The response may also be analysed by considering the response of the interferometer to the sidebands produced by the phase modulation (Monchalin and Héon 1986). These workers have treated the general case and performed experimental verification with a confocal interferometer.

Since a change of integral multiples of 2π in the extra phase shifts makes no difference to the result, the response is evidently periodic as a function of frequency with a spacing v_s, the separation of orders. Examples of the frequency dependence of the amplitude W_{D0} of the response are shown in figure 3.40 below. Usually the only region of practical interest is where $f \ll v_s$; then a simpler formula is applicable:

$$W_{D0} = \frac{2\pi a}{\lambda} \frac{2\beta}{(4 + \beta^4)^{1/2}} W_I$$

(3.69)

where

$$\beta = 2\pi f \tau_D^* = \frac{2\varepsilon}{\Delta} = \frac{2f}{\delta v_w}.$$

(3.70)

The maximum response occurs for $\beta = \sqrt{2}$, corresponding to a frequency given by

$$f_{\text{max}} = \frac{1}{\pi\sqrt{2}\tau_{\text{D}}^*} = \frac{\delta v_{\text{w}}}{\sqrt{2}}. \tag{3.71}$$

It will be noted that, to this approximation, the response curve depends only on the resolution of the interferometer and not on the reflectivity and mirror separation separately.

As Monchalin *et al* (1989) have pointed out, there is an advantage in using the beam reflected back towards the source instead of the transmitted beam. This gives a better response at frequencies above the passband of the interferometer. This beam may be considered as the result of interference between two components, (a) an initial reflection, and (b) a sum of multiple reflections which has virtually the same amplitude and phase as the transmitted beam. With the assumptions detailed above, we find that the amplitude of the intensity fluctuation of this back-reflected beam resulting from the sinusoidal displacement of the surface of amplitude a and frequency f is

$$W_{\text{D0}}^* = \frac{4\pi a\beta}{\lambda}\left(\frac{1 + \beta^2}{4 + \beta^4}\right)^{1/2} W_{\text{I}} \tag{3.72}$$

where β is given by equation (3.70) above. At low frequencies the amplitudes of the fluctuations of the transmitted and back-reflected beams are equal and have opposite phase as we expect, but at high frequencies W_{D0}^* approaches the value $4\pi a W_{\text{I}}/\lambda$, giving an essentially flat response at frequencies above δv_{w} until the order spacing v_{s} is approached, when equation (3.72) is no longer adequate.

The behaviour at high frequencies may be interpreted in terms of 'optical sideband stripping' described by Monchalin *et al* (1989). Because of the narrow passband of the interferometer, the component (b) that has passed through the interferometer has been stripped of the sidebands separated from v by the acoustic frequency f. The component (a), however, still has these sidebands and hence interference with the correct phase shift produces intensity fluctuations which are not dependent upon frequency. In principle, there is an increase of a factor of two in the maximum attainable sensitivity in this mode of operation but in practice this is likely to be lost by the optics required to separate out the back-reflected beam from the incident beam.

Another way of producing much the same result is to split off a part of the input beam and recombine it with the transmitted beam with the appropriate phase shift in an auxiliary interferometer. The Fabry–Perot interferometer can then be tuned to the peak of the response curve. This possibility has also been discussed by Monchalin *et al* (1989).

Finally in this section, we note that the output from a Fabry–Perot interferometer is evidently very sensitive to the spacing of the mirrors. Very small displacements of these may thus be detected. The multiple reflections across the cavity give an increase in sensitivity of perhaps a hundred fold over that of a simple interferometer. This may be used as the basis of an ultrasonic detector of extraordinary sensitivity in which the specimen acts as one of the mirrors (Thomson *et al* 1973). Assuming the use of a 1 mW helium–neon laser, 99% reflectivity, an appropriate silicon diode photo-detector and a system bandwidth of 1 Hz, they estimate a theoretical sensitivity limit of 10 attometres (i.e. 10^{-17} m). The method has, however, limited practical application because of the extreme requirements of surface flatness and reflectivity, but it could be useful for some special investigations.

3.11.3 The confocal Fabry–Perot interferometer

An important consideration in the use of interferometers or spectrometers is the light-gathering power for extended sources. In the plane Fabry–Perot interferometer, this has similar limitations to that of a simple long-path interferometer due to the dependence of the difference in optical path length on the inclination of a ray as shown in figure 3.35. In the terminology of Fabry–Perot spectroscopy, the 'étendue' is limited. This quantity is the product of the aperture area of the interferometer and the spread of inclinations (expressed as a solid angle) that may be accepted without reducing resolution. For a plane Fabry–Perot interferometer with mirrors of a given diameter, the product of the étendue and resolving power is constant. Thus light-gathering power must be sacrificed for extra resolving power or vice versa. To obtain adequate étendue and resolving power together, impracticably large mirror diameters may be required (see §3.11.5).

An alternative Fabry–Perot arrangement, that with a confocal configura-tion, has however a very much less severe restriction on light-gathering power and offers a more efficient instrument when light from an extended source is being analysed, permitting the use of a larger illuminated spot when measuring the velocity of a surface without reduction of sensitivity. A confocal Fabry–Perot interferometer consists of two coaxial partially transmitting mirrors of equal radii of curvature R_M separated by a distance R_M. The optical properties of this arrangement have been analysed by Hercher (1968). The path of a typical ray making two alternate transits to and fro across the confocal cavity is shown in figure 3.38. It will be seen that alternate transits are not equivalent and thus the effective length of the cavity determining resolution and the separation of orders is twice the actual separation of the mirrors. Unless the inclination angle is very small, the paths on successive double transits are not exactly coincident due to spherical aberration. However, to a good approximation the paths always cross the confocal plane at the same points and thus coherent interference of the emergent beams is obtained. For an exactly confocal separation of the mirrors, the optical path length around the cavity is very insensitive to the angle of

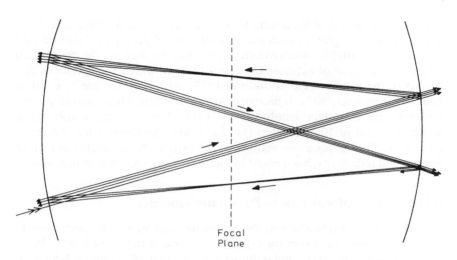

Figure 3.38   Light beams in a confocal Fabry–Perot interferometer.

inclination and distance from the axis, and thus light may be accepted over a wide input angle.

The application of a confocal Fabry–Perot interferometer to the detection of ultrasonic movements of a surface has been demonstrated by Monchalin (1985b) and by Bruinsma and Vogel (1988). Monchalin's arrangement is shown in figure 3.39. Light from a 5 mW helium–neon laser is directed to the sample collinearly with the detection system. Use of the polarizing beam splitter (PBS) and quarter-wave (QW) plate ensures that all the laser light is directed to the sample whilst the majority of the light scattered back from the sample goes to the interferometer. The output of the interferometer, which contains

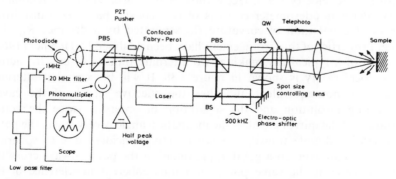

Figure 3.39   Confocal Fabry–Perot interferometry applied to the measurement of ultrasonic movements of a surface (Monchalin 1985b).

the signal proportional to the velocity of the surface, is recorded by the avalanche photodiode.

To ensure that this confocal Fabry–Perot interferometer is maintained at the most sensitive part of the response curve, a stabilization system is incorporated. A sampling beam coming directly from the laser is passed through the interferometer. This has a polarization orthogonal to the signal beam and is separated from it by a polarizing beam splitter. The transmission of this unshifted beam is maintained at half its peak value by the feedback which controls the separation of the mirrors by means of a piezoelectric displacer.

Also provided is a calibration system in which a simulated velocity signal is obtained by frequency modulation of the illuminating beam by means of the electro-optic phase shifter. A comparatively low frequency of 500 kHz is used so that the calibration signal may be separated by a filter at the output.

3.11.4 Theory of the confocal Fabry–Perot interferometer

The resolution, frequency response and sensitivity of a confocal Fabry–Perot interferometer may be calculated in the same way as for a plane Fabry–Perot system, and the formulae (3.57)–(3.71) in §§3.11.1 and 3.11.2 may be applied (at least for $R \simeq 1$) with some modifications as follows.

(i) Because alternate transits around the interferometer are not equivalent, the effective cavity spacing is doubled. Thus the mirror spacing h must be replaced by $2R_M$ for the confocal instrument. Thus the separation of orders corresponding to equation (3.58) is now

$$v_s = c/4R_M. \tag{3.73}$$

(ii) There are twice as many reflections in a double transit around the interferometer, i.e. four as against the two previously assumed. Thus R should be replaced by R^2. Hence the finesse is now given by

$$\mathscr{F} = \frac{\pi R}{1 - R^2}. \tag{3.74}$$

(iii) Interference only occurs between beams arising from alternate transits around the interferometer cavity. There are thus two output beams whose amplitudes are governed by multiple interference between beams arising from odd and even numbers of transits respectively. When the number of transits is large, each of these evidently has half the amplitude calculated for the output beam of the plane Fabry–Perot interferometer, and hence a quarter of the intensity. Both beams may now be collected by a detector, but since they are not coherent their intensities rather than their amplitudes add. This results in a general reduction in output by a factor of two compared with the plane Fabry–Perot interferometer.

In particular, we note that with the assumptions as used in the derivation of equation (3.69), the response to a small sinusoidal displacement of the reflecting or scattering surface is

$$W_{DO} = \frac{\pi a}{\lambda} \frac{2\beta}{(4 + \beta^4)^{1/2}} W_I \qquad (3.75)$$

where

$$\beta = 2\pi f \tau_D^* = \frac{8\pi R_M f R}{c(1 - R^2)} = \frac{8 R_M f}{c} \mathscr{F}. \qquad (3.76)$$

The maximum value of W_{DO} is $\pi a W_I / \lambda$.

It may be noted that, to a first approximation when $R \simeq 1$, the modifications (i) and (ii) above compensate, so that plane and confocal interferometers with the same mirror separation and reflectivity have virtually the same resolution and frequency response.

The use of the reflected beam from the interferometer gives a better response at frequencies above δv_w (the optical passband of the interferometer) as discussed in the last section. This arises from the interference of the initial reflection at the first surface with reflections which have passed around the interferometer and thus have been stripped of the sidebands generated by the acoustic frequency (Monchalin et al 1989).

In the confocal interferometer there are two reflected output beams not mutually coherent, corresponding to odd and even numbers of crossings of the interferometer cavity. Only the latter is coherent with the initially reflected beam, and can thus produce the desired output by interference. The former has the properties of a transmitted beam and produces a fluctuation which is out of phase at low frequencies and tends to zero at high frequencies. It is, however, generally impracticable to separate the two beams whilst maintaining the desired étendue of the interferometer. We thus suppose that they both reach the detector. The amplitude of fluctuation of the total back-reflected light due to a small sinusoidal movement of the surface of amplitude a and frequency f is then (with $R \simeq 1$ and the other assumptions detailed in §3.11.2)

$$W_{DO}^* = \frac{2\pi a \beta}{\lambda} \left(\frac{1 + \beta^2}{4 + \beta^4} \right)^{1/2} W_I. \qquad (3.77)$$

The frequency dependence is in fact the same as for the reflected beam from a plane Fabry–Perot interferometer (equation (3.72)) but the magnitude is reduced by a factor 2. W_{DO}^* reaches the value $2\pi a W_I / \lambda$ at high frequencies. The example of the frequency variation shown in figure 3.40 (curve CF1*) has been calculated using a more general formula, but this makes little difference until the frequency approaches v_s.

3.11.5   Light-gathering power

Although the maximum of the response is only half that of a plane Fabry–Perot interferometer, the confocal device is preferred for measurements on rough surfaces because of its much greater 'étendue', i.e. ability to accept input beams appreciably inclined to the axis without loss of resolution. A confocal interferometer can accept light rays that intercept both mirrors up to a mirror aperture radius r_M given by (Hercher 1968)

$$r_M^4 = \frac{R_M^3 \lambda}{\mathscr{F}} = \frac{R_M^3 \lambda (1 - R^2)}{\pi R}. \tag{3.78}$$

The étendue (Et) is thus

$$\mathrm{Et} = A\Omega = \frac{\pi^2 R_M \lambda}{\mathscr{F}} = \frac{\pi R_M \lambda (1 - R^2)}{R} \tag{3.79}$$

where A is the aperture area of the interferometer and Ω the solid angle in which the direction of the incident light may lie without significant loss of resolution. For example, a confocal interferometer with a mirror radius (and hence separation) of 100 mm and reflectivity of 98% can employ a mirror diameter of 3.2 mm, giving a theoretical étendue of 6×10^{-3} mm^2.

When the interferometer is used to analyse light scattered from an illuminated spot on a rough surface and delivered to the interferometer by a suitable lens system, the restriction imposed by the étendue may be written as

$$\mathrm{Et} = \frac{\pi}{4} \left(\frac{D}{F}\right)^2 \frac{\pi}{4} d^2 \tag{3.80}$$

where D/F is the angular aperture of reception and d is the diameter of the illuminated spot. Thus in the above example $D/F = 0.1$ (i.e. $f/10$) is the maximum useful reception aperture with a spot diameter of approximately 1.0 mm.

We may note that the étendue of a plane Fabry–Perot interferometer is (Hercher 1968)

$$\mathrm{Et} = \frac{\pi r_M^2 \lambda}{h \mathscr{F}} = \frac{r_M^2 \lambda (1 - R)}{h \sqrt{R}} \tag{3.81}$$

where h is the separation of the mirrors of aperture $2r_M$. It follows from equations (3.79) to (3.81) that to achieve the same étendue as a confocal interferometer of the same spacing and mirror reflectivity R (and hence approximately the same line width), a diameter about five times the separation of the mirrors would be required. This is clearly impractical in most circumstances. It follows therefore that the confocal type of interferometer is preferred whenever étendue considerations are important.

It will be seen that the étendue of a confocal interferometer increases with the mirror separation (and hence with radius of curvature), provided that the mirror diameters are also increased in accordance with equation (3.78). The required diameters remain quite small and practicable. It is thus an advantage to use a large mirror separation for ultrasonic studies on rough surfaces even though the highest resolution may not be required. This may enable a low reflectivity to be used, thereby reducing the effective number of transits and hence the accuracy necessary in figuring the mirrors. The low finesse is of no great disadvantage in this application. The greatest separation reported in use so far is 500 mm (Monchalin and Héon 1986). With a 90% mirror reflectivity, their interferometer has a bandwidth of 10 MHz: the useful mirror diameter is $2r_M = 17$ mm, giving a theoretical étendue of 0.2 mm^2.

3.11.6 Comparison of the sensitivities of Fabry–Perot and other types of interferometer

The sensitivity of Fabry–Perot interferometers is ultimately limited by photon noise considerations as with the other types of interferometer, and the theoretical minimum detectable displacement may be calculated in the same way using equation (1.8). We shall now extend the discussion of sensitivity to include a comparison of the different types of interferometer. Since the frequency responses are different, it is not possible to make a comparison which is valid for all frequencies and types of signal. Our calculations are based on the assumption of a sinusoidal surface displacement of frequency f. It should be remembered that this may not accurately reflect the relative merits of each interferometer for displacement pulses or other waveforms.

To facilitate comparison we express the sensitivity as follows:

$$\frac{1}{\delta a_N} = \text{SF}\pi \left(\frac{\eta W_0}{hc\lambda\Delta f} \right)^{1/2}. \tag{3.82}$$

Here δa_N is the amplitude of sinusoidal oscillation which gives an RMS output signal equal to the RMS photon noise, i.e. a signal to noise ratio of unity. This is a measure of the minimum detectable displacement. The right-hand side of equation (3.82) has been written as the product of two terms. The first, SF, is the 'sensitivity factor', which is characteristic of an interferometer operated under prescribed conditions. The second term is dependent only on the available laser power (W_0), optical wavelength λ, detector efficiency η, and the required bandwidth Δf, and is the same for all of the interferometer types to be considered. Interferometers may thus be compared on the basis of the sensitivity factor SF.

(a) Reflecting surfaces

For mirror-like reflecting surfaces, there are no great differences between the maximum sensitivities of the various interferometric techniques. For reference-beam interferometers (RBI), we deduce from equation (3.13) that SF(RBI) = 2.

Comparison with long-path interferometers (LPI) has already been made in §3.10.3, from where we may deduce that $\text{SF}(\text{LPI}) = 4\sin(\pi f\tau_\text{D})$. Although the maximum value is twice that of the reference-beam instrument, the advantage is limited to sinusoidal displacements of a frequency exactly matched to the delay time of the interferometer.

Since étendue is not a consideration for a mirror-like reflecting specimen surface, we may choose a plane Fabry–Perot interferometer for comparison. This has a slightly better performance under these conditions than the confocal type. Using equations (1.8) and (3.69) (which assume that $R \approx 1$ and $f \ll v_\text{s}$), taking $W_\text{I} = W_0$ and noting that the mean light power at the detector is $\frac{1}{2}W_0$, we deduce that for transmission

$$\text{SF}(\text{FPI}) = \frac{2\sqrt{2}\beta}{(4 + \beta^4)^{1/2}} \tag{3.83}$$

where β is given by equation (3.70). The maximum value of this expression is $\sqrt{2}$, equal to that for the reference-beam technique without balanced detection. An attractive option is to use the back-reflected output which gives a higher sensitivity and better frequency response. From equation (3.72) the $\text{SF}(\text{FPI})$ at high frequencies is found to be $2\sqrt{2}$ (using the approximations of §3.11.2). Separation of the output beam from the input should present no difficulty, because polarizing optics can be used since light polarization is usually preserved under these conditions.

There are thus only small differences (6 dB) in sensitivity for reflecting surfaces. The choice of interferometer type will depend rather upon the required ultrasonic frequency, type of signal waveform, degree of background vibration, cost and convenience. Reference-beam interferometry must often be a good choice here because of its compactness, relatively low cost and flat frequency response.

(b) Rough surfaces

When making measurements on rough surfaces, the differences between the different types of interferometer become more apparent and the situation more complicated. To make a useful comparison, we must make assumptions about typical measurement conditions and practical interferometers. In our idealized example, we assume that the laser light is scattered by a perfectly diffusing surface (see §3.3.2) whose small sinusoidal ultrasonic movement is being detected, and that there are no light losses in the optics.

Confocal Fabry–Perot interferometers and long-path-difference interferometers of the type discussed in §3.10.5 generally have adequate étendue, and the limiting factors are the aperture available for the reception of the scattered light and the practicability of long interferometer paths. With an aperture of diameter D at the focal distance F from the surface, the light power entering the interferometer is $W_\text{I} = W_0(\text{D}/2\text{F})^2$. We thus derive from equations (3.75) and (1.8) the sensitivity factor for the confocal instrument under these conditions (and assuming $R \approx 1$ and $f \ll v_s$ as before):

$$\text{SF}(\text{CFPI}) = \frac{D}{2F} \frac{2\beta}{(4 + \beta^4)^{1/2}}. \tag{3.84}$$

The corresponding value for the long-path-difference interferometer obtained from equation (3.53) is

$$\text{SF}(\text{LPI}) = \frac{2D}{F} \sin(\pi f L/c) \tag{3.85}$$

where L is the optical path length difference in the interferometer.

The possible value of D/F depends greatly on the particular application, but a typical and perhaps rather generous value of 0.1 (e.g. diameter of 100 mm at a distance of 1 m) will be assumed. On this basis, sensitivity factors are shown as a function of frequency in figure 3.40 for four examples of confocal Fabry–Perot interferometer and two long-path-difference inter-

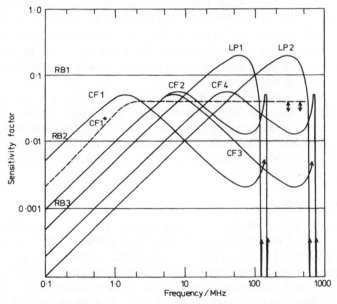

Figure 3.40 Comparison of the photon-noise-limited sensitivities of various interferometers specified in table 3.1 for the detection of ultrasonic displacements of frequency f. The sensitivity factor (SF) is defined in the text (equation (3.82)). The maximum aperture available for light reception is assumed to be $f/10$ ($D/F = 0.1$). An example of the use of the back reflection from a confocal Fabry–Perot interferometer (i.e. with 'sideband stripping') is shown; curve CF1*. 50% light utilization is assumed in this case.

ferometer (table 3.1). The smaller of these latter is the fairly compact interferometer postulated in §3.10.5 having an optical path difference of 500 mm. The larger has a path difference of 2.5 m, making a rather cumbersome, but still quite feasible, instrument. The Fabry–Perot interferometers have mirror separations either of 100 mm (representative of most instruments) or 500 mm (as described by Monchalin and Héon (1986)). The curves for reflectivities (assuming no absorption) of 90 and 98 % are given for both mirror separations. These have been plotted using a more general formula based on equation (3.68) (which does not assume $f \ll v_s$) to give a fair representation of the behaviour over the frequency range covered by the graph. Also included is an example of using the back-reflected beams as described in §3.11.4 to improve the frequency response by using 'sideband stripping' (Monchalin *et al* 1989). Because of the difficulty of separating the return beam from the input beam, we have assumed a 50 % loss in the optics.

The factor limiting the sensitivity of the reference-beam system is generally the size of the illuminated spot. From equation (3.29) we deduce that

$$\text{SF}(\text{RBI}) = \frac{2\sqrt{2}\lambda}{\pi r_0}. \tag{3.86}$$

The maximum value of SF(RBI) is when the spot size is diffraction limited after expanding the illuminating beam to fill the available aperture, assumed to be $D/F = 0.1$. To avoid distortion of the Gaussian profile, we have assumed that the beam is expanded to a diameter $D/\sqrt{2}$ at the $1/e^2$ intensity points.

Table 3.1 Specifications of interferometers compared in figure 3.40.

Confocal Fabry–Perot interferometers

Mirror separation = mirror radius (mm)		Reflectivity (%)	Spacing of orders (MHz)	Line width (MHz)	Maximum spot diameter (mm) ($\lambda = 0.5\ \mu m$)
CF1	500	98	150	1.9	2
CF2	500	90	150	10.1	5
CF3	100	98	750	9.6	1
CF4	100	90	750	47.8	2

Long-path-difference interferometers

LP1 $L = 2.5$ m
LP2 $L = 500$ mm

Reference-beam interferometers

RB1   Illuminated spot radius $r_0 = 9\lambda$ (maximum possible with aperture limitation)
RB2 $r_0 = 90\lambda$
RB3 $r_0 = 900\lambda$

The illuminated spot diameter $(2r_0)$ at the $1/e^2$ intensity points is then

$$2r_0 = \frac{2F}{D} \frac{2\sqrt{2}\lambda}{\pi} = \frac{40\sqrt{2}\lambda}{\pi} \simeq 18\lambda \qquad (3.87)$$

where equation (1.14) has been used with z large. Various factors may prevent a perfectly focused spot being obtained, and values of the sensitivity factor corresponding to spot diameters of 180λ ($\simeq 0.1$ mm, which is perhaps more typical) and 1800λ ($\simeq 1$ mm) are also included in figure 3.40 for the reference-beam interferometer. Note that reference-beam signals are always subject to random variations due to speckle (§3.3.1). The sensitivities given are average values. The problem can be avoided with the other types of interferometer.

It should perhaps be remarked here that the sensitivity factors quoted for the reference-beam technique assume that the interferometers are always at their most sensitive point, as occurs when they are stabilized (§3.7). This is not so when heterodyne or quadrature techniques are used, generally leading to a reduction in the sensitivity factor by at least $\sqrt{2}$.

While the spot size is unlikely to be a limitation with a long-path-difference interferometer of the appropriate design (see §3.10.4), the finite étendue of a confocal Fabry–Perot interferometer may sometimes be important. The maximum acceptable spot sizes may be calculated using equations (3.79) and (3.80), and are listed for the examples given in table 3.1.

The comparison (figure 3.40) shows that the sensitivities of LP1, CF2 and CF3 are rather similar in the frequency range most used in non-destructive testing, i.e. 1–10 MHz. However, to be competitive in this range, the long-path-difference interferometer does have to be rather bulky. Because of the folded path, a Fabry–Perot interferometer is a much more compact and feasible proposition for lower frequencies. The fabrication of an interferometer of the high resolution required does, however, require high-procision optical work that may not be easy to achieve. If a small spot size can be obtained and fluctuations due to speckle can be tolerated, the reference-beam technique in one of its many forms may be the most practical system. This will almost always be true below 1 MHz. This technique does have the advantage of having the flat frequency response for displacement that is important for the accurate reproduction of waveforms. Above 100 MHz the long-path-difference interferometer seems best for sinusoidal waveforms over a limited frequency range, but the oscillatory frequency dependence may be a problem when detecting over a wide passband, and the instrument may have to be used well below its maximum sensitivity. Then a confocal Fabry–Perot interferometer would have comparable sensitivity and a better frequency response especially if operated in back reflection (i.e. with sideband stripping). It could take the form of a very small unit in this frequency range and be readily available.

REFERENCES

Aharoni A, Tur M and Jassby K M 1987 *Proc. Ultrasonics International 87, London* (London: Butterworths) p 623

Bondarenko A N, Drobot Y B and Kruglov S V 1976 *Sov. J. NDT* **12** 655

Bruinsma A J A and Vogel J A 1988 *Appl. Optics* **27** 4690

Buchave P 1975 *Disa Information* no 18 15

Cooper J A, Crosbie R A, Dewhurst R J, McKie A D W and Palmer S B 1986 *IEEE Trans. Ultrasonics Ferroelectric Freq. Contr.* **UFFC-33** 462

De la Rue R M, Humphryes R F, Mason I M and Ash E A 1972 *Proc. IEE* **119** 117

Drain L E 1980 *The Laser Doppler Technique* (New York: Wiley) p 42

Drain L E and Moss B C 1972 *Opto-electronics* **4** 429

Drain L E, Speake J H and Moss B C 1977 *Proc. 1st. Eur. Congr. on Optics applied to Metrology* SPIE vol 136 52

Gillard C W 1970 *US patent* no 3503 012

Goodman J W 1984 *Topics in Applied Physics* vol 9 *Laser Speckle and Related Phenomena* ed J C Dainty (Berlin: Springer) p 38

Hercher M 1968 *Appl. Optics* **7** 951

Hernandez G 1988 *Fabry–Perot Interferometers* (Cambridge: Cambridge University Press)

Jackson D A and Paul D M 1971 *J. Phys. E: Sci. Instrum.* **4** 173

Jette A N, Morris M S, Murphy J C and Parker J G 1977 *Mater. Eval.* **35** no 10 90

Kaule W 1977 *US Patent* no 4046 477

—— 1983 *US Patent* no 4388 832

Kim H C and Park H K 1984 *J. Phys. D: Appl. Phys.* **17** 673

Kline R A, Green R E Jr and Palmer C H 1978 *J. Acoust. Soc. Am.* **64** 1633

Massey G A 1968 *Proc. IEEE* **56** 2157

Mezrich R, Vilkomerson D and Etzold K 1976 *Appl. Optics* **15** 1499

Monchalin J P 1985a *Rev. Sci. Instrum.* **56** 543

—— 1985b *Appl. Phys. Lett.* **47** 14

Monchalin J P and Héon R 1986 *Mater. Eval.* **44** 1231

Monchalin J P, Héon R, Bouchard P and Padioleau C 1989 *Appl. Phys. Lett.* **55** 1612

Moss B C 1987 Private communication

Mueller R K and Rylander R L 1979 *J. Opt. Soc. Am.* **69** 407

Palmer C H and Green R E Jr 1977a *Appl. Optics* **16** 2333

—— 1977b *Mater. Eval.* **35** no 10 107

Paul M, Betz B and Arnold W 1987 *Appl. Phys. Lett.* **50** 1569

Reibold R and Molkenstruck W 1981 *Acustica* **49** 205

Smeet G and George A 1978 *Rev. Sci. Instrum.* **49** 1589

Thomson J K, Wickramasinghe H K and Ash E A 1973 *J. Phys. D: Appl. Phys.* **6** 677

Vaughan J M 1989 *The Fabry–Perot Interferometer: History, Theory, Practice and Applications* (Bristol: Adam Hilger)

Watrasiewicz B M and Rudd M J 1976 *Laser Doppler Measurements* (London: Butterworths) p 19

White R G and Emmony D C 1985 *J. Phys. E: Sci. Instrum.* **18** 658

Whitman R L, Laub L J and Bates W J 1968 *IEEE Trans. Sonics and Ultrasonics* **SU-15** 186

4   Applications of Laser Interferometry to Ultrasonic Displacement Measurement

In this chapter, we consider the applications of laser interferometry of various types to ultrasonic measurements and visualization methods which do not also involve the use of laser-generated ultrasound. Consideration of the many important applications in which both techniques are combined is postponed until Chapter 6, after laser generation of ultrasound has been explained.

4.1   THE MEASUREMENT OF ACOUSTIC FIELDS

4.1.1   The thin-membrane technique

In a transparent medium, it is possible to make direct measurements of the displacement of a point in an acoustic field by optical means, provided a reflecting, partially reflecting or scattering surface is suitably placed in the medium where the measurement is required.

For experiments in a water tank, a reflecting surface that follows the ultrasonic displacement may be conveniently provided by the immersion of a plastic membrane or pellicle preferably having a thin aluminium or gold coating to give a good light return. Measurements of the movement of this reflecting surface may be made by a laser interferometer through a window in the tank as illustrated in figure 4.1.

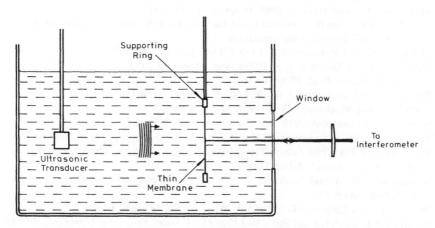

Figure 4.1   Measurement of acoustic fields in a water tank by the displacement of a thin membrane (pellicle) observed by a laser interferometer.

The membrane is usually stretched over a frame large enough to avoid disturbance of the acoustic field and the plane of the surface would normally be at right angles to the direction of propagation of the acoustic wave. The thickness of the membrane can be made very small compared with the acoustic wavelength. Since the acoustic properties of the plastic are also a fair match to those of water, the reflection of ultrasound is small and the membrane can faithfully transmit the acoustic displacement in the liquid. The plotting of ultrasonic fields using thin pellicles has been described by Mezrich *et al* (1975, 1976) and Vilkomerson *et al* (1976).

For a sinusoidal plane wave, the ratio of the displacement amplitude of the surface of a membrane of thickness Δ to that in the absence of the membrane may be expressed as follows (assuming that there is no absorption):

$$\frac{A}{A_0} = \left[1 + \frac{1}{4} \left(\frac{v'\rho'}{v\rho} - \frac{v\rho}{v'\rho'} \right)^2 \sin^2 \left(\frac{2\pi\Delta}{\Lambda'} \right) \right]^{-1/2} \tag{4.1}$$

where v and v' are the ultrasonic velocities in the liquid and the membrane respectively, ρ and ρ' are the respective densities and Λ' the acoustic wavelength in the membrane material. For thin membranes, the fractional reduction in the displacement amplitude can be approximated thus:

$$\frac{\delta A}{A_0} = \frac{1}{2} \left(\frac{v'\rho'}{v\rho} - \frac{v\rho}{v'\rho'} \right)^2 \frac{\pi^2 \Delta^2}{\Lambda'^2}. \tag{4.2}$$

Strong polyester film is available with thicknesses of only a few micrometres. This is very thin compared with the acoustic wavelength in water which is typically 0.4 mm at 5 MHz. Thus we may estimate that the use of a film 10 μm thick would lead to a reduction in displacement amplitude of the order of only 0.7%, quite negligible for most purposes. However, this does not take account of the metal coating and, for accurate work, a measurement of the transmission coefficient through the membrane would be desirable.

Since there is a good return from the metallized film, the use of a reference-beam type of interferometer is preferred. Sufficient power would normally be available from a small helium–neon laser. Depending on the amplitude of the ultrasonic displacement, one of the signal-processing techniques described in §§3.6–3.9 would be appropriate. Most often the displacements are small compared with the wavelength of light and a compensated interferometer as described in §3.7 would be the most suitable, or the computer evaluation technique described in §3.9.4 could be used.

To avoid the effect of acoustic reflections and reverberations around the tank, ultrasonic pulses are preferred to continuous wave (cw) excitation, the measurement time being restricted to the actual passage of the ultrasonic pulse through the membrane. A pulse may be either a few cycles of damped oscillation produced by a step (or 'edge') excitation of a piezoelectric transducer, or a larger number of cycles of a definite frequency, i.e. 'pulsed or gated cw' (or 'tone burst'), simulating true continuous-wave excitation. Either the peak displacement or velocity amplitude alone may be recorded or the full pulse shape may be captured on a transient recorder.

Although measurements of single pulses are usually quite satisfactory, considerable enhancement of signal to noise ratio may be obtained by signal averaging. The reverberation of ultrasound usually dies away in a few milliseconds and hence a hundred or so repeat measurements can be made every second and the signals averaged with suitable equipment. However, during signal averaging the membrane must not move more than a fraction of an acoustic wavelength. Otherwise the time delay between the excitation of the transducer and the movement of the membrane would not be sufficiently constant to ensure consistent triggering of the transient recorder in relation to the signal. This would result in the broadening of the averaged signal. Quadrature interferometers and some compensated interferometers can record low-frequency displacements directly and, if necessary, a continuous correction could be made to the delay time before triggering the transient recorder. It is also important in signal averaging that all the pulses are exactly reproduced. Thus in pulsed cw the phase of the cw should have a constant relation to the start of the pulse or the pulse envelope, i.e. the ultrasonic frequency should be an integral multiple of the repetition frequency.

4.1.2   Correction for acousto-optic interaction

In the above technique, the illuminating laser beam and the reflected beam both traverse the medium through which the ultrasound is travelling. Thus in addition to the optical path length change induced by the motion of the reflecting surface, a change also results from the variation of refractive index with pressure changes associated with the acoustic wave. This effect has been discussed in detail in §2.1.3. Thus the interference signal is not simply related to the displacement of the surface by equation (3.1) where the wavelength involved is that of the light in the medium. This is only true at very low frequencies where the acoustic wavelength is long compared with the dimensions of the tank. For ultrasonic pulses a correction is normally necessary.

This correction may be computed simply if we assume that the ultrasonic wave does not spread out laterally, i.e. it approximates to a plane wave. This is usually true in pulse experiments if the length of the pulse is small compared with the diameter of the ultrasonic beam at the pellicle.

We consider the measurement of the displacement of a membrane surface S in figure 4.2 by interferometry through a length $\mathscr{L}$ of liquid (or other medium) through which a longitudinal ultrasonic pulse is travelling. We suppose that at a time t, the surface is displaced by a small distance x from its initial position O. Due to the passage of the ultrasonic wave, the density

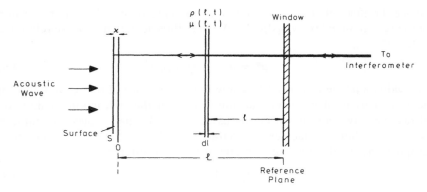

Figure 4.2   Calculation of the correction for refractive index variations of the medium through which light passes due to the passage of plane acoustic waves.

and refractive index deviate from their initial values ρ_1 and μ_1 by small amounts $\delta\rho$ and $\delta\mu$ which depend on position and time. Thus the effective optical path length in the liquid from the reference plane W (which could be at the entrance window as shown in figure 4.2) to the surface is given by

$$\text{OP}(t) = \int_0^{\mathscr{L}+x} (\mu_1 + \delta\mu)\, \mathrm{d}l. \tag{4.3}$$

The increase in path length due to the movement of the surface is thus

$$\delta\text{OP}(t) = \mu_1 x + \int_0^{\mathscr{L}+x} \delta\mu\, \mathrm{d}l. \tag{4.4}$$

Now consider the mass of material contained within the surface and the reference plane within a cylinder of cross-sectional area A whose axis is along the direction of propagation. This is given by

$$M = A \int_0^{\mathscr{L}+x} (\rho_1 + \delta p)\, \mathrm{d}l. \tag{4.5}$$

Since we have assumed a plane wave, there is no movement transverse to the axis of the cylinder and thus the mass of material contained therein must remain constant. Hence

$$\rho_1 x + \int_0^{\mathscr{L}+x} \delta\rho\, \mathrm{d}l = 0. \tag{4.6}$$

Assuming that refractive index is linearly related to density (quite justifiable for the small fluctuations involved), equations (4.4) and (4.6) give

$$\delta\text{OP}(t) = x(\mu_1 - \rho_1\, \mathrm{d}\mu/\mathrm{d}\rho). \tag{4.7}$$

Hence the effect of acousto-optic interaction is to reduce the apparent value of the refractive index by $\rho_l\, d\mu/d\rho$. We may thus define an effective refractive index μ^*:

$$\mu^* = \mu - \rho\, d\mu/d\rho. \tag{4.8}$$

To evaluate μ^* we thus need to know the dependence of refractive index on density. This is of considerable importance in the diffraction of light by ultrasonic waves and has been discussed in §2.1.3. The simplest assumption possible is that the increase in refractive index over that in vacuum is proportional to the amount of material per unit volume:

$$(\mu - 1) \propto \rho. \tag{4.9}$$

This leads to the result $\mu^* = 1$. When no other information is available, this is probably the best value to assume although theoretically it is arguably better to use the Lorentz theory discussed in §2.1.3 (equations (2.5) and (2.6)), which leads to the result

$$\mu_L^* = 1 - \frac{(\mu - 1)^2}{6\mu}\,(\mu^2 + 2\mu - 2). \tag{4.10}$$

If, however, the piezo-optic constant of the material is known, μ^* may be calculated. Values of the adiabatic constant $\rho(\partial\mu/\partial\rho)_S$ are given for some liquids in table 2.1. For water at 20 °C, for a wavelength 633 nm, $\mu = 1.332$ and the best value of $\rho(\partial\mu/\partial\rho)_S$ from table 2.2 is 0.323 and thus $\mu^* = 1.009$. The estimated uncertainty of 1 % implies 0.3 % in μ^*. For silica a value of $\mu^* = 1.035$ may be deduced from the information given in table 2.3 using equation (2.25).

4.1.3 The total reflection technique

An alternative method of measuring ultrasonic fields is based on the observation of the displacement of a free surface of the medium at which the ultrasonic waves are reflected. This has the following advantages.

(i)   It does not require the medium to be optically transparent.
(ii) Correction for acousto-optic interaction is not necessary.
(iii) At normal incidence, the displacement amplitude is twice that of the ultrasonic wave.

A disadvantage is that access to the exact position where the ultrasonic field is measured, to insert a transducer for calibration for example, cannot be gained without some modification to the geometry.

Measurements are normally made at an air interface. Air is so poorly matched acoustically to most solid and liquid media that the interface may, for all practical purposes, be treated as an unrestrained surface giving total internal reflection of acoustic waves. The free surface of a liquid, e.g. water,

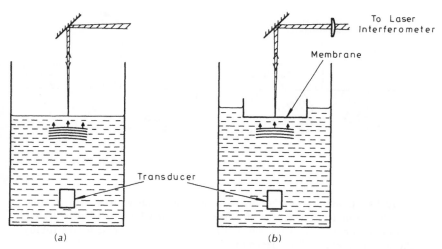

Figure 4.3 Two methods for the measurement of acoustic waves by observation of the displacement of a liquid–air interface at which the waves are reflected.

may be used, as in figure 4.3(*a*), the 2% or so specular reflection giving quite a good optical interference signal. An acoustic wave incident on the surface is totally reflected with a 180° phase change. Thus, for normal incidence, the displacement of the surface is exactly twice that associated with the incident wave.

Free liquid surfaces are, however, particularly prone to perturbation by vibration and air movements and it is more practical to use a thin membrane as shown in figure 4.3(*b*). Furthermore, if this has an aluminized surface, a better optical reflection is obtained. The error due to finite thickness of the membrane can be more significant than in the transmission method described in §4.1.1. The fractional increase in displacement due to a thickness Δ of the membrane (assumed to be small compared with the acoustic wavelength) is given by

$$\frac{\delta A}{A_0} = 2\left(\frac{v'\rho'}{v\rho} - 1\right)^2 \frac{\pi^2\Delta^2}{\Lambda'^2} \tag{4.11}$$

where v and v' are the acoustic velocities in the medium and the membrane respectively and ρ and ρ' their densities. Λ' is the acoustic wavelength in the medium material. For a polyester film 10 μm thick, the deviation would be approximately 1.3% at 5 MHz. This is not really significant but there is a possibility of appreciable error if air bubbles are present under the membrane affecting the acoustic coupling.

The reflection technique is very suitable for the measurement of ultrasonic waves in solids, especially metals where the free surface may be polished to give excellent interferometer signals. For normally incident longitudinal

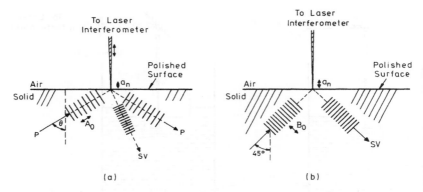

Figure 4.4 Examples of the measurement of acoustic waves in a solid from the normal displacement of a free surface at which they are reflected. (*a*) Longitudinal waves with mode conversion, (*b*) transverse waves totally reflected at an angle of incidence of 45°.

waves, the surface displacement is twice that associated with the ultrasonic wave but for oblique incidence, as illustrated in figure 4.4(*a*), the ratio of the normal surface displacement amplitude a_n to the longitudinal displacement amplitude A_0 in the wave differs from the value $2\cos\theta$ which is valid at a liquid surface. This is because some of the energy of the incident wave is mode converted to a shear wave. The displacement of the surface may be calculated by adding the contributions of the incident and two reflected waves (i.e. P and SV) using the standard treatment of acoustic reflection at the surface of a semi-infinite medium (Achenbach 1973, Graff 1975). The displacement is given by

$$\frac{a_n}{A_0} = \frac{2\cos\theta(k^2 - 2\sin^2\theta)k^2}{4\sin^2\theta\cos\theta(k^2 - \sin^2\theta)^{1/2} + (k^2 - 2\sin^2\theta)^2} \tag{4.12}$$

where θ is the angle of incidence and k the ratio of longitudinal to shear wave velocity, which is related to the Poisson's ratio of the solid, v. Thus

$$k^2 = \frac{2(1 - v)}{1 - 2v}. \tag{4.13}$$

The ratio of amplitudes given by equation (4.12) is plotted in figure 4.5(*a*) for the cases of aluminium ($k = 2.05$) and steel ($k = 1.84$).

Usually the direction of incidence of the ultrasonic wave is known but it could in principle be determined from the time delay or phase lag of the response at two or more points on the surface. Ideally these measurements should be made simultaneously, which requires the use of more than one interferometer, but they could be made sequentially if the generation of the acoustic wave is sufficiently reproducible.

We note that shear waves polarized in the plane of incidence (SV waves) also give rise to a normal displacement of the surface if they have oblique

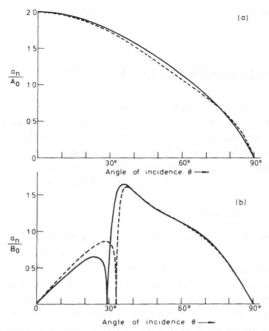

Figure 4.5 Ratio of the amplitude of normal surface displacement to that of an incident longitudinal wave (upper graph) or transverse wave (lower graph) for reflection at a free surface of an isotropic solid. Calculation for ratios of longitudinal to transverse wave velocities of 1.84 (broken curve, e.g. steel) and 2.05 (full curve, e.g. aluminium).

incidence. The relationship between the amplitude of the normal surface displacement a_n and that of the transverse displacement in the shear wave B_0 is complicated by mode conversion when the angle of incidence θ is less than the critical angle $\theta_c = \sin^{-1}(1/k)$ and by phase shifts above. In the latter case, unless $\theta = 45°$, some ultrasonic energy is also propagated along the surface at the compression wave speed in an SP wave. Figure 4.4(b) illustrates the simple case of $\theta = 45°$ when the shear wave is totally reflected from the surface and the surface displacement is normal and its amplitude is $\sqrt{2}$ times that of the transverse displacement of the shear wave. In general the ratio of displacements is given by

$$\frac{a_n}{B_0} = \frac{4 \sin \theta \cos \theta (1 - k^2 \sin^2 \theta)^{1/2}}{4 \sin^2 \theta \cos \theta (1 - k^2 \sin^2 \theta)^{1/2} + k(1 - 2 \sin^2 \theta)^2} \qquad \text{if } \theta \leqslant \theta_c$$

or

$$\frac{a_n}{B_0} = \frac{4 \sin \theta \cos \theta (k^2 \sin^2 \theta - 1)^{1/2}}{[16 \sin^4 \theta \cos^2 \theta (k^2 \sin^2 \theta - 1) + k^2 (1 - 2 \sin^2 \theta)^4]^{1/2}} \qquad \text{if } \theta > \theta_c$$

$$(4.14)$$

This relationship is illustrated in figure 4.5(*b*) for the cases of aluminium and steel.

4.2 SCANNED LASER INTERFEROMETRY

A key feature of laser techniques for the measurement of ultrasonic surface displacement is that no contact is required and thus a change in measurement position on the surface may be made very quickly simply by the movement of a light beam. Combined with the high resolution obtainable from the laser spot, this enables a detailed picture of the ultrasonic movements over a surface to be built up very rapidly by optical scanning.

4.2.1 Angular scanning

The usual method of steering a light beam is by the use of one or more moveable mirrors. The scheme for an $x-y$ scan over a surface shown in figure 4.6 employs two separate deflecting mirrors. This is mechanically simpler than using one mirror incorporating two axes of rotation, but the second mirror does have to be large enough to accommodate the small deviations of the beam as the first mirror scans. The distance between the mirrors should of course be as small as practicable. With a mirror scanning system, it is generally an advantage to use an interferometer arrangement like those in figures 3.6 or 3.8 where the returned light is received along the same path as the outgoing beam, as this reduces the size of the deflecting mirrors required. This also gives the maximum range of distance from the inter-

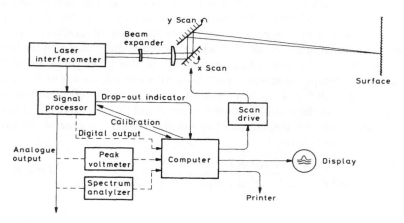

Figure 4.6 Plotting the acoustic or ultrasonic displacement of a surface by laser interferometry using angular scanning of a laser beam.

ferometer over which a satisfactory signal can be received for a given focal setting.

As the angle of incidence at the specimen surface is not constant, it is not possible to make use of specular reflection with this method of scanning. It is thus most suitable for rough surfaces, and the sensitivity of the interferometer is limited to that obtainable from this type of surface (§3.3.5), though a big increase may be obtained by the use of retroreflecting paint (§3.3.2) if this is permissible. In any case, the returned light will suffer random phase and intensity fluctuations due to speckle (§3.3.1) and thus the signal intensity from an interferometer of the reference-beam type will vary considerably as scanning proceeds. When using a compensated interferometer to measure small displacement amplitudes, calibration as described in §3.7.5 will be necessary at every point scanned. This may be done automatically under computer control just before and/or just after a measurement at a point of the scan, the value of the parameter V_0 being stored and used to scale the output of the interferometer at that point.

When the intensity fluctuations are dominated by speckle, there will always be some points at which the signal is insufficient to obtain a measurement of ultrasonic displacement. In principle this difficulty may be overcome by 'diversity reception' using one or more extra interferometer channels receiving scattered light at different angles. It is then very unlikely that there will be a 'drop-out' in all channels simultaneously. In practice, recourse is rarely made to this, the usual procedure being to take a sufficient excess of measurement points that a few missing ones do not matter. An alternative is to move the scan very slightly from the specified position when a 'drop-out' is sensed. This will usually restore the signal to a usable amplitude. Speckle is not generally a problem for the velocity or long-path-length-difference interferometer (§3.10) or with Fabry–Perot interferometers (§3.11) since these can usually average over many speckles. However, as the output of these interferometers is usually dependent on the light return as well as velocity, their sensitivity can vary over the surface scanned and frequent calibration is desirable.

4.2.2 Parallel scanning

The main disadvantage of angular scanning is that it is not suitable for specularly reflecting surfaces which are otherwise desirable for high sensitivity. The direction of movement to which the interferometer is sensitive also varies slightly across the scan. In addition, variations in optical path length may give rise to difficulties when employing lasers which are not operating in a single longitudinal mode. Parallel scanning overcomes these difficulties.

For mirror surfaces, the angle of incidence of the laser beam to the surface should be constant during the scan so that light is returned efficiently to the interferometer, either along the same path as the incident beam or with a constant deflection angle. This is also a consideration for scanning with the knife-edge technique discussed in §2.6.4. This 'parallel scanning' may be achieved by moving the specimen surface transversely on a slide, by the sideways movement of a mirror and lens combination or by rotating a glass block (Massey 1968), but a scanning system in which a mirror is tilted is much faster and more convenient. A scheme for producing a parallel scan in this way using one lens is shown in figure 4.7. The axis of rotation of the mirror passes through one focus of the lens and the sample surface is in the other focal plane. Thus in addition to producing the parallel scan, the lens also focuses the laser beam onto the surface for maximum resolution and minimum sensitivity to tilt of the surface. The lens must be at least as large as the surface area to be scanned. This is a limitation of this system but in ultrasonics there is usually no problem, as samples are rarely more than a few centimetres across.

To ensure the accurate return of the reflected light to the interferometer for all points on a wide scan, it is important that the lens is well corrected for spherical aberration to focus a parallel beam towards the mirror. For this reason, a doublet lens has been shown in figure 4.7. Ideally, the lens should also be corrected for focusing the laser beam onto the surface but this is not so critical except for high-resolution studies.

When x and y deflections are both required, it is probably more convenient to use one scanning mirror rather than two, although providing the two axes of rotation on a single mirror entails some mechanical complication. Since

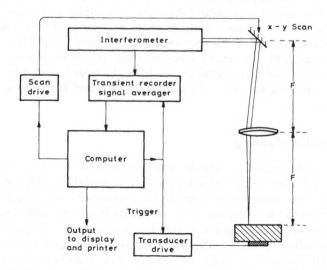

Figure 4.7 Parallel scanning of a surface to plot ultrasonic displacements by laser interferometry.

only one mirror can be accurately at the focus of the lens, the use of separate mirrors for the x and y scans entails an extra lens or lenses to provide an additional focal point. This method is illustrated in figure 2.25.

With this type of scanning over a reasonably good reflecting surface, the signal strength should be fairly constant; nevertheless for accurate work it is desirable that the calibration be checked at each point on the scan.

When signals are repetitive, or when an accurate triggering pulse is available, as in the case when the ultrasound is produced by a piezoelectric transducer or a laser pulse, the signal to noise ratio of the interferometer may be increased by signal averaging as described in §4.1.1. It is often quite possible to repeat a measurement at each point in a scan a hundred or so times and perform a calibration there within a second. In general it is better to do the averaging in a purpose-built transient recorder/signal averager than in the computer since the summations are performed on line and delays associated with transfers to the computer are avoided. Ideally, the complete waveforms of pulses recorded at each scan position should be transferred to the computer memory for subsequent processing to produce tabulated information or stereographic or contour plots. For cw or pulsed cw excitation, however, the only quantities of interest may be the amplitude and phase of the signal and only these need be stored. In this case, processing by a phase-sensitive lock-in detector is an alternative to the use of a transient recorder.

Many ultrasonic displacement patterns obtained by parallel scanned laser interferometry are given in §4.5. Results are illustrated for scans over transducer faces (§4.5.1), pellicles in water (§4.5.2) and specimen surfaces (§4.5.4).

4.2.3 Ultrasonic imaging

Scanned laser interferometry may be used to generate images of objects insonified in a water tank. The arrangement is shown diagrammatically in figure 4.8. The object is placed in front of an ultrasonic transducer and the acoustic field mapped by parallel scanning over a reflective surface placed just beyond the object or at an image formed by an acoustic lens.

In experiments at an acoustic frequency of 10 MHz (Mezrich *et al* 1976), the surface scanned was of a 6 μm thick aluminized pellicle suspended in water, the 'thin membrane' of the technique described in §4.1.1. A conventional reference-beam interferometer with a 'wiggler' mirror (§3.7.1) was used to ensure that consistent results were obtained with or without significant background vibration. This method gives only the peak amplitude of the ultrasonic movement at the point scanned but this is sufficient for ultrasonic imaging. Many of the types of compensated or heterodyne interferometer described in §§3.7 and 3.8 could evidently be used to give phase or waveform information. A system employing a Bragg cell frequency shifter was used in early experiments by Massey (1968).

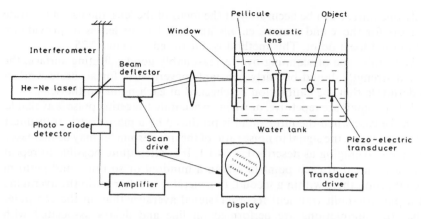

Figure 4.8 Ultrasonic imaging of an object using laser interferometry scanning a pellicle in the image plane (Mezrich *et al* 1976).

Biological specimens may be investigated and, if an argon laser of moderate power is used, the ultrasonic power requirement is low enough to allow imaging to be done *in vivo* (Mezrich *et al* 1976).

A high-resolution system or 'scanning laser acoustic microscope' (SLAM) using interferometric detection has been described by Rudd *et al* (1987). This operates at a frequency of approximately 100 MHz, corresponding to an acoustic wavelength of about 15 μm. The beam from a 150 mW argon laser is expanded so that it may be focused to a spot only 10 μm in diameter by a 50 mm focal length lens. The reflective surface is formed by a 6 mm thick transparent plastic cover-plate mirrored on the side next to the specimen, i.e. facing the acoustic waves. The material has a high acoustic absorption at the operating frequency, thus reducing problems from unwanted ultrasonic reflections.

The long-path-difference or time delay type of interferometer (§3.10) is normally used in this system as it is practicable at this frequency and is insensitive to background vibration, although active stabilization of the optical delay path is necessary. A single longitudinal mode laser is needed with this type of interferometer.

The knife-edge technique (§2.6.2) could also be used, but was found not to be generally satisfactory as it depends on the tilting of the surface rather than displacement or velocity, and tilt may not always be very significant in this arrangement. A scheme more suited to the knife-edge technique was discussed in §2.6.4. There is of course no reason why reference-beam interferometry could not be applied to acoustic microscopy, provided it is adequately compensated or stabilized against background vibrations as described in §3.7.

A different type of laser acoustic microscope based on the thermoelastic generation of ultrasound by a mode-locked laser is discussed in §6.3.

4.2.4 Measurement of vibration

It is of interest to note that the ability of a laser interferometer to measure surface movement from a considerable distance is also of great value at frequencies in the acoustic range and for low-frequency vibration. The main interest here is in studies of modes of vibration and the generation and transmission of sound. The main advantages of laser interferometry are firstly that the laser 'probe' spot has effectively zero mass so that the dynamics of the system being studied are not affected, and secondly that the measurement point may be moved rapidly without the need for the removal and reattachment of a transducer. A map of vibration levels may thus be built up rapidly by scanning the laser beam over the surface. As the surfaces studied are not usually specular reflectors or specially prepared, angular scanning (figure 4.6) is normally appropriate, particularly in view of the large size of many of the objects studied.

Since laser interferometry with a reference beam is intrinsically sensitive to displacement, its performance relative to techniques measuring velocity or acceleration improves as the frequency is decreased. The reduction in bandwidth compared with that needed for ultrasonic measurements at 1 MHz or more is also helpful in obtaining a good signal to noise ratio at low frequencies. Thus, although the surfaces of objects are often poor reflectors, there is usually adequate sensitivity for vibration measurement even from a nominally matt black surface. If necessary, the surface may be treated with retroreflecting paint (see §3.3.2). This can often increase the signal strength by two orders of magnitude. It should be remarked that the use of long-path-difference or time delay interferometry (sensitive to velocity) is impractical at low frequencies.

Since vibration displacements are generally greater than one wavelength, signal processing by frequency tracking (§3.8.3) or cycle counting (§3.9.3) is preferred. The outputs of these processors are not dependent on signal amplitude (provided a given threshold is exceeded) and calibration at each point on a scan is not necessary. Since the surfaces are normally optically rough, signal amplitudes are dominated by speckle, and drop-outs will occur. These need to be sensed and appropriate action taken, as discussed in §4.2.1. Here we ignore special problems that may arise at very high vibration levels where tilting of the surfaces and/or sideways displacement by more than one diameter of the focused spot can give rise to difficulties with signal processing.

When the excitation of the object is periodic, the waveform of the vibration at a point on the surface may be analysed for amplitude, phase and harmonic distortion and, with this information for a complete scan over a surface, a picture of the surface movement during a cycle of excitation may be built up. By processing an analogue displacement or velocity signal through a phase-sensitive 'lock-in' amplifier, movements not synchronous with the excitation may be excluded. The effect of background vibration may thus be eliminated.

An ideal application is to the study of the movement of loudspeaker diaphragms so important for high-fidelity reproduction. Figure 4.9(a) is a photograph showing the plotting of the movements of a loudspeaker by scanned laser interferometry. Figure 4.9(b) shows the scanning mirror system. Displacement patterns may be obtained for any specified excitation frequency which may show up unexpected modes of vibration. The response to pulse excitation may also be plotted if required.

For random or quasi-random vibration, such as that generated by engines or machinery, the analogue or displacement signal can be spectrally analysed to determine the dominant frequencies or noise in designated passbands. At most typical frequencies this may be done on line and a spectrum obtained and stored for each point on the scan. This information may then be used to construct vibration amplitude maps for each frequency or frequency band. The time spent by the laser interferometer at each point on the scan must be sufficient to allow the spectrum analyser to respond for the narrowest passband used.

Unfortunately, multi-point correlation information is not obtained since it is not possible to measure movement at all points on the surface simultaneously. However, two (or more) point correlations can be obtained by using two (or more) interferometers focused simultaneously onto different points on the surface.

An example of the application of scanned laser interferometry to the vibration of a large object is shown in figure 4.10. Contours of the vibration level of the side of a van are shown superimposed on a photograph of that object.

Mention should be also made of the capability to measure vibration (or ultrasonic movement) at large distances. Assuming that there are no problems from atmospheric absorption or refractive index gradients, the sensitivity of reference-beam interferometry for a given laser power and type of rough surface depends on the focused spot size only (equation (3.26)). If this can be kept small, no loss in sensitivity with distance need occur. However, to obtain a small spot, a large beam diameter at the projection lens is required and there are practical limitations to this. A diameter of about 100 mm is about the maximum that can be accommodated in readily available high-quality lenses and at the He–Ne laser wavelength of 633 nm this gives a spot diameter of 0.80 mm when focused at a distance of 100 m. From equation (3.26) the theoretical signal to noise amplitude ratio for a white diffusing surface is then about 400:1, assuming a 5 mW laser and a bandwidth of 10 kHz. Even with a poorly diffusing surface, the signal to noise ratio should be quite adequate for frequency tracking. For long-distance work, it is desirable to employ a single (longitudinal) mode laser because it is clearly

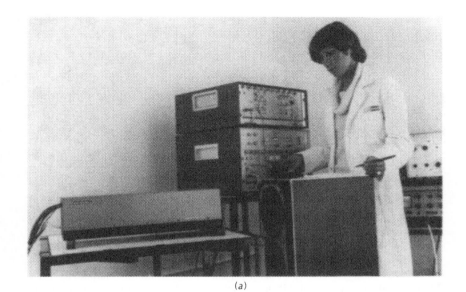

(a)

(b)

Figure 4.9 (*a*) Plotting patterns of loudspeaker movement by laser interferometry. (Reproduced by courtesy of B and W Loudspeakers). (*b*) The scanning mirror system.

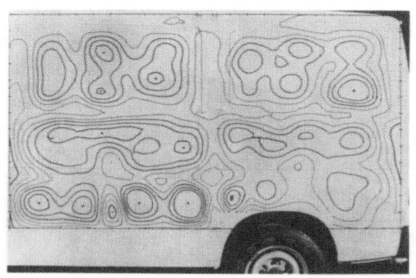

Figure 4.10 Contours of vibration levels of the side of a van plotted by scanning laser interferometry. The RMS velocity varies from below $1.8 \, \text{mm s}^{-1}$ over areas marked $-$ to over $14 \, \text{mm s}^{-1}$ over areas marked $+$, the contours being at intervals of 3 dB. (Reproduced courtesy of MIRA.)

impracticable to equalize the signal and reference-beam paths even approximately. However, the use of a small helium–neon laser may be satisfactory, provided the path length difference is not close to an odd multiple of the cavity length (see §3.5.1).

4.3 FULL-FIELD VISUALIZATION OF SURFACE DISPLACEMENT

An apparently attractive alternative to scanning is the use of whole surface interferometry which offers the ability to record small surface displacements over the whole surface simultaneously. One technique using diffraction was described in §2.6.1. This is only suitable for surface waves on smooth reflective surfaces and so has a rather limited application. Visualization by optical interferometry is possible by the techniques of holographic and speckle pattern interferometry which are particularly appropriate for rough surfaces. There are, however, practical limitations to these techniques especially for the detection of displacements small compared with the wavelength of the laser illumination. The techniques do, however, provide ways of obtaining a complete qualitative picture of any pattern of surface movement in one exposure.

Holography and speckle pattern interferometry may be used in a variety of ways and have many applications. The reader is referred to textbooks for a detailed description of the theory and practice of these techniques (Kock 1975, Vest 1979, Francon 1979, Jones and Wykes 1983). Here, we briefly discuss their application to dynamic surface displacement measurements, particularly their potential use at ultrasonic frequencies.

4.3.1 Holographic interferometry

In the well known technique of holography, the wavefront of light reflected from (or transmitted through) an object illuminated by a laser is recorded in amplitude and phase as a system of fine fringes on a high-resolution photographic plate by interference with a reference beam derived from the same light source. A typical arrangement is shown in figure 4.11. The wavefront may be reconstructed by illuminating the processed photographic plate (or hologram) with a reference beam in exactly the original geometry. A reconstructed three-dimensional image may then be seen by viewing through the plate.

When a double exposure hologram of an object is made with a slight distortion of the object or movement between exposures, fringes are observed crossing the reconstructed image. These follow contours of equal displacement, differing by $\frac{1}{2}\lambda$ (assuming normal incidence of the laser illumination and holographic recording). These fringes are the result of the interference between the two holograms, the phase differences being introduced by the displacements of various parts of the surface.

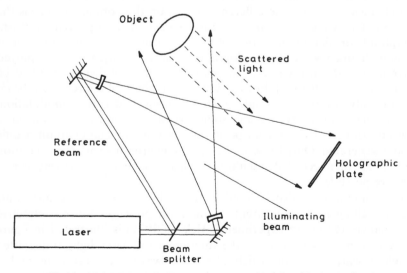

Figure 4.11 Diagram illustrating the principle of holography.

Now if a hologram is made of a vibrating object, it is effectively a multi-exposure one and the interference effect is averaged over the range of displacements associated with the vibration. In general, this leads to a reduction in intensity in the reconstructed image by destructive interference which varies in an oscillatory manner with the amplitude of vibration. For sinusoidal motion, it may be shown that the intensity is reduced (from that with zero vibration) by a factor $[J_0(4\pi a/\lambda)]^2$ where a is the amplitude of vibration and J_0 is a Bessel function of zero order. Thus on reconstructing the object from the hologram, fringes of equal vibration amplitude will be observed corresponding to the maxima and minima of this function. A complete picture of the pattern of vibration amplitude may thus be obtained with one exposure. The normal limit of sensitivity is about $\frac{1}{2}\lambda$, but by careful measurement of fringes a resolution of about a tenth of a fringe, i.e. ~ 30 nm, is possible. Large vibration amplitudes of more than a few wavelengths can lead to difficulties as the fringe contrast deteriorates.

These simple time-averaged holograms clearly cannot give information on the time variation of displacement but, in the case of vibration in response to a periodic excitation, one way of determining the phase of the displacement is by stroboscopic holography where the laser illumination is flashed in synchronization with the excitation. This evidently improves the fringe contrast as well. The time history of displacements through a cycle may thus be built up from several holograms.

Improved sensitivity to very small displacements may be achieved by holographic subtraction (Harihan 1973). In this method, a double exposure hologram is made, once with the vibration or ultrasonic movement present and once with the surface stationary but with the phase of the reference beam changed by 180°. The difference hologram thus obtained cancels the background showing up small surface movements with improved contrast.

Amplitude or phase modulation of the reference beam at the vibration or ultrasonic frequency during a holographic exposure may also be employed to give increased sensitivity and to enable the phase of displacement to be determined (Aleksoff 1971). For example, if the phase of the reference beam is periodically reversed by single sideband suppressed carrier modulation, i.e. frequency shifting by the ultrasonic excitation frequency, a difference hologram between displacement positions separated by half of an ultrasonic period is generated. Only those movements synchronous with the excitation frequency are then recorded and ultrasonic displacement sensitivities down to $\lambda/100$ are possible.

The technique may also be used in transmission to visualize ultrasonic waves travelling in a transparent medium via pressure-induced refractive index changes (§2.1). An experiment (Aleksoff 1971) is illustrated in figure 4.12. A hologram is formed by the interference of laser light transmitted through a water tank in which ultrasonic waves are travelling and a reference-beam frequency shifted by a Bragg cell (see §2.5.3). The ultrasonic

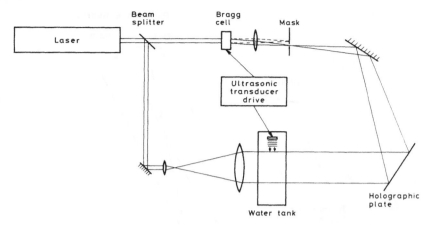

Figure 4.12 Holographic method for visualizing ultrasonic waves in a water tank (Aleksoff 1971).

waves can be seen as stationary fringes on reconstructing the image from the hologram.

4.3.2 Double pulsed holography

Information about surface movement may also be obtained from double pulsed holograms where two exposures by a pulsed laser are made at a closely spaced interval, possibly only a small fraction of a microsecond when visualizing ultrasound. The pattern of displacement during that interval is indicated by fringes. This method avoids a major difficulty in holography, that is its susceptibility to background vibration.

The use of double exposure holography for the visualization of surface waves has been described by Wagner (1986). Comparatively large amplitude acoustic waves are necessary to obtain readily interpretable fringes on a simple holograph. About half a fringe may be discerned, giving a resolution of approximately 120 nm with an argon ion or frequency-doubled YAG laser.

To resolve the smaller amplitudes normally associated with ultrasonic frequencies, the technique of heterodyne holographic interferometry has been devised (Dändliker 1980). In this, a double exposure hologram is made using different reference beams for the two exposures, as illustrated in figure 4.13. The idea is to change the relative phase of the reconstructed images by adjusting the phases of the reference beams used for viewing the holograms. If the phase difference between these is varied continuously, i.e. there is a frequency shift, amplitude fluctuations due to interference will be observed at each point on the reconstructed image of the surface. The phase of these fluctuations varies linearly with the displacement of the point on the surface between holographic exposures. These fluctuations may be recorded by a

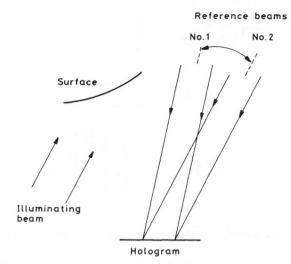

Figure 4.13 Making a double exposure hologram with two reference beams.

photodetector and, since they are sinusoidal, it is possible to measure small differences in phase, for example by mixing in a phase-sensitive detector. A sensitivity down to about 0.85 nm has been demonstrated (Wagner 1985a).

The frequency shifting of a reference beam that is required in this technique is usually obtained by the use of a Bragg cell or combination to give a more convenient lower frequency as described in §2.5.3. The fast switching of the reference beam between holographic exposures may be accomplished by electro-optic modulators. Alternatively, a triple exposure technique is available that avoids the necessity for this operation (Wagner 1985b).

Although this technique requires the scanning of at least one detector over the reconstructed image, information from all over the surface is obtained in one very short (double) exposure rather than from a multiplicity of measurements as in single-point scanning interferometric techniques. It is thus particularly useful when the repetition of an event being studied is difficult or undesirable. However, for the study of randomly occurring events, some sufficiently fast auxiliary technique must be available for triggering the pulsed laser.

A completely different type of application of double exposure holography to ultrasonic measurement has been demonstrated by Reibold (1980) who used the technique to measure the static distortion of a liquid surface due to the radiation pressure of an ultrasonic beam. The two exposures were made by a pulsed laser, both with and without an ultrasonic beam vertically incident on the surface. Fringes on the reconstructed image were used to observe the distortion and hence calculate the distribution of radiation pressure and accurately deduce the power in the ultrasonic beam.

4.3.3 Speckle pattern interferometry

The technique of speckle pattern interferometry is much less demanding of the resolution of the recording medium and of background vibration, though some precautions are still necessary. It is a more practicable proposition than holography in most circumstances. The formation of speckles by the scattering of coherent light from a rough surface is discussed in §3.3.1, and there are many ways in which they may be used to observe both static and dynamic displacement.

Many techniques involve the photography of speckles but the moderate resolution requirements permit the use of television equipment giving rise to the technique of electronic speckle pattern interferometry (ESPI) (Butters and Leendertz 1971) which can give an immediate display of a time-averaged vibration pattern and which has found considerable application to the rapid examination of surface movement.

A diagram of an arrangement for speckle pattern interferometry sensitive to out of plane movement of the surface (like the majority of the interferometric techniques we have discussed) is shown in figure 4.14. The surface is illuminated by light from a CW laser from which a reference beam is split by the beam splitter BS1. This is expanded and superimposed on the image of the surface on the TV camera tube by means of the beam combiner BS2. The variable attenuator VA enables the ratio of the intensities of the image and reference beam to be adjusted to an optimum value. A speckled image is formed in the camera due to its finite aperture and the coherence of the light, and the reference beam is arranged to form a coherent beam apparently diverging from the centre of the aperture. There is then spatial coherence (see §3.2.3) over each speckle between the reference beam and the beam forming the image, and strong interference can occur. Since the phases of

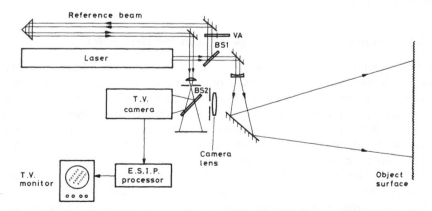

Figure 4.14 Arrangement for electronic speckle pattern interferometry.

the speckles are arbitrary, there is a random mixture of constructive and destructive interference and, in the absence of surface movement, strong contrast between speckles is observed. Vibration, however, will average the interference of each speckle over a range of phases and the contrast is evidently reduced. Note that it is essential for the camera to be able to resolve the speckles to observe this effect as the intensity averaged over many speckles is not affected by much. The required speckle contrast signal may be obtained simply by processing the TV video output through a high-pass filter and rectifier. This may then be displayed on a TV monitor screen to show fringes of equal vibration amplitude. For sinusoidal movement, their position is governed by the $(J_0)^2$ function given above for the time-averaged holograms. As in that case, the phase is not obtained and the fringe contrast decreases with increasing vibration amplitude and precautions are usually necessary to reduce the effect of background vibration. Nevertheless, time-averaged ESPI is a powerful tool for obtaining an overall picture of surface vibration very quickly.

Phase information may be obtained and fringe contrast may be improved by stroboscopic illumination if this can be synchronized to an exciting waveform. A simpler alternative is to phase modulate the reference beam, with a vibrating mirror for example. Then those parts of the surface vibrating in phase with this will appear stationary and maintain high speckle contrast. In this way the phase and amplitude of vibrations may be mapped over a surface.

The sensitivity of ESPI is normally limited to $\sim \frac{1}{2}\lambda$ but modifications to achieve a much higher sensitivity are possible. By phase modulating the reference beam at a frequency slightly different from the excitation frequency and observing flicker, Høgmar and Løkberg (1977) have been able to measure displacement amplitudes down to 10^{-2} nm. However, the technique implies a very restrictive bandwidth and it is doubtful if it would be applicable to ultrasonic studies.

4.3.4   Comparison with single-point laser interferometry

It will be seen that, although holographic and speckle pattern interferometry provide immediate spatial coverage of a surface, obtaining detailed time-dependent information is difficult. At present there is no satisfactory method of obtaining continuous space and time coverage simultaneously. It is expected that single-point interferometry with scanning will prove more practical for detailed vibration and ultrasonic studies except perhaps for surface acoustic waves. It has fundamentally more sensitivity for a given bandwidth since the laser light is concentrated in a small spot rather than being spread over the whole surface. This is particularly important in ultrasonic studies. Obtaining sufficient intensity to illuminate a large object is often a problem in holography and speckle interferometry.

The smallest displacements detectable by these techniques are of the order of 0.1–1 nm. This is comparable with single-point (reference-beam) interferometry from a rough surface with a small to medium power laser and a bandwidth of 10 MHz (see §3.5.5). However, the effective bandwidth is very much smaller. In time-averaged holography or speckle interferometry, either the time dependence of the displacement is not obtained or a phase-sensitive technique, relative to an excitation waveform, is used, implying a very narrow bandwidth.

In pulsed holography, a 'snapshot' picture of surface displacement is obtained and this cannot be repeated many times in quick succession due to the limited energy storage capacity of the laser. The information obtained is thus really of a different type from that obtainable from single-point scanning. When acoustic fields produced by infrequent events need to be studied and a suitable trigger is available, this method could clearly give results not obtainable in other ways.

With the exception of pulsed holography, these techniques are susceptible to background vibration. Although this is a problem in the measurement of very small ultrasonic displacements by single-point reference-beam interferometry, it may be overcome by the use of compensated interferometers (§3.7) or by the techniques described in §§3.8 or 3.9, without the need for mechanical antivibration mounts.

4.4   THE MEASUREMENT OF SURFACE WAVES

Laser techniques are particularly suitable for observing surface waves, as the focused spot that may be obtained from a laser can be made smaller than the ultrasonic surface wavelength. This is true even up to frequencies as high as 500 MHz where the wavelength is typically a few micrometres, thus allowing the resolution of individual waves. Interferometry has, however, not been so widely used as might have been expected since a shorter wavelength means that there is a greater surface tilt for a given wave amplitude, and the knife-edge technique discussed in §§2.6.2 and 2.6.3 provides a practical and frequently simpler alternative. If a reflecting surface can be provided and the optimum spot diameter is used, the techniques have very similar sensitivity (see §2.6.3). Interferometry can, however, also work with optically rough surfaces, though with much reduced sensitivity and (in reference-beam techniques) with speckle problems.

Surface waves are usually continuous waves or consist of long bursts of sinusoidal oscillation. Very narrow bandwidth detection is frequently used. With good surfaces, amplitudes less than 0.01 pm may be detected in a bandwidth of 1 Hz with a low-power helium–neon laser. With wide-bandwidth detection it is, however, possible to study sharp surface-wave pulses produced, for example, by pulsed laser excitation (Scruby and Moss 1985). Some of these measurements are discussed in §5.10.1.

As with the knife-edge technique, interferometry can measure the surface-wave displacement amplitude at a point and its phase relative to an excitation waveform fed to a transducer. Determination of the direction of travel and standing-wave patterns needs measurements to be made at more than one point. With a fast scanning system, direction may be deduced from the apparent 'Doppler' shift of frequency, as discussed in §2.6.4.

4.4.1 The effect of surface distortion on interferometry

The tilting of the surface by a surface wave which is the basis of the knife-edge technique is undesirable for interferometry. In reference-beam techniques it leads to a reduction of coherence and a consequent reduction in interference signal. However, in practice this is a small effect and is not usually a problem. The maximum tilt occurs at the zero crossing point of the sine wave, as shown in figure 4.15(a). On a reflecting surface, the return beam is deflected by an angle $4\pi a/\Lambda$ where a is the sinusoidal displacement amplitude and Λ the surface wavelength. Back in a conventional reference-beam interferometer, this leads to a displacement of the return beam relative to the reference, a

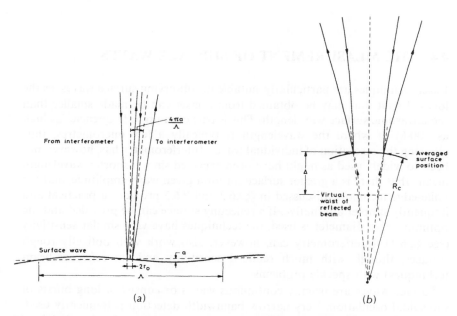

Figure 4.15 (a) The deflection of the returned beam due to surface tilt in laser interferometry of surface waves. (b) Effect of surface curvature on reflection at the waist of a Gaussian laser beam.

reduction in overlap and hence leads to reduced interference. Assuming Gaussian beam optics (§1.2.9), the reduction factor may be calculated:

$$\text{RF} = \exp\left[-2\left(\frac{2\pi^2 a r_0}{\Lambda\lambda}\right)^2\right]. \tag{4.15}$$

Here $2r_0$ is the beam waist diameter on the surface (to $1/e^2$ intensity points). It is of course desirable to keep r_0 as small as possible to reduce this effect but, for a typical $r_0/\Lambda = 0.1$, a displacement amplitude $a = \frac{1}{10}\lambda$ would produce a reduction of approximately 8% which should be acceptable. The interferometer signal is small in this region anyway.

If the spot size is not very much smaller than the surface wavelength, the interferometer response must average the surface displacement over the diameter of the focused beam. This effect is a maximum near the displacement peak, where the curvature of the surface perturbs the returned beam, as shown in figure 4.15(b). This may be described as a shift of the apparent waist of the return beam by

$$\Delta = \frac{2\pi^2 r_0^4}{\lambda^2 R_c} \tag{4.16}$$

where R_c is the radius of curvature of the surface at its maximum (assumed to be much greater than $2\pi r_0^2/\lambda$). This leads to an extra optical path difference:

$$\Delta\text{OP} = \frac{r_0^2}{4R_c} \tag{4.17}$$

when averaged over the interfering beams in the far field in the interferometer. As might be expected, this result is exactly that that may be derived from averaging the surface displacement over the waist of the beam weighted by the beam intensity. Using the result

$$\frac{1}{R_c} = \frac{4\pi^2 a}{\Lambda^2} \tag{4.18}$$

for a small-amplitude sinusoidal wave, we obtain for the apparent reduction in surface displacement as measured by the interferometer

$$\delta a = \frac{\pi^2 r_0^2 a}{2\Lambda^2}. \tag{4.19}$$

For $r_0/\Lambda = 0.1$, this is a reduction of approximately 5%. This is fairly acceptable but it is undesirable for r_0/Λ to be much above this value, though it may be unavoidable at very high frequencies.

At the peak of the wave, the curvature also theoretically leads to some reduction in coherence due to the slightly changed divergence of the returned beam. However, this is a very small effect that may be neglected in most circumstances.

4.4.2 Measurement of surface-wave velocity

The determination of the wavelength and direction of a travelling wave
requires phase measurement as a function of position on the surface. De la
Rue *et al* (1972) employed the single-sideband heterodyne technique discussed
in §3.8.2. Their experimental arrangement with a double-pass Bragg cell was
very similar to that shown in figure 3.26. The determination of phase requires
the production of a signal of the ultrasonic frequency proportional to the
displacement amplitude with phase unperturbed by low-frequency vibration.
This is obtained by mixing one sideband with the carrier (i.e. shift frequency)
component of the output of the detector. A diagram of the signal-processing
system is shown in figure 4.16. The signal is then down converted to a
convenient frequency around 5 kHz and mixed with the transducer excitation
waveform in a phase-sensitive detector, producing an output proportional
to the component of the sinusoidal motion in phase with the transducer. A
very narrow passband can be used with this type of detector. To compensate
for changes of sensitivity due to variations in light level, etc, the output may
be normalized against the original signal level obtained directly from the
photodetector as shown.

The output may be fed to a recorder and the laser spot moved slowly
across the surface. For a simple travelling wave, a sinusoidal trace is then

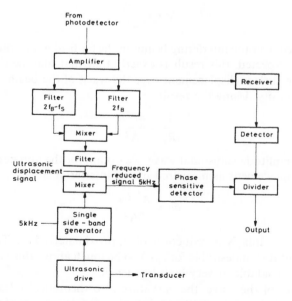

Figure 4.16 Signal processing system for the measurement of surface-
wave amplitude and phase using single sideband heterodyne interferometry (De la Rue *et al* 1972).

obtained from which the surface wavelength may be determined. Knowing the excitation frequency, this gives the phase velocity. The group velocity may be obtained by varying the frequency and making measurements at two fixed locations, of the frequency interval between output zeros. It may be shown that if the velocity does not vary too rapidly with frequency, the group velocity v_g may be expressed in the form

$$v_g = L_{AB} \frac{\Delta f_A \Delta f_B}{\Delta f_A - \Delta f_B} \tag{4.20}$$

where Δf_A and Δf_B are the frequency intervals at the two locations and L_{AB} is the separation of the locations.

Wickramasinghe and Ash (1975) have used the above technique for the measurement of the variation of surface-wave phase velocity with direction of propagation on lithium niobate and quartz. Very high accuracy is possible with measurements of this type. The measurement of phase to $\pm 3°$ reported by Murray and Ash (1977) should enable an accuracy of one part in 10^5 to be achieved at frequencies around 60 MHz with a 40 mm long sample, provided sufficient precision in mechanical scanning can be arranged. Absolute accuracies of five parts in 10^5 have been obtained in measurements on surfaces of lithium niobate (Murray and Ash 1977); the relative accuracy is higher. Such precise measurements can provide valuable information about surface layers and surface preparation as well as improving our knowledge of elastic constants and enabling crystal orientation to be determined accurately. Accurate measurements of surface-wave velocities on silicon (Ameri *et al* 1980) have shown differences of about one part in 1000 between samples of p and n types, and a 0.3 μm thick layer of oxide results in a 0.6 % decrease in surface-wave velocity.

4.4.3 Studies of surface-wave scattering

The heterodyne technique was also used by Bouchard and Bogy (1985) in their study of the scattering of Rayleigh waves by a plate on the surface of a steel block, a situation resembling the attachment of a piezoelectric transducer to a surface. A sketch of the experiment is shown in figure 4.17. Bursts of compression waves produced by the piezoelectric transducer were converted to Rayleigh waves by a perspex wedge angled correctly. These were reflected by an aluminium plate attached to the surface of the block by coupling oil or grease. The frequency of the transducer could be varied between 0.8 and 1.4 MHz. Signals were generated from the incident Rayleigh waves as they passed the focused laser spot (300μm diameter) and from the reflected waves on their way back. The ratio of the amplitudes of these signals was noted.

As in the experiments described in the last section, the method of mixing one sideband with the carrier was used to reproduce the sinusoidal ultrasonic

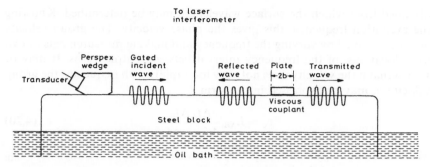

Figure 4.17 The measurement by laser interferometry of the reflection of surfaces waves on a steel block by a plate coupled to the surface (Bouchard and Bogy 1985).

displacement waveform. The lower sideband and carrier were selected from the output of the commercial interferometer which had been shifted by 40 MHz by a Bragg cell. This was done by a bandpass filter centred on 39 MHz with a width of 2.3 MHz to accommodate the frequency range required. It was important that the gate period be long enough for the response to be considered a steady state one as this type of processor does not reproduce transients, but not so long that the incident and reflected waves overlap. A gate period of 30 μs and pulse interval of 16 ms were found to be suitable.

To obtain reproducible results, the orientation of the plate had to be set carefully and good coupling maintained. Results obtained with grease coupling are shown in figure 4.18. As expected it was found that the fraction

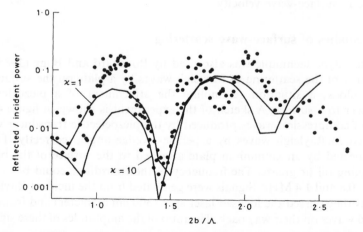

Figure 4.18 Results of measurements of the reflection coefficient of surface waves by a plate coupled to a surface. The coefficient as a function of plate width is compared with theory (Bouchard and Bogy 1985).

of the power returned depended only on the ratio of the width of the plate to the Rayleigh wavelength. Two theoretical curves are shown for comparison; these are calculated for different values of the couplant viscosity parameter (Angel and Bogy 1981).

The reflection of Rayleigh waves from a surface slot (simulating a crack) has been studied in a similar way by Jungerman *et al* (1984a,b). The reflection coefficient as a function of the ratio of the slot depth to the wavelength is shown in figure 4.19 where the results are compared with the theory of Hirao and Fukuoka (1982).

An alternative way of observing the perturbation of surface waves by a crack is to plot the ultrasonic displacement amplitude as a function of position on the surface around the crack when the incident gated cw pulse is centred over the crack. An image of a fatigue crack in an aluminium sample obtained in this way is shown in figure 4.20. The defect is 450 μm long, half-penny shaped and 200 μm deep. The ultrasonic frequency is 5.5 MHz. The resolution obtainable from a laser interferometer is clearly essential in obtaining images of this type.

The type of interferometer used in these experiments was a long-path-difference or velocity interferometer (§3.10) using a fibre-optic delay line due to Bowers (1982). In this way long delays can be achieved in a compact interferometer, e.g. 1 μs with a 200 m fibre length. This enables this type of interferometer to operate down to frequencies as low as 200 kHz. A diagram of the arrangement is shown in figure 4.21. Near-infrared light from the laser diode is pigtailed directly into a single-mode fibre. The light is then split, one half being delayed by τ_D in the fibre loop, before being recombined

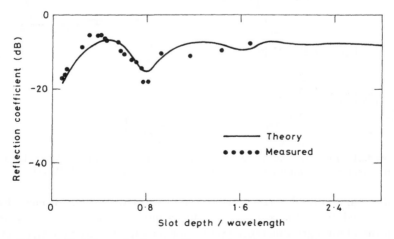

Figure 4.19 The reflection coefficient of Rayleigh waves by a surface slot measured by laser interferometry as a function of slot depth is compared with theory (Jungerman *et al* 1984a).

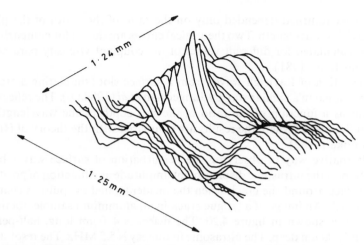

Figure 4.20 Acoustic displacement image of a fatigue crack in an aluminium sample. Vertical displacement is plotted as a function of position. The defect is 450 μm long, half-penny shaped and 200 μm deep. The 5.5 MHz transducer is to the left (Jungerman *et al* 1984a).

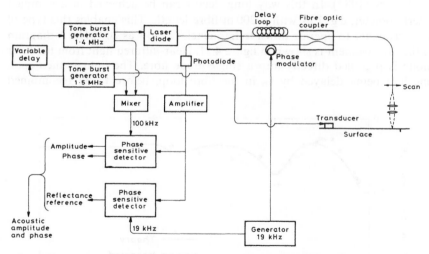

Figure 4.21 Investigating surface waves with a long-path-difference laser interferometer incorporating a fibre-optic delay (Jungerman *et al* 1984a).

with the minimally delayed portion and directed to the surface. For light returned from the surface, the fibre-optic couplers linked by the fibres form a long-path-difference interferometer with a delay time τ_D exactly the same as for the outgoing beam. Light leaving the long-path arm thus consists of two components having delays of τ_D and $2\tau_D$, whilst in the short arm the delays are 0 and τ_D. Because the coherence length of the semiconductor laser is short, only interference between light components suffering almost the same delay, i.e. τ_D, is observed. There is, however, a small phase difference between these components since the surface moves slightly between their arrivals and this gives rise to the surface displacement signal, as discussed in §3.10.2.

Laser diodes may be readily modulated and advantage is taken of this to modulate the output at a frequency close to the ultrasonic frequency, the consequent stroboscopic effect producing a low-frequency output. This improves the detector sensitivity and enables the signal processing to be performed at the much lower difference frequency.

The advantages of this method are that long path differences can be accommodated in a small space; it is very insensitive to fluctuations in fibre path length due to pressure or temperature, etc, and a laser diode of poor temporal coherence can be used. Sensitivity to periodic displacements down to 0.03 pm is possible with a passband of 0.1 Hz. It works quite well with rough surfaces but this is due to the small spot size essential for efficient coupling to the single-mode fibre, rather than to an ability to handle inclined beams discussed in §3.10.4. In this respect it is similar to reference beam interferometry and it is also subject to signal intensity fluctuations due to speckle. To allow for this, a continuous calibration system is incorporated via a low-frequency phase modulation of the fibre-optic interferometer.

The application to surface-wave measurement of a more conventional heterodyne interferometer incorporating fibre-optical coupling has also been described by Jungerman et al (1982). It is interesting to note that the fibre-optic output may be coupled directly to the surface without the use of a lens. The fibre end must then be very close to the surface and this is not normally a very practical arrangement.

An interesting possibility is the reconstruction of images of surface defects from acoustic holograms of surface-wave fields (Wickramasinghe and Ash 1973, 1974). Such a one-dimensional hologram may be generated from laser probe measurements which give the phase and amplitude of the surface-acoustic-wave field along a line intercepting surface waves scattered by one or more defects. Although reconstruction of the (acoustic) image by an optical method has been demonstrated, computing techniques are more flexible and satisfactory. The theory of these has been outlined by Wickramasinghe and Ash (1974). Although the computation is considerably simpler than for image reconstruction from 2D acoustic holograms, it is usually complicated by the anisotropy of the substrate. Ameri et al (1980) have described a method, using this principle, of locating and characterizing surface defects using scans across three lines intercepting a surface-wave field.

4.4.4 Differential interferometry

A favourable factor in the application of laser interferometry to the observation of surface waves which have a short wavelength is that the net movement of the surface need not be measured. It is sufficient to record between points close together, separated only by a fraction of a wavelength. This enables the effect of background vibration to be eliminated making compensation for this problem (discussed in §3.7) unnecessary. By using interference between light reflected or scattered from adjacent points on the surface, ideally separated by a half a surface acoustic wavelength, the effects of background optical path variations are cancelled out.

This principle is used in the scheme of Palmer *et al* (1977) shown in figure 4.22. After expansion, the beam from a helium–neon laser is split by a grating (or holographic beam splitter, H). The resulting beams are then focused and the ± 1 diffraction orders selected by the slits. These are converted to parallel beams again before being focused by the variable focus lens down to two adjacent spots P and Q on the surface (at the beam waists). Light returned from the spots is made into parallel beams by passing through the variable focus lens and directed towards the photodetector by the mirror M (which is out of the plane of the incident beams). The beams arising from each spot are inclined to each other by a small angle and a system of parallel interference fringes is formed whose spacing is $\lambda F/\Delta_{PQ}$ where Δ_{PQ} is the separation of the spots on the surface and F the focal length of the lens. These fringes shift sideways as the optical path difference changes due to the difference in movement of the surface at P and Q. This movement is translated into intensity fluctuations at the detector by passage through a matching grating, G. For maximum sensitivity, the grating should be displaced relative to the fringes by a quarter of their spacing.

If the separation of the spots has its optimum value of half a wavelength, the sensitivity of this technique is comparable to that of conventional reference-beam techniques and, assuming both techniques are operated optimally, its sensitivity is also very similar to that of the knife-edge technique

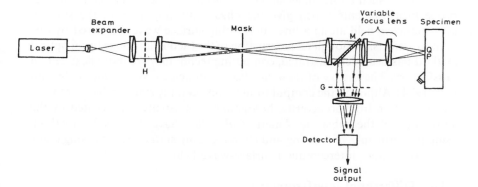

Figure 4.22   The measurement of surface waves with a differential laser interferometer (Palmer *et al* 1977).

(§§2.6.2 and 2.6.3) to which it is in many respects rather similar. It is best suited to sinusoidal waves and sensitivity to displacement amplitudes down to 0.06 pm has been reported for waves on a glass surface (Palmer *et al* 1977) with a 1 mW helium–neon laser and a bandwidth of 1 Hz. (For ease of detection and obtaining a narrow bandwidth, the laser light was chopped at 480 Hz and phase-sensitive detection used.)

Simplification of this scheme is possible by the use of the same grating for splitting of the incident beams and gating the fringes (Jablonowski 1978). This is illustrated in figure 4.23. The use of a grating consisting of alternating reflecting and clear strips enables the return signal to be obtained by reflection off the grating, thus improving optical efficiency. It is not necessary to separate diffraction orders but there could be complications of interpretation in some cases due to there being more than two spots. There also appears to be some advantage in using a rotating grating (Jablonowski 1978).

The technique has also been extended to measurements in two orthogonal directions on the surface simultaneously by the use of two pairs of spots arranged in a cross and separated by polarization (Turner and Claus 1981). In this way the true amplitude and direction of propagation of an arbitrary surface wave may be determined.

In a rather similar 'twin path' technique of Yoneda *et al* (1981), the two beams of the differential interferometer are separated and recombined by a phase grating. However, one beam is reflected from the surface being studied whilst the other is reflected from an adjacent mirror which is attached to a piezoelectric compensator to enable the phase of the interference to be adjusted for maximum sensitivity. This interferometer thus measures the actual displacement of a point on the surface and not the difference between two adjacent points as do the other differential interferometers discussed

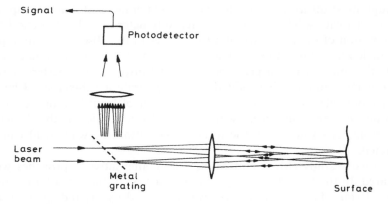

Figure 4.23 A simple scheme for detecting surface waves based on the principle of differential interferometry (Jablonowski 1978).

above. Insensitivity to background vibration is achieved through the close proximity of the two beam paths.

4.5 CALIBRATION OF TRANSDUCERS

Laser interferometry has a very important contribution to make to the calibration of ultrasonic transducers. Although ultrasonic transducers are now very widely used, they are not always calibrated as comprehensively as they could be. This is partly because a full calibration (i.e. sensitivity defined in absolute units as a function of frequency and position in the beam) is difficult and costly. Calibration is particularly difficult in the case of contact transducers for use in the inspection of solids.

Various techniques are available, including reciprocity methods, standard test blocks, and water-tank procedures. All have drawbacks of one kind or another, and there is still a need for a rapid, reliable and flexible technique that will give an accurate value for transducer sensitivity over a wide frequency range. The laser interferometer offers absolute measurement over a wide frequency band, high spatial resolution for beam plotting, rapid scanning, freedom from couplant (eliminating thereby one of the most important sources of error) and non-contact sensing (so that the transducer and its wavefield are undisturbed).

In the following sections we discuss the potential of interferometry for the calibration of ultrasonic transducers in a variety of situations, i.e. free-standing (coupled into air), coupled into a fluid (water), and in direct contact with a solid.

4.5.1 Transducers coupled into air

Although most ultrasonic transducers are used either with liquid couplant, or in contact with solids, it may be of interest to determine the characteristics of the motion of the front face of a transducer in the absence of any liquid or solid loading. One possible reason is to check for any manufacturing defects that might cause a perturbation in the displacement pattern across the front face, and which would therefore generate a sub-standard beam. Another possible use for accurate measurement of front-face motion is in the design of new transducers with special characteristics. If the displacement profile is known, then the form of the ultrasonic wave field can be calculated.

In an early study (Moss and Milne 1980), alternative approaches were compared. In the first, the average response across of the face of a nominal 1 MHz, 10 mm diameter transducer was measured by expanding the beam of the interferometer up to approximately 15 mm (figure 4.24). The front face of the piezoelectric element was polished to reflect the maximum light. One

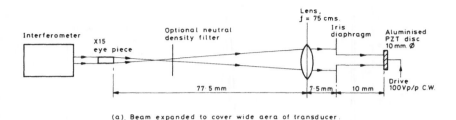

(a). Beam expanded to cover wide aera of transducer.

(b). Beam focused to point at centre of transducer.

Figure 4.24 Optical system for measuring displacements over (*a*) a large area, and (*b*) a very small area of the front face of a piezoelectric transducer (Moss and Milne 1980).

serious practical problem was the high sensitivity of the apparatus to minor misalignments and to any loss of flatness of the transducer surface. The displacement waveform was then subjected to frequency analysis (figure 4.25). There is a broad resonance at 1 MHz. The diameter of the interferometer beam was reduced, first to 3 mm by means of an aperture, and finally to 1 mm by means of an eye-piece lens. As figure 4.25 shows, the effect of reducing the probe beam size is to detect more structure in the spectrum. The additional peaks in the spectrum that are observed, especially at 1 mm diameter, are almost certainly due to various radial and other coupled resonance modes of the piezoelectric element. The effect of enlarging the probe area is to average out the displacements caused by these modes so that their amplitude in the spectrum is reduced.

Effects such as the excitation of radial resonance modes can seriously impair the performance of a piezoelectric element. These can be studied in more detail by scanning the interferometer beam (diameter approximately 0.1 mm) across the surface of the transducer. By way of example, figure 4.26 shows data obtained from a 0.5 MHz transducer excited with 2 V cw at a range of frequencies. These data were obtained before the implementation of a digital recording system, and the scan plots are taken from an ultraviolet recorder. The change in radial-mode pattern as the frequency is varied around the nominal thickness resonance of 0.5 MHz can be clearly seen.

Transducers may alternatively be excited by edge pulses. These are generally used to obtained broader bandwidth performance. Figure 4.27 shows a pen-plotter scan of the surface displacements across the front face

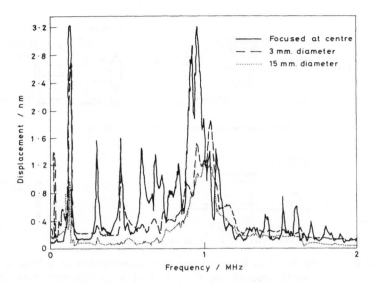

Figure 4.25 Comparison of frequency spectra of the displacement of the front face of a 1 MHz transducer, for laser interferometer beams of different diameter. When focused at the centre of the transducer, sharp thickness and radial resonances are observed, which are progressively averaged out as beam diameter is increased (Moss and Milne 1980).

of a 2.5 MHz piezoelectric transducer, excited by an edge-pulse ultrasonic driver. Note that there is no phase information in a plot such as this: the interferometer is only being used to measure peak displacement amplitude as a function of position. Nevertheless it can be seen that edge-pulse excitation of this particular piezoelectric element produces a reasonably uniform displacement across the central region of the face. Although there is an annulus of slightly increased displacement amplitude around the edge of this region, this transducer is emulating the ideal 'piston' source reasonably well.

Experimental data such as these are virtually impossible to obtain by any other method. A knowledge of the mode pattern across the face of a transducer should prove very useful, both in explaining the intricacies of the resulting ultrasonic beam, and in enabling new types of transducer to be developed in a systematic manner. The sensitivity of the interferometer technique in detecting unwanted perturbations to the motion of the front face is well illustrated in figure 4.28, where an angular scan across the face of a 2 MHz transducer is presented. The returned light level was not sensitive to angle of incidence of the interferometer beam because the front face was matt black. With adequate beam expansion, the focused-beam technique returned sufficient light to give scans with acceptable signal/noise, even with this type of surface. The structure outside the central pattern of rings should be ignored:

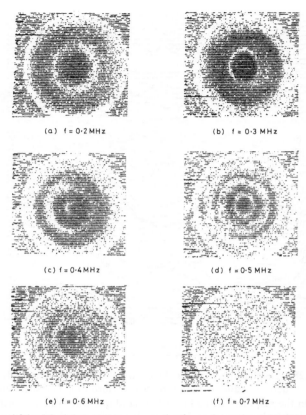

(a) f = 0·2 MHz   (b) f = 0·3 MHz

(c) f = 0·4 MHz   (d) f = 0·5 MHz

(e) f = 0·6 MHz   (f) f = 0·7 MHz

Figure 4.26 Modal patterns across the front face of a 0.5 MHz transducer are shown as a function of frequency of excitation, recorded by an ultraviolet recorder (Moss and Milne 1980).

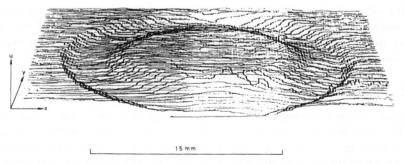

15 mm

Figure 4.27 Displacement (u) across the front face of a 15 mm diameter, 2.5 MHz transducer, when driven by edge pulse (Moss and Milne 1980). The data outside the 15 mm diameter of the active element have no meaning and should be ignored.

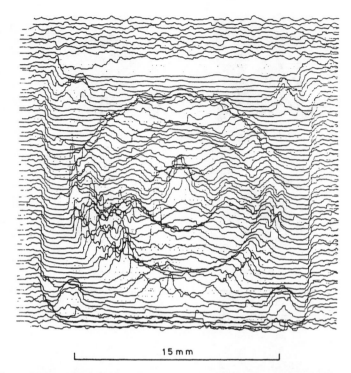

Figure 4.28 Displacement across the front face of a 2 MHz transducer. The ring structure is due to radial modes of vibration. The scan shows a defect in the response in the lower left-hand quadrant (later identified with an electrical connection). The transducer was attached to a square plate with four screw holes at the corners (Moss and Milne 1980). The data outside the plate have no meaning and should be ignored.

this transducer was attached to a plate with four holes at the corners. Figure 4.28 shows very clearly that the otherwise symmetrical pattern of modal rings is disturbed at the bottom left. This 'defect' was later found to correlate with the electrical connection to the front face.

4.5.2   Transducers coupled into liquids

Many ultrasonic transducers are immersed in liquids such as water when in use. This may be either in order to provide reliably constant coupling during the scanned inspection of a solid in a water tank, or else when making measurements on fluids, or aqueous materials (many biological materials for instance exhibit similar ultrasonic behaviour to water). It is thus important to know the characteristics of their ultrasonic wavefield under these conditions. In addition to the conventional beam-plotting method in which small spheres are used as targets, and reflected ultrasonic amplitudes are measured, the laser interferometer can be used to obtain a direct measurement of the field. Because it is by far the most important fluid in this context, we shall restrict our discussion to water. Small corrections have to made for other fluids, depending upon the density and ultrasonic velocity.

Water is transparent to light; it is therefore necessary to introduce a thin reflecting membrane (pellicle) into the water. Experience has shown that metallized Melinex or Mylar film is close to ideal. The film is stretched over a frame to give a good planar reflector. Such films are very thin (typically in the range $2-5\,\mu\mathrm{m}$) and light. As calculated in §4.1.1, they barely perturb the acoustic wavefield as it propagates across the pellicle. Measurements (Bacon 1988) have in fact shown that acoustic reflection coefficients of 1 % or less are typical below 10 MHz, and therefore negligible except for the most accurate of measurements.

Using the experimental arrangement shown in figure 4.1, the pellicle technique was used to determine the linearity of various ultrasonic transducers (Moss and Milne 1980). Figure 4.29 shows the results for a 1 MHz probe,

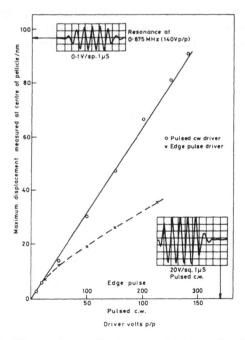

Figure 4.29 Ultrasonic amplitude as a function of pulsed cw and edge-pulse drive voltages for a nominal 1 MHz transducer. A laser interferometer was directed at the centre of a pellicle 60 mm from the front face of the transducer (Moss and Milne 1980).

the pellicle being positioned 60 mm from the probe, and the laser beam being directed accurately along the axis of the transducer. The transducer exhibits an excellent linear response under gated cw excitation. Under pulsed excitation its response deviates somewhat from linearity. This is because the output waveform from the edge-pulse excitation unit changes markedly with output voltage, rather than because of any non-linearity in the response of the piezoelectric element.

The laser interferometer measurement system shows its true potential when scanned measurements are made. Thus for the data of figure 4.30 a 12 μm thick aluminized Melinex pellicle is placed at right angles to the beam from a 1 MHz probe in water, and the probe excited at its resonance with a gated sine wave. For these measurements the probe was scanned horizontally so that the ultrasonic beam moved across the pellicle. The probe–pellicle distance was 50 mm, such that the scan was made through the 'focus' of the probe (i.e. the point of maximum ultrasonic intensity). The data show that the probe produces a symmetrical beam that is approximately Gaussian in section. In figure 4.31, the scan has been made along the beam axis by varying the distance between the same 1 MHz probe and pellicle. This figure is interesting in that it also shows the results of a conventional probe beam plot, in addition to the theoretical intensity calculated assuming the transducer acts as an ideal 'piston' source. As figure 4.31 shows, the agreement between both experimental techniques and the theory is remarkably good.

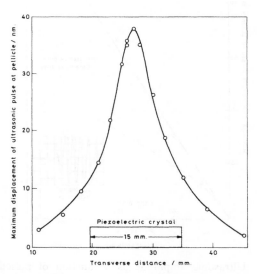

Figure 4.30   A transverse scan across the beam from a 1 MHz transducer at focus (50 mm, i.e. maximum amplitude), under pulsed cw excitation at a resonance of 0.9 MHz (Moss and Milne 1980).

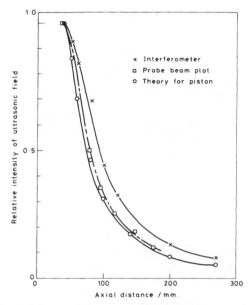

Figure 4.31 Axial scan of the same 1 MHz transducer under edge-pulse excitation, showing comparison of interferometer with conventional water-tank data, and theory for an ideal piston (Moss and Milne 1980).

The technique described above of scanning the transducer mechanically by stepping motors has two major disadvantages. Firstly, it is very slow, especially if a full raster scan is desired. Secondly, the motion of the probe through the tank disturbs the water, setting up standing waves. A better alternative is to scan the beam of the interferometer. The angular scanning arrangement used earlier is unsatisfactory in the present case because of the high reflectivity of the pellicle. Any deviation from normal incidence causes an unacceptable loss of signal. Thus a parallel scanning system was developed (as described in §4.2.4) to ensure constant normal incidence of the light beam at the pellicle. The beam from a 5 MHz annular (axicon) transducer was directed perpendicularly at the pellicle at ranges of 25, 50 and 100 mm. Figure 4.32 shows the results of a raster scan across the pellicle by the interferometer. Figure 4.33 displays the data at 25 mm in the form of a contour plot. From this and the previous figure it can be seen that this transducer produces an exceptionally finely focused beam at a range of 25 mm in water. The beam is still reasonably narrow at 50 mm.

4.5.3 Hydrophones

All users of ultrasonics recognize the importance of transducer calibration in terms of the temporal and spatial characteristics of the ultrasonic field.

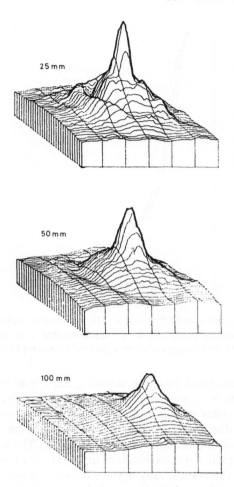

Figure 4.32 Raster scan of the ultrasonic field generated by a 5 MHz
axicon (annular) transducer at ranges of 25, 50 and 100 mm, respectively.
The vertical scale is displacement amplitude.

However, in medical ultrasonics it is also necessary to determine the *absolute*
strength of the field for reasons of clinical safety and to enable meaningful
biomedical research. Miniature hydrophones are most commonly used to
make these measurements, but since these are not themselves absolute devices,
they must first be calibrated absolutely. There are a number of methods in
current use, but they are mostly difficult and time consuming if an accurate
result is to be achieved. The most common methods are the reciprocity and
total power planar scanning techniques. The laser interferometer offers a
method of hydrophone calibration that is highly reproducible, accurate and
rapid to use.

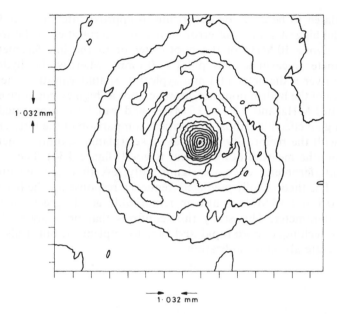

1·032 mm

1· 032 mm

Figure 4.33 Data of figure 4.32 at 25 mm range plotted as contour plot. First contour is 2 nm, interval is 0.5 nm, and hence the maximum amplitude at the focus of the beam is approximately 9 nm.

The principle of the interferometric method for hydrophone calibration is similar to that for immersion transducers. The hydrophone is placed in a large water tank opposite an ultrasonic transmitter. The transmitter is excited with gated (pulsed) continuous waves from a suitable signal generator. The output from the hydrophone is fed into a digital recording system. The hydrophone is then carefully replaced by a reflecting pellicle (described earlier) at exactly the same location in the tank. The interferometer beam is focused in the plane of the pellicle, and can be scanned across it. Lateral movement of the beam can be achieved either by rotating a refracting glass block, or by means of rotating mirrors, as described in §4.2.2. The output signal from the interferometer can also be fed into the same digital recording system to enable a direct comparison to be made.

The above method is a substitutionary one: an even more direct and rapid calibration can be made in the case of a membrane hydrophone, e.g. of polyvinylidene difluoride, by using the hydrophone itself as the pellicle, since it has gold-plated active elements. While this method gives accurate results at typically 3 MHz and below, small errors can accrue above this value due to the finite thickness of the gold layer.

Since the requirement is for absolute accuracy, Bacon (1988) carried out a very detailed analysis of systematic and random sources of error in hydrophone calibration. Below 10 MHz both systematic and random errors are calculated to be 2.5% or less, rising to approximately 6% at 15 MHz. One of the biggest sources of error is due to uncertainties in the frequency response above 10 MHz of the photodetector in the interferometer. The interferometer measures at a point <0.1 mm, whereas the hydrophone measures over a diameter of 1 mm typically. Spatial variations across the active area of the hydrophone are, however, not as high as might be expected (0.06% at 1 MHz and 1% at 15 MHz) and can be compensated for by calculating an effective radius. In a critical comparison of the interferometric method with the reciprocity and total power (planar scanning) methods of calibration, the measure of agreement is good (figure 4.34). The error bars on the interferometer data are significantly less than for the other two methods. The theoretically predicted frequency response for the hydrophone between 0.5 and 15 MHz is also in remarkably good agreement with the laser interferometer. This study thus confirms that not only is the interferometric technique more rapid and easy to implement, but it also gives a more accurate absolute calibration.

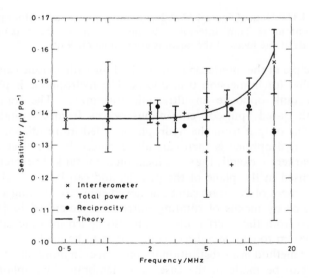

Figure 4.34 Sensitivity of hydrophone as a function of frequency, as determined by interferometer (×), total power, i.e. planar scanning (+), and reciprocity (●) techniques. Error bars are given for the interferometer at all frequencies (bold symbols), for planar scanning at 2.25 and 10 MHz, and for reciprocity at 1, 5 and 10 MHz. The interferometer data are in good agreement with theoretical frequency response (solid curve) (Bacon 1988).

4.5.4   Transducers in contact with solids

Contact piezoelectric transducers are used in non-destructive testing to inspect a wide variety of engineering structures. The accuracy with which any defects might be located and sized depends in part upon knowing the ultrasonic beam profile, i.e. its orientation to the surface and its width. It is also important to ensure there are no significant sidelobes, or unwanted beams, which might otherwise give spurious defect indications. Hence each transducer needs to be calibrated before service.

A common method is to plot the ultrasonic beam from the transducer in pulse-echo mode using a small spherical reflector in a water tank. The beam profile in a solid such as steel or aluminium is deduced by means of a refraction correction, which takes into account the different wave speeds in water and the solid. It is clearly important to know whether this indirect method correctly predicts the beam in the solid. Indeed, one consideration that is invariably omitted when making the correction is the change of loading of the front face of the transducer when switching from water to solid. This would have at least a small influence on beam profile.

An alternative calibration method is to use a test block of the correct solid material, and to measure the amplitude of the echo from a standard side-drilled hole as the transducer is scanned across the block. This has the obvious advantage that the transducer is correctly coupled into the solid, and should therefore be good for estimating the sensitivity of the probe to defect echoes at the same range as the hole. It is, however, less well suited to general beam-profile measurement because of couplant variations as the transducer is scanned across the block. The configuration is difficult for calculating beam angles accurately, so that it is more difficult to generalize the results to other geometries and other types of defect. Finally, both these techniques use the transducer in pulse-echo mode, and therefore do not test its performance as a transmitter and receiver separately, as would be needed for pitch-and-catch and transmission configurations.

(a) Calibration of ultrasonic transmitters

Ideally, beam profiles are best measured when the transducer is in contact with a sample of the material to be inspected, using a receiver which does not disturb the ultrasonic field in any way, and which can be readily scanned without fluctuations in sensitivity. Because it is impossible to scan a receiver inside a solid sample, a surface measurement has to suffice. Only non-contact sensors will leave the ultrasonic field undisturbed. While electromagnetic/electrodynamic probes are cheaper, they do not have the high spatial resolution (<0.1 mm) and bandwidth ($\sim$ 100 MHz) of the laser interferometer that are needed for accurate beam measurement. Furthermore, interferometry is capable of absolute measurement (referred to the wavelength of light).

While this is unimportant for most non-destructive testing applications, absolute accuracy is of relevance for calibration studies.

Measurement of the beams generated by compression-wave contact transducers is generally more straightforward than shear-wave transducers. Figure 4.35 shows data from 5 and 10 MHz probes in normal contact with a polished 10 mm aluminium plate. The data are obtained by scanning the laser interferometer in a raster across the face of the block opposite the transducer. Although the interferometer only measures perpendicular surface displacement, no correction was deemed necessary to account for the

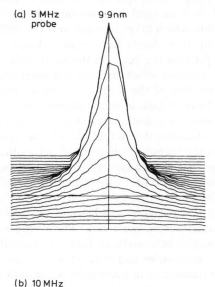

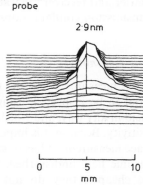

Figure 4.35 Compression-wave beam generated by 5 and 10 MHz contact probes as measured by a laser interferometer on the far side of a 10 mm aluminium plate (Moss and Scruby 1988).

variations in angle of incidence of the ultrasonic beam, because the beam widths are small (a few millimetres). A Michelson system (Chapter 3) was used for these measurements in combination with a 5 mW helium–neon laser. With a bandwidth of 100 MHz, this interferometer would have too high a noise threshold for such measurements (of the order of 1 nm). However, rapid on-line averaging of typically 200 waveforms per measurement point reduced the noise by a factor of 14 approximately to an acceptable level.

The most popular method for ultrasonic inspection is almost certainly to use an angled shear-wave probe in pulse-echo mode. It is more difficult to use the interferometer to calibrate such a transducer because there is no simple relationship between shear-wave amplitude (B_0) and perpendicular surface displacement (a_n). Assuming that the radiation from the probe can be approximated to plane waves, the relationship is given by the inverse of equation (4.14), i.e.

$$\frac{B_0}{a_n} = \frac{4\sin^2\theta\cos\theta(1 - k^2\sin^2\theta)^{1/2} + k(1 - 2\sin^2\theta)^2}{4\sin\theta\cos\theta(1 - k^2\sin^2\theta)^{1/2}} \tag{4.21}$$

where θ is the angle between the incident shear wave and the surface normal. The expression only holds when θ is less than the critical angle for shear waves. Outside this range the expression is complex, and the following relationship for amplitude modulus becomes more appropriate:

$$\frac{B_0}{a_n} = \frac{[16\sin^4\theta\cos^2\theta(k^2\sin^2\theta - 1) + k^2(1 - 2\sin^2\theta)^4]^{1/2}}{4\sin\theta\cos\theta(k^2\sin^2\theta - 1)^{1/2}}. \tag{4.22}$$

We note that there are problems for the calibration of shear-wave probes. Thus at 0° the normal displacement is zero, so that only angled probes can be calibrated by this method. However, the displacement is also zero at the critical angle. This is particularly serious for broad-beam or shallow-angled probes which encompass this angle. It does not present any significant problems for the most common angles of 45 or 60°. Indeed, the factor simplifies to a constant $\sqrt{2}/2$ at the former angle.

Figure 4.36 shows the experimental arrangement for calibrating shear-wave transducers by means of a laser interferometer. The interferometer beam executes a parallel scan in a raster across the surface of the specimen. The results for a 5 MHz probe fitted with a 45° perspex shoe are shown in figure 4.37. The signal went below the noise threshold at the edges of the scan. A small number of scans made at a higher gain did not indicate the presence of any sidelobes at the edges of the beam. Nevertheless, it must be borne in mind that the laser interferometer is a relatively sensitive receiver of ultrasound, even with appreciable signal averaging, so that it might not detect very weak sidelobes in the way a piezoelectric transducer would.

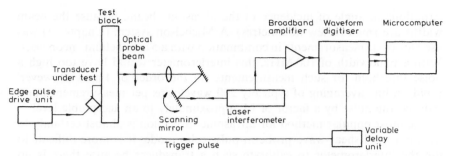

Figure 4.36 Experimental arrangement for calibrating ultrasonic fields generated by shear-wave transducers. The interferometer beam executes a raster scan across the specimen by means of a parallel scanner (Moss and Scruby 1988).

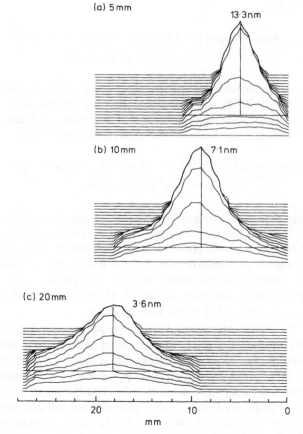

Figure 4.37 Comparison of beams from a 5 MHz 45° shear-wave transducer measured on the far surface of 5, 10 and 20 mm aluminium plates (Moss and Scruby 1988).

The data of figure 4.37 can be used to calculate the width of the beam in two perpendicular sections. If the beam widths are then plotted as a function of depth through the specimen, a fuller picture of ultrasonic propagation from the transducer through the aluminium plate is obtained (figure 4.38). The laser interferometer can thus be used to find the apparent source of the beam within the transducer itself, in addition to the actual beam angle (41° in this case, rather than 45° as quoted by the manufacturer) and width.

Another application of the interferometer is to compare the beams generated by different transducers. In figure 4.39 three different probe beams are compared. It can be seen that the 2 MHz probe produces the broadest beam, while the 5 MHz probe gives the highest peak displacement amplitude. Note especially that the interferometer gives signal strengths in absolute units.

This study demonstrates how well suited laser interferometry is to the calibration of ultrasonic transducers. The most important advantage has been discussed in some detail earlier, i.e. the non-contact nature of the interferometer in no way disturbs the ultrasonic wave field. However, the study also illustrates the speed with which a two-dimensional scan can be obtained—a matter of minutes, depending upon the resolution desired—and the amount of signal averaging needed.

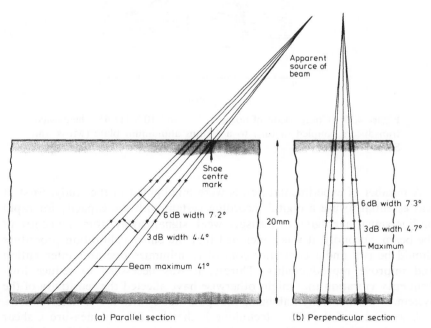

Figure 4.38 Reconstruction of a beam from a 5 MHz 45° shear-wave transducer using the data of figure 4.37 (Moss and Scruby 1988).

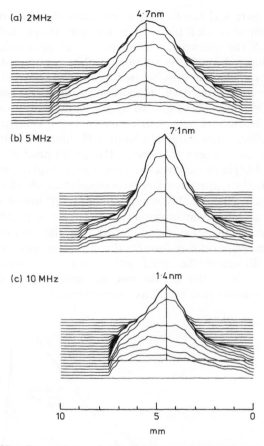

Figure 4.39 Comparison of nominal 2, 5 and 10 MHz 45° shear-wave transducers coupled in turn to a 10 mm aluminium plate (Moss and Scruby 1988).

A number of experimental points were highlighted in the study. First, it was essential to use a digital recording system with the capacity for rapid on-line signal averaging to measure weak signals away from the centre of the beam. Secondly, it was concluded that the entire calibration procedure should be run under computer control to minimize operator intervention and improve reproducibility. Thirdly, it was necessary to reduce low-frequency vibration that might otherwise have affected the sensitivity of the system, by mechanically stabilizing the apparatus by means of a massive table.

A weakness of the above technique is that it can only measure a shear amplitude indirectly via the normal surface displacement, which requires a complex correction factor. An alternative approach is to direct the interferometer beam obliquely at the specimen surface, so that parallel as well as perpendicular motion of the surface is measured. Since mode conversion occurs at the surface of an elastic solid, measurements should be made at two different orientations to enable resolved components to be deduced. This is discussed in more detail in §4.7 below.

(b) Calibration of ultrasonic receivers

The above measurements have all concentrated on the performance of ultrasonic transducers or transmitters. From a practical point of view it is equally important to assess their sensitivity as ultrasonic receivers. One way the laser interferometer can be used for this is by a substitution method. Thus the ultrasonic field produced by some source is measured first by the piezoelectric transducer under test. The transducer is then removed and the field measured by the interferometer. The problem with this method is that the piezoelectric transducer loads the surface of the specimen when in contact with it whereas the interferometer does not. Thus making the substitution slightly perturbs the sound field, so that the calibration may be inaccurate.

Alternatively, a special test block can be constructed to permit the interferometer beam to measure the ultrasonic field simultaneously with the contact probe (figure 4.40). The block consists of two glass prisms joined together by a drop of glycerol, and then firmly clamped to ensure good

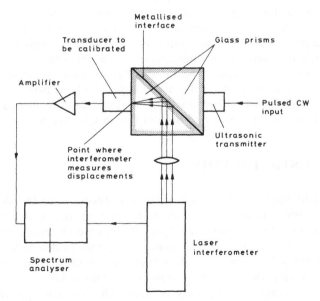

Figure 4.40 Special double prism calibration block (Moss and Milne 1980).

transmission of sound across the interface. The compression-wave speed in the glass was measured to be 5970 m s^{-1}, which is similar to steel, and therefore a reasonable substitute for calibration purposes. The transducer under calibration and the driver were then clamped with silicone grease to the left-hand and right-hand prisms, respectively. The laser interferometer beam was focused on to the centre of the transducer by reflection from the interface between the prisms. The hypotenuse of the left-hand prism was also coated with a thin layer of aluminium to increase optical reflectivity without affecting ultrasonic transmission. It was often found that the transducer front face reflected insufficient light: this was corrected by aluminizing this face of the glass block also. Another problem was that the interface disturbed the ultrasonic field as it passed across it and distorted the measured signal.

The experimental arrangement of figure 4.40 was used to obtain the absolute spectral response of various transducers in the receiver mode. The approach was to drive the transmitter with gated cw of variable frequency, and measure the output waveform of both the interferometer and the transducer under calibration. The next stage was to divide the Fourier transform of the transducer waveform by the Fourier transform of the interferometer waveform. The result gave the sensitivity of the transducer in V nm^{-1} as a function of frequency.

One problem in both the above cases is that a piezoelectric transducer measures the average surface displacement (or velocity) over the area of contact, whereas the interferometer measures displacement at a single point. To make a realistic comparison requires a scan to be made over an area corresponding to the probe face, followed by spatial averaging.

An alternative method for characterizing the performance of ultrasonic receivers is to measure their response to a standard ultrasonic source, i.e. to perform the inverse measurement to the calibration of ultrasonic transmitters. One of the most attractive methods is to employ a laser-generated ultrasonic source. It is more appropriate to discuss this in Chapter 6.

4.6 ACOUSTIC EMISSION

An important application of ultrasonic transducers is to the detection of acoustic emission. Before commencing this section, the reader may find a brief description of this technique helpful. When solid materials are subjected to increasing applied stress, there comes a stage when deformation and/or fracture of the solid may occur. If a crack, for instance, starts to grow at some point within the solid, the stress and strain fields in the immediate vicinity of the crack are suddenly relieved. This sudden appearance of a localized change of stress acts as a source of stress waves (also known as elastic waves or acoustic emissions) which radiate out from the crack, distributing the new stress state throughout the solid specimen. If the strain energy dissipated is large enough, and also occurs sufficiently abruptly, then these elastic waves can be detected by ultrasonic transducers attached to the solid. The technique of acoustic emission has developed a wide range of applications, centred on the detection and monitoring of defects in a wide range of materials and engineering structures (Williams 1980, Scruby 1987).

While conventional contact transducers are adequate for most applications of acoustic emission, there are a small number of specialized applications where a non-contact sensor would be of considerable advantage. As in the case of ultrasonics, these include use in hostile, high-temperature environments, or where high spatial and temporal resolution are desired.

4.6.1 Detection of deformation and crack growth

Most acoustic emission measurements of crack growth employ narrow-band piezoelectric transducers. While these have high sensitivity and are suitable for locating the approximate position of the defect, they give little information about its characteristics, since the source characteristics tend to be spread over a much wider frequency range than a resonant transducer. Some work has been carried out employing broadband capacitive transducers to characterize crack growth (Wadley and Scruby 1983). Capacitive sensors are, strictly speaking, also non-contact devices, since the sensing plate is separated from the specimen by an air gap of several micrometres. Although they have adequate bandwidth characteristics (of the order of 20 MHz) for acoustic emission studies, they require very careful sample preparation and mounting, thus restricting them to laboratory applications. Unless a very small plate (or ball) is used as the active element, they also tend to be too directional for general application.

More recently, some excellent point-contact piezoelectric transducers have become available for use in acoustic emission systems (Proctor 1982, Scruby 1985). These overcome most of the disadvantages of capacitive transducers, but at the expense of reduced bandwidth (typically 3 MHz). Furthermore, because of their lower cost and ease of mounting, they are suitable for industrial application, and also for use in multi-channel arrays. They still nevertheless suffer from the main disadvantages of contact devices.

The laser interferometer offers both high bandwidth and high spatial resolution in a non-contact form, and should therefore be of value for laboratory studies of the characteristics of the acoustic wave fields generated by growing cracks, etc. The most comprehensive study of optical detection of acoustic emission has been carried out at Johns Hopkins University. Most of the work has been carried out using a Michelson interferometer, with some of the features discussed in Chapter 3 to improve performance as a non-destructive probe, such as beam expansion and stabilization against background vibration by means of feedback to a piezoelectric mirror.

Historically, one of the first laboratory applications of the acoustic emission technique was to the deformation-induced twinning of various metals such as tin. It was appropriate therefore that the first Johns Hopkins study (Palmer and Green 1977, Kline *et al* 1978) should demonstrate the optical detection of acoustic emission signals from twinning. For all their measurements, the output from the Michelson interferometer was compared with the output from a conventional piezoelectric transducer (figure 4.41). Although the theoretical background noise level for the 1 mW helium–neon laser used and their chosen detection bandwidth of 10 kHz–1 MHz was 4 pm, they found that in practice their detection threshold was 50 pm (0.5 Å). They attributed most of this noise to the laser itself. The optical and piezoelectric waveforms (figure 4.42) were rather similar (manifesting the acoustic response of the specimen), but closer examination showed that some resonances due to the piezoelectric element were absent in the optical waveform, while the latter contained more high frequencies (figure 4.43).

The second study however was of stress-corrosion cracking (SCC) in E4340 steel (figure 4.44). This was a wise choice of problem to study with the relatively insensitive interferometer, because SCC is known to generate large-amplitude emissions in high-strength steels such as 4340. The early experiments were carried out using the same video-tape recording system as the twinning measurements, which restricted the upper band-pass limit to 1 MHz. The two optical waveforms shown in figure 4.45 were obtained by directing

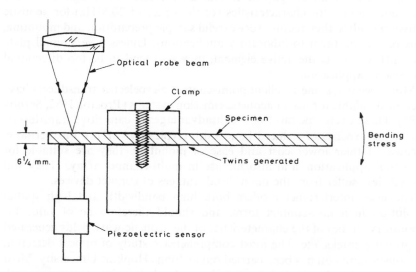

Figure 4.41   Method for clamping and bending specimens so as to induce deformation twinning. The interferometer beam is directed at a point on the specimen directly opposite the piezoelectric transducer (Palmer and Green 1977).

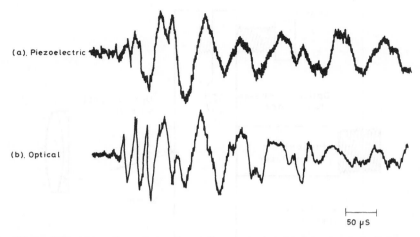

(a). Piezoelectric

(b). Optical

50 µS

Figure 4.42 Acoustic emission signals from a single twinning event in cadmium, enabling optical and piezoelectric data to be compared (Palmer and Green 1977).

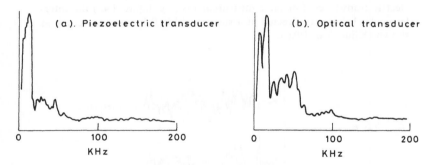

(a). Piezoelectric transducer

(b). Optical transducer

0   100   200
KHz

0   100   200
KHz

Figure 4.43 Frequency spectra of waveforms shown in figure 4.42 (Palmer and Green 1977).

the interferometer beam at a point on the polished specimen surface that was approximately 0.5 mm from the crack tip, thus greatly increasing the effective sensitivity of the probe compared with the piezoelectric transducer, which was located approximately 20–25 mm from the crack. The detection bandwidth was also increased to 5 MHz and the videorecorder replaced by a 10 MHz digitizer. When the optical and piezoelectric waveforms are compared the most apparent feature is a sharp first arrival on the optical channel that is absent on the piezoelectric. The most likely reason is that this short-duration arrival is smoothed out as it propagates across the 12.5 mm front face of the latter. The shortest duration first arrival observed in this

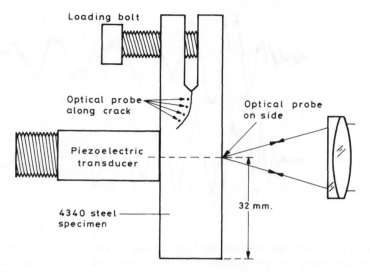

Figure 4.44 Method for inducing stress corrosion cracking in 4340 steel. The laser interferometer was first focused directly opposite the piezoelectric transducer. For later measurements (e.g. figure 4.45) the interferometer beam was directed at a surface within 0.5 mm of the crack as shown (Kline *et al* 1980).

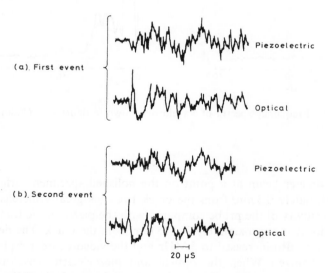

Figure 4.45 Two consecutive emission events due to stress corrosion cracking of 4340 steel, enabling optical and piezoelectric data to be compared. Note the short duration initial 'spike' on optical traces absent from piezoelectric (Kline *et al* 1980).

study (figure 4.46) had a rise time of 200 ns and an amplitude of approximately 1 nm (10 Å).

In a more recent paper, Kline *et al* (1981) carried out a more extensive analysis of the optical waveforms from scc in 4360 steel, and concluded that the amplitudes of the first (compressional) arrivals lay in the range 1–10 Å (100 pm–1 nm), and rise times lay in the range 0.2–1.4 μs. These characteristics are consistent with the results of other studies of the acoustic emission from brittle crack growth using calibrated capacitive transducers, in which compression-wave amplitudes lay in the range 5–500 pm, and rise times in the range 30–600 ns (Wadley *et al* 1981). Thus the interferometer data represent the large-amplitude, longer rise-time end of the distribution detected by the more sensitive capacitive transducer. While lacking the high spatial resolution of optical probes, capacitors can readily be made with the capability of detecting displacements ~ 1 pm over a bandwidth of 20 MHz.

Although the poor sensitivity of an optical detection technique can often be at least partially compensated for in ultrasonic applications by the judicious use of signal averaging, the stochastic and irreproducible nature of acoustic emission waveforms precludes the use of this method for reducing background noise. Poor sensitivity is likely therefore to prevent optical transducers from making a significant impact on most of the field of acoustic emission, except possibly for an application such as welding, as discussed in the next section.

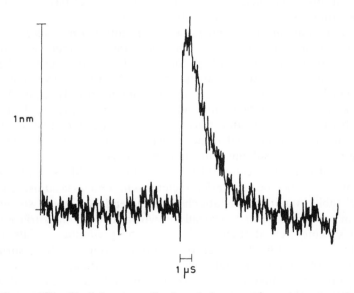

Figure 4.46 Single large-amplitude emission event from stress corrosion cracking. Rise time estimated to be approximately 200 ns (Kline *et al* 1980).

Nevertheless, because of their excellent broadband and high-resolution response, optical transducers can be of tremendous value for a few special applications, such as the experimental confirmation of theoretical predictions. Thus Kline *et al* (1981) also show that the optically detected waveforms from scc in steel, and from a simulated emission source (glass capillary fracture), are in good agreement with the results of wave propagation theory.

4.6.2 Detection of defects during welding

The acoustic emission technique has been proposed for the detection of defects as they are formed during the fusion welding of structural materials such as steel. Typical fabrication defects that are occasionally formed include solidification cracks and slag inclusions. These defects may be detected after the weld head has passed. As the temperature falls, thermoelastic and (in ferritic steels) phase transformation stresses tend to open up any incipient defects, with the consequent emission of bursts of energy which can be detected by acoustic transducers attached to the material being welded. The motivation for the early detection of these defects is that the quicker the weld can be repaired the better, both in terms of final weld quality and in terms of reduced costs. The economic advantage of a technique which detects these defects immediately after their formation increases rapidly with thicker section multi-pass welds, which are very expensive to repair. Thus a number of organizations (Wehrmeister 1977, Bentley *et al* 1982) have investigated weld fabrication monitoring by acoustic emission.

There are two major disadvantages with conventional piezoelectric transducers for this application. Firstly, the temperatures experienced by the transducers are likely to be too high for normal operation, unless they are placed so far away from the hot metal (e.g. by the use of waveguides) that they are insufficiently sensitive. Secondly, they are not readily scanned across a hot metal surface without creating concern for the maintenance of adequate coupling between specimen and transducer. A sensor that can be scanned in synchronization with the weld head is clearly ideal. An optical transducer offers the potential for solving both these problems.

In a trial of optical detection of acoustic emission (Moss *et al* 1982), a compensated reference-beam interferometer was mounted via adjustable jacks on a steel table. A mirror was attached to the end of a boom so that the light beam could be directed vertically downwards onto the surface of the steel plates being welded (figure 4.47). For the preliminary trials, it was found that insufficient light was returned from the rough oxidized surface of the specimen plates for adequate sensitivity. Two attempts were made to overcome this. In the first, the steel plates to be welded were coated with a thin layer of reflecting paint (which contains tiny glass spheres). This improved matters, but not sufficiently to give unequivocal emission signals associated with the formation of weld defects. The second, more drastic, method was to attach a small mirror to the surface of the plate. Under these conditions

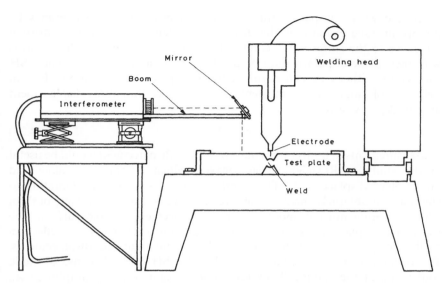

Figure 4.47 Diagram of apparatus to detect acoustic emission signals from defect formation during arc welding. For some of the measurements, a mirror was attached to the surface to increase the signal to noise ratio.

there was sufficient light returned to detect acoustic emission signals associated with the formation of weld defects such as slag inclusions. Indeed, because the acoustic background noise level was so high during welding, from a combination of the arc itself and phase transformations and cracking of the flux, the piezoelectric detection system used in parallel to the interferometer was only marginally more sensitive. The weld defect emission signals detected by the interferometer correlated well with the signals detected by the piezoelectric transducer.

In reality, the use of a reflecting device such as a mirror would be out of the question. However, for these tests the laser used in the interferometer was a low-power helium–neon one. If this were replaced by a more powerful laser, e.g. an argon laser, the sensitivity could be considerably increased. There would also need to be development of the system to deal with a continuously moving specimen surface. Both the quadrature and long-path interferometers would be better systems for such applications.

4.7 THE MEASUREMENT OF TRANSVERSE AND VECTOR DISPLACEMENT

4.7.1 The requirements for vector surface displacement measurement

While the main interest in using laser interferometry is likely to be in the measurement of normal surface displacements due to the incidence of ultrasonic waves, in-plane (transverse) displacements may sometimes be important. One example is the detection of plane shear waves at normal incidence to the surface. A second is the oblique incidence of shear waves with their plane of polarization perpendicular to the plane of incidence (SH waves). In neither case is a normal surface displacement produced. For an SH wave of amplitude B_0 incident at angle θ, the transverse displacement amplitude, a_t, is given by

$$a_t = 2B_0. \tag{4.23}$$

Measurements of both components of surface displacement would be useful in the study of Rayleigh surface waves, which comprise in-plane and out-of-plane displacements. They would also be useful when attempting to characterize obliquely incident compression and shear (SV) waves, both of which generate normal and transverse displacements in the surface. The transverse displacements can be calculated from the plane-wave reflection coefficients in a similar manner to the normal surface displacements. The equations are more complicated than for SH waves because of mode conversion at the surface. Thus for a compression wave of amplitude A_0 incident at angle θ, the transverse displacement, a_t, is the sum of the contributions from the incident wave plus the two (P and SV) reflected waves, using the reflection coefficients (Achenbach 1973, Graff 1975)

$$\frac{a_t}{A_0} = \frac{4k^2 \sin\theta \cos\theta(k^2 - \sin^2\theta)^{1/2}}{4\sin^2\theta \cos\theta(k^2 - \sin^2\theta)^{1/2} + (k^2 - 2\sin^2\theta)^2}. \tag{4.24}$$

Similarly, for a shear (SV) wave of amplitude B_0 incident at angle θ, the transverse displacement, a_t, is given by

$$\frac{a_t}{B_0} = \frac{2k \cos\theta(\cos^2\theta - \sin^2\theta)}{4\sin^2\theta \cos\theta(1 - k^2 \sin^2\theta)^{1/2} + k(1 - 2\sin^2\theta)^2}$$
$$\text{for } |\theta| \leqslant \sin^{-1}(1/k) \tag{4.25}$$

and

$$\left|\frac{a_t}{B_0}\right| = \frac{2k \cos\theta(\cos^2\theta - \sin^2\theta)}{[16\sin^4\theta \cos^2\theta(k^2 \sin^2\theta - 1) + k^2(1 - 2\sin^2\theta)^4]^{1/2}}$$
$$\text{for } |\theta| > \sin^{-1}(1/k). \tag{4.26}$$

This amplitude is complex for angles greater than critical θ. The second equation therefore gives the modulus of the amplitude, but it should be remembered that the phase of the wave also varies in this regime. This is similar to the behaviour of a_n in equation (4.14).

Because the phase of a_n and a_t is zero for the case of an incident compression wave, these displacement components may be combined into a vector surface displacement amplitude. The magnitude of the vector can be obtained as follows:

$$|a| = (|a_n|^2 + |a_t|^2)^{1/2}. \tag{4.27}$$

The orientation (ϕ) of the vector displacement for an incident compression wave is obtained from the ratio of the tangential to perpendicular displacements, thus:

$$\tan \phi = a_t/a_n$$

$$= \frac{2 \sin \theta (k^2 - \sin^2 \theta)^{1/2}}{k^2 - 2 \sin^2 \theta}. \tag{4.28}$$

This equation is used to plot the orientation of the displacement vector as a function of angle of incidence (figure 4.48(a)). For angles less than about

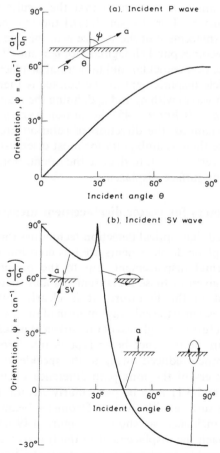

Figure 4.48 Orientation of surface displacement vector as a function of angle of incidence of (a) compression and (b) shear waves. ($k = 2$, e.g. aluminium.)

45°, the vector displacement is approximately parallel to the incident compression wave. However, as the incident angle approaches 90°, the displacement is no longer parallel, tending to an orientation of 60°. In this regime a correction is needed to deduce the orientation of the incident ultrasound from the surface displacement.

The situation for SV waves is more complicated, due to the complex reflection coefficient and critical angle effects. The ratio of the tangential to perpendicular displacements in this case is given by

$$\frac{k(\cos^2 \theta - \sin^2 \theta)}{2 \sin \theta (1 - k^2 \sin^2 \theta)^{1/2}}. \tag{4.29}$$

The inverse tangent of this ratio is plotted in figure 4.48(b). Both for $\theta = 0$ and at the critical angle, a_t is zero so that the resultant displacement is tangential to the surface. For $\theta < \sin^{-1}(1/k)$ the two components are in phase so that the displacement of the surface is in the direction $\tan^{-1}(a_t/a_n)$. The denominator becomes purely imaginary for $\theta > \sin^{-1}(1/k)$. Thus outside the critical angle, the perpendicular and tangential displacements are 90° out of phase. The particle displacement at the surface is therefore elliptical (cf the surface Rayleigh wave) with a_n and a_t defining the major and minor axes, except at 45° where $a_t = 0$. For $\theta > 45°$ the motion is again elliptical but, since the sign of a_t has changed, the direction of rotation around the ellipse is reversed. Thus, since there is ambiguity for most orientations in addition to phase changes, it is very difficult to deduce the orientation of an incident SV wave from the vector surface displacement.

4.7.2 Interferometers for vector displacement measurement

It must be recognized that optical detection techniques for the detection and measurement of in-plane displacements are less convenient and satisfactory than those for normal displacement. Up to the present, relatively little attention has been given to these measurements and there is little practical experience reported in the literature. It is clearly not possible to use well-polished surfaces, since the sideways motion of these cannot be discerned optically. Some structure or roughness of the surface is necessary, as illustrated in figure 4.49, and in the reference-beam type of interferometer at least, the basic sensitivity is much reduced owing to the speckle limitations discussed in §3.3. Interferometers of the long-path-difference or Fabry–Perot types described in §§3.10 and 3.11 may be particularly advantageous.

By observing the surface with the interferometer beam at a very oblique (shallow) angle of incidence, as shown in figure 4.49(a), the instrument's response can approach the displacement in the transverse direction, though it is clearly not possible to eliminate sensitivity to normal displacement altogether. Monchalin (1985) uses an oblique angle of 39° for the laser beam to sense the shear waves generated by a 2.5 MHz shear-wave transducer.

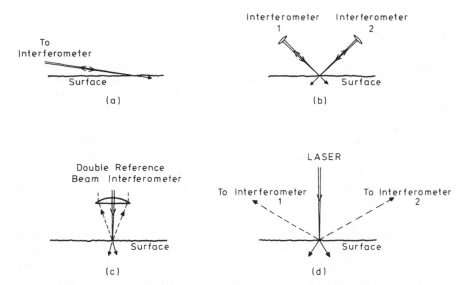

Figure 4.49 Various arrangements for measuring the transverse displacement of a surface. In (*b*), (*c*) and (*d*) two components of displacement are measured simultaneously. The arrows indicate the directions of movement measured by each channel.

The shear waves are nominally at normal incidence to the surface, so that there should be no normal component of surface displacement to cause complications. In practice, for a beam of finite width, it is difficult to prevent some conversion to normal displacement, particularly at the edges of the beam.

To obtain transverse displacement unmixed with any normal component, it is necessary to use a twin interferometer arrangement as illustrated in figures 4.49(*b*), (*c*) or (*d*). In the symmetrical arrangements shown, subtraction of the displacements measured by the two interferometers gives a result proportional to the transverse displacement. The normal component of displacement may be obtained by summation.

Unless one is fortunate with an existing surface finish, some form of surface treatment will be necessary. Roughening on a micrometre scale can give a satisfactory diffuse optical reflection without adversely affecting the acoustic properties. A thin layer of matt white paint will also give satisfactory results. A paint that will give a dramatic increase in signal strength is the retroreflecting type described in §3.3.2. Whilst this is satisfactory at low frequencies, considerable caution should be exercised in its use at ultrasonic frequencies above about 1 MHz, since commonly available retroreflecting paints contain spheres of glass (or other transparent material) of diameter $30\text{--}50\,\mu\text{m}$. This means that a coat of paint must inevitably be quite thick and can thus significantly modify the surface acoustic behaviour. Furthermore, since the back-scattered light intensity from spheres can be extremely sensitive to their diameter (Drain *et al* 1983) it is possible that distortion of the spheres by the ultrasonic field may modulate the returned light. For ultrasonic work, smaller spheres would be more satisfactory though optically less efficient.

Another specialized finish giving an efficient light return in a specified non-specular direction is a diffraction grating. The required line spacing for $45°$ angles of incidence and reflection would be $\lambda/\sin 45° = 0.9 \ \mu m$ for the helium–neon wavelength. A smaller angle would give a more practical grating spacing, but would also give reduced sensitivity to transverse displacement. It may be a feasible proposition to produce a grating finish by a photographic or holographic technique directly onto the surface or by transfer of a suitably exposed gelatine film, a technique successfully used in the measurement of strain (Post 1985).

When two separate reference-beam interferometers are used for two component displacement measurements, as shown in figure 4.49(b), it is desirable that they both observe exactly the same point on the surface. It is thus possible that light from one interferometer could interfere with the operation of the other. Optical isolation may be obtained through directional scattering, polarization or colour. Retroreflecting paint gives efficient directional discrimination since the scattered light is strongly concentrated in the direction of return along the incident beam. In addition, the direction of polarization is usually retained, enabling further discrimination by this means. With matt white paint, or a similar optically rough surface, these mechanisms are ineffective and the use of lasers of different colours could be considered. Helium–neon lasers of alternative wavelengths are available (see table 1.1). Another possibility is the use of an argon ion laser operating in the multiline mode. The main blue and green wavelengths may be separated out and used independently.

It would also be possible to split the beam from one laser for use in two interferometers, arranging the optical paths so that cross interference effects are suppressed by unfavourable path length differences in relation to the coherence length of the laser. But the use of two independent lasers of the same nominal wavelength would probably not be satisfactory because of beating effects when their frequencies drift close to each other.

These problems could be avoided by the use of one illuminating beam only, as in figures 4.49(c) and (d). In the double-reference-beam interferometer (c), the return paths are separated as far as possible so that the directions of motion to which the two channels are sensitive are inclined to each other as much as possible. An optical arrangement for such a double interferometer is shown in figure 4.50. With the single projection lens system shown, the practical limit would be about $30°$. Scattered light may also be received in two different directions by long-path-difference or Fabry–Perot interferometers (figure 4.49(d)). These can be completely separate from the light source, and the angle between the displacement directions greater, thus improving sensitivity to transverse displacement.

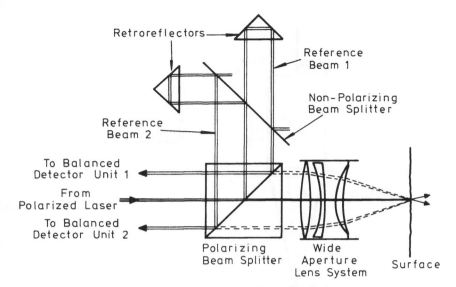

Figure 4.50 A possible double interferometer arrangement for measuring two components of surface displacement simultaneously.

We note that, when using two interferometer channels, their calibration is normally different and may vary randomly with speckle effects. Calibration thus needs careful monitoring if reliable separation of the displacement into normal and transverse components is to be made. If compensated interferometers are used, separate compensation will be required for each interferometer or channel. One practical advantage of using two interferometer channels is that full vector displacement measurement is possible, by combining normal and tangential components, as discussed at the beginning of this section.

4.7.3 Transverse displacement measurement by the differential Doppler technique

An alternative approach to the measurement of transverse displacement is the differential Doppler technique (Drain 1980). This is the method most commonly used in laser Doppler studies of fluid flow. It is particularly effective when the density of scattering centres is low, but satisfactory results may be obtained from rough surfaces giving random scattering. In this technique, illustrated in figure 4.51, the laser beam is split into two accurately parallel beams which are then brought to a focus in a small crossing region where a system of interference fringes is formed. Scattered light may be received in any direction. If the surface is moving transversely with velocity u, the received light fluctuates with a frequency f_D given by

$$f_D = \frac{2u}{\lambda} \sin(\tfrac{1}{2}\alpha) \tag{4.30}$$

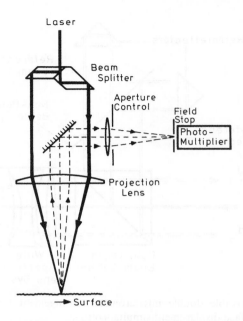

Figure 4.51 Differential Doppler system for measuring transverse displacement.

where α is the crossing angle of the intersecting beams and λ is the optical wavelength.

The best aperture for the detection of the signal depends on the structure of the surface. With a low density of scattering centres, a wide aperture is best but with a small-scale roughness, as with a matt surface, there may be no advantage in increasing the aperture over that for coherent detection from the illuminated spot, so that the amount of useful light is similar to that available in laser interferometry with a reference beam. In these circumstances, the signal may be considered to be derived from the coherent beating of the contributions to the scattered light from the two incident beams, the frequencies being different because the Doppler shift depends on the angle of scattering.

As the surface moves transversely for a short distance the output of the detector varies sinusoidally about a mean level. For greater displacements, however, we note that the amplitude of the sinusoidal variation and current mean level vary randomly on a scale of the diameter of the illuminated spot on the surface. Observation of movements of a fraction of a cycle are evidently possible as in laser interferometry. For the small movements involved here, the calibration factor remains effectively constant.

The differential Doppler technique has the advantage that it directly measures transverse displacement but is not as sensitive as interferometry is to normal displacement. Compared with laser interferometry, there are also a number of unsatisfactory features. The absence of a reference beam means that effective amplification by coherent mixing is not available and the use of photomultipliers or avalanche photodiodes is necessary to obtain photon noise limited operation.

Usually reflection or scattering at a rough surface does not retain polarization. Thus balanced detection is not possible, resulting in increased sensitivity to any modulation of the laser. It also makes compensation for low-frequency vibration more difficult to implement. However, because the length period of the sinusoidal fluctuations of the signal is greater, low-frequency vibrations are more easily tolerated and compensation may not be necessary. Sensitivity is correspondingly reduced. Quadrature detection is not possible but directional discrimination may be made by frequency shifting one or both of the intersecting beams. This is the customary method in laser anemometry applications.

With a suitable surface finish, polarization might be retained, partially at least, and this would enable balanced and/or quadrature detection to be implemented. This is the case for laser anemometry using scattering from particles in a near-forward direction (Hiller and Meier 1972). Retroreflecting paint is evidently not useful in the differential Doppler technique unless the crossing angle is very small. A grating finish matched to the fringe spacing could be very efficient. Then the scheme is virtually equivalent to a Moiré fringe technique.

4.8 SENSITIVITY COMPARISON WITH OTHER ULTRASONIC DETECTION TECHNIQUES

The principal advantage of the laser interferometric technique for the detection of ultrasound is its unique measurement-at-a-distance capability, and for this some sensitivity may have to be sacrificed. However, even when access to the surface is available, the use of laser interferometry does have a number of useful features (listed in §3.1.1) which may make it desirable in these circumstances.

The other principal techniques available for ultrasonic detection and measurements are

(i)   capacitive transducers
(ii) electromagnetic acoustic transducers (EMATS)
(iii) piezoelectric transducers.

Whilst (i) and (ii) may be, in the strictest sense, non-contact techniques, the transducers have to be very close to the surface (within a fraction of a millimetre in the case of a capacitive transducer). In contrast, the laser interferometer can operate from several metres or more if necessary. In most techniques, the sensitivity increases with the area of the surface sampled. In laser interferometry this is not generally the case, the sensitivity with mirror surfaces being limited by the laser power available, and with rough surfaces the sensitivity actually increasing as the spot size is reduced. Thus if the ultimate in spatial resolution is required, laser interferometry excels on the basis of sensitivity per unit area. However, this requirement is unusual, a sampling area of about $10\,mm^2$ usually being small enough for ultrasonic measurements (except for surface waves). It will also be noted that for (i) and (ii) an electrically conducting medium or surface coating is required.

Capacitive transducers are perhaps the most closely comparable in performance with laser interferometers. A transducer of this type (Scruby and Wadley 1978) consists of a conducting plate placed very close to the surface to form an electrical capacitor, whose capacitance depends upon the separation of the plate from the surface. For a constant charge, the voltage across the capacitor varies with the plate separation and thus gives a measure of the movement of the surface. The plate is normally connected to a voltage supply through a resistor whose value determines the frequency response and potential signal to noise ratio.

Assuming that impedance can be matched to a low-noise amplifier, the noise will be given by the thermal (Johnson) noise from the resistor multiplied by a factor b_N, the noise figure of the amplifier. If low-frequency response is not required, a low-value resistor may be used to give an improved signal to noise ratio. On this basis we deduce that, if a frequency response down to f_0 is required, the RMS noise equivalent displacement (at frequencies well above f_0) is given by

$$u_N = \frac{2\xi b_N}{V_c} \left(\frac{kT\Delta f}{2\pi f_0 C_0} \right)^{1/2} \tag{4.31}$$

where ξ is the separation of the plate from the surface, V_c the applied voltage and C_0 the capacitance. k is Boltzmann's constant and T the absolute temperature. Typically for a 6 mm diameter plate, $V_c = 50\,V$, $\xi = 5\,\mu m$ and $C_0 = 50\,pF$. For $f_0 = 1\,MHz$ and a bandwidth $\Delta f = 10\,MHz$, and taking optimistically $b_N = 1$, we find that $u_N = 2.3\,pm$.

Scruby *et al* (1978) report measurements using a capacitive transducer with a 4 μm air gap, in which the minimum acoustic signal they could detect above the electrical noise was 1 pm. The bandwidth was of the order of 10 MHz. A 5 μm air gap would have a correspondingly larger noise threshold. These data are consistent with the above calculation.

A noise level of 2.3 pm for the capacitor is very comparable with the ultimate photon noise limited value of 3.2 pm for an interferometer with a 2 mW helium–neon laser source given in §3.2.2. This is a fair comparison under similar conditions, since the surface would have to be highly polished to obtain the quoted performance for both techniques.

EMATS work by measuring the voltage generated by the movement of conducting material in a magnetic field. They are thus basically sensitive to velocity rather than displacement. The voltages actually induced in the material are very low: for example, a sinusoidal movement of amplitude 1 nm at 1 MHz would generate 1.6 μV in a 5 mm conducting length in a magnetic field of 0.05 T. However, the internal impedance is low, and if efficient coupling and impedance matching to a low-noise amplifier can be made, a good signal to noise ratio is theoretically possible. For a 5 mm square sensitive area of aluminium surface, the resistance (in the skin depth at 1 MHz) is approximately 300 $\mu\Omega$. If the thermal noise in this resistance is the limiting factor, a noise-equivalent displacement of 5 pm is deduced at 1 MHz in a field of 0.05 T for a bandwidth of 10 MHz. However, it would be very difficult to achieve this sensitivity and bandwidth in practice.

Piezoelectric devices of course need good mechanical contact, but given this they can achieve very good electrical coupling to the acoustic field. With perfect coupling from ultrasonic waves in aluminium say, a noise-equivalent displacement of the order of 0.01 pm may be possible at 1 MHz over an area of 20 mm^2, assuming a bandwidth of approximately 100 kHz. Their sensitivity is therefore very much higher than any of the non-contact transducers, but the ultrasonic field is perturbed by their presence, and the response is a function of the efficiency of the mechanical coupling. It is also very difficult to achieve a wide and flat frequency response. If the bandwidth of such a piezoelectric device could be increased from 100 kHz to 10 MHz, to make a fairer comparison with the interferometer and capacitor, then the noise-equivalent displacement would increase to 0.1 pm. The piezoelectric transducer is therefore still a good order of magnitude more sensitive than the other devices.

Several groups have compared the performance of interferometric, capacitive, EMAT and piezoelectric transducers (Hutchins et al 1986b, Dewhurst et al 1987) using a broadband laser-generated pulse as a reproducible standard acoustic source. Dewhurst et al (1987) quantify their results in terms of relative sensitivities. Thus they show that, for an identical compression-wave step from a thermoelastic laser source, the signal to noise ratios for the interferometer, capacitor, EMAT and piezoelectric transducer (a wideband commercial 5 MHz probe) are 16, 50, 6.6 and 90, respectively. Thus the capacitor and interferometer are 5 and 15 dB less sensitive respectively than the piezoelectric transducer. The EMAT is much less sensitive than any of the other transducers.

Since the gains of their interferometer and capacitor are also given, it is possible to calculate the RMS noise-equivalent displacements, i.e. 8 pm (interferometer), 2 pm (capacitor), 15 pm (EMAT) and 1 pm (piezoelectric). The value for the capacitor is in good agreement with the calculation above, but the noise level in their interferometer is apparently somewhat higher than the above figure of 3.2 pm. However, for the increased power (5 mW) and increased bandwidth (130 MHz) of their system, the calculated noise level should be 7 pm, which agrees well. The high background noise from the EMAT reflects the difficulty in making such a device of high sensitivity. It is not clear why the piezoelectric transducer had a higher noise than expected. It might not have been ideally matched electrically at its output, or else the detection system might not have been input noise limited for this device.

These comparisons were made so that the wavefronts were parallel to the front face of each transducer. For some applications (e.g. to acoustic emission) it cannot be guaranteed that the ultrasonic waves will emanate from a source on the transducer axis. The performance of devices with a large front-face area deteriorates markedly in terms of high-frequency response for oblique incidence. The optical probe area is so small that the interferometer does not suffer from this problem.

The above results confirm that a laser interferometer has a similar sensitivity to the capacitor under good experimental conditions (e.g. good surface reflectivity). Even though its bandwidth is much higher, it is still more sensitive than the EMAT. Indeed, since the capacitor is not practical for industrial measurement, we may conclude that interferometry offers not only high bandwidth, high spatial resolution, remote operation and ease of scanning, but also a very competitive sensitivity compared with other non-contact ultrasonic sensing methods.

The sensitivities of various types of reference-beam and velocity interferometer have been calculated (tables 4.1 and 4.2) for reflecting and rough, i.e. perfectly diffusing, surfaces. The type of signal processing used in the reference-beam instrument (e.g. heterodyne or homodyne, method of stabilization) should make little difference (not more than a factor of two) to the theoretical sensitivity, provided all the available light is used. Of the sensitivity values that are published in the literature for practical reference-beam interferometers on polished surfaces, several approach reasonably close to the theoretical limit. Thus the system of Dewhurst et al (1987) has a sensitivity of $\sim 2 \times 10^{-15}$ m mW$^{1/2}$ Hz$^{-1/2}$, and that of Hutchins et al (1986) $\sim 7 \times 10^{-15}$ m mW$^{1/2}$ Hz$^{-1/2}$. In these units, the theoretical limit is 1.4×10^{-15} m mW$^{1/2}$ Hz$^{-1/2}$.

Although there are few quantitative data in the literature on the absolute sensitivities of EMAT and piezoelectric transducers, some values have been estimated for purposes of comparison with interferometry. Thus it can be seen from table 4.1 that a reference-beam interferometer is ~ 30 dB less sensitive than a broadband piezoelectric transducer on highly reflecting specimen surfaces. In principle a sensitivity increase of ~ 27 dB could be obtained for the interferometer by replacing the 2 mW He–Ne laser by a 1 W argon laser to make it comparable with the broadband transducer. However, one major problem is that the photodetectors would rapidly become saturated at such high light levels and a means of overcoming this would need to be found. For interest, the sensitivity of the knife-edge technique (with optimized spot size) has also been included in table 4.1, since this

Table 4.1 Comparison of sensitivity (expressed as RMS background noise displacement) of the reference-beam interferometer (RBI) with the knife-edge and other non-contact (capacitive and EMAT) and contact (piezoelectric) transducers for a highly reflecting surface. Assumptions in the calculations and further information can be found in the sections noted and other references given. Values for velocity types of interferometer are not given since they offer no significant sensitivity advantage for reflecting surfaces, differing by ± 6 dB under ideal conditions for each type of instrument.

Ultrasonic receiver	Reference (section)	Bandwidth (MHz)	Sensitivity (pm)
2 mW He–Ne RBI	3.2.2, 3.3.5	10	3
2 mW Knife-edge	2.6.3	10	4
Capacitor	4.8 and [1]	10	2
EMAT	4.8 and [2]	∼1	∼15
Damped PZT	4.8 and [3]	10	∼0.1
Resonant PZT	4.8 and [3]	0.1	∼0.01

[1]   Scruby *et al* (1978)
[2]   Dewhurst *et al* (1987)
[3]   Speake (1978)

Table 4.2 Comparison of sensitivity (expressed as RMS background noise displacement) of the examples of the reference-beam (RBI), long-path-difference (LPDI) and confocal Fabry–Perot (CFPI) interferometers discussed in §3.11.6 and figure 3.40 with other types of non-contact (EMAT) and contact (piezoelectric) transducers. The specimen is assumed to have an ideal rough surface in the form of a perfect diffuser. The sensitivities of the interferometers depend upon the assumptions made for key variables as referenced.

Ultrasonic receiver	Reference (section)	Bandwidth (MHz)	Sensitivity (pm)
1 W argon RBI[1]	3.3.5	10	15
1 W argon LPDI[2]	3.10.5	10	15
1 W argon CFPI[3]	3.11.6	10	3
EMAT	4.8 and [4]	∼1	∼15
Damped PZT	4.8 and [5]	10	∼0.1
Resonant PZT	4.8 and [5]	0.1	∼0.01

[1]   Spot diameter $= 180\lambda = 90$ μm
[2] $L = 500$ mm, $D/F = 0.1$
[3] $r = 500$ mm, $R = 90\%$, $D/F = 0.1$
[4]   Dewhurst *et al* (1987)
[5]   Speake (1978)

detection method is applicable to surface waves and obliquely incident bulk waves at a reflecting surface.

Turning to table 4.2, the data available for other types of transducer suggest that on rough surfaces it should be possible to attain a sensitivity comparable with an EMAT by using either a reference-beam or long-path interferometer of the right design and under good conditions. For the examples shown, the confocal Fabry–Perot system offers a small sensitivity advantage over both the EMAT and the other types of interferometer. However, no interferometer system comes even close to competing with contact piezoelectric probes on such surfaces, other than by using exceptionally powerful lasers.

The table confirms the results of Chapter 3 that the reference-beam and long-path-difference interferometers have similar intrinsic sensitivities, and that the confocal Fabry–Perot system has slightly higher sensitivity to 10 MHz ultrasonic waves under typical experimental conditions. An important advantage of both types of velocity interferometer is their relatively large étendue, which means that they do not rely on light returned from individual speckles, nor do they require such a sharply focused spot as the reference-beam instrument, so that they are far more flexible as probes for rough surfaces under industrial measurement conditions. They are also likely to be appreciably less sensitive to other potential environmental hazards such as convection currents and steam clouds. Note again that their sensitivity increases in proportion to frequency below the maximum, whereas the response of the reference-beam instrument is frequency independent. Above the maximum, the sensitivity of both velocity interferometers falls away, which is somewhat of a disadvantage. However, this can be eliminated in the confocal Fabry–Perot system by means of optical sideband stripping (Monchalin *et al* 1989) as discussed at the end of Chapter 3.

While there is clearly considerable benefit in using long-path and confocal Fabry–Perot interferometers for many of the industrial applications of laser ultrasound to be discussed in Chapter 6, and where reception at higher frequencies is desired, figure 3.40 shows that the reference-beam instrument is to be preferred for lower frequencies (especially below 1 MHz), and when a highly reflecting surface is available as in many laboratory measurements and calibration studies. To summarize, the choice of interferometer system depends not only on the sensitivity required, but also on a large number of other factors, such as the type of surface to be monitored and the frequency range of the ultrasound that is expected.

REFERENCES

Achenbach J D 1973 *Wave Propagation in Elastic Solids* (Amsterdam: North-Holland)
Aleksoff C C 1971 *Appl. Optics* **10** 1329
Ameri S, Ash E A, Murray D and Wickramasinghe H K 1980 *Rev. Cethedec (France)* **17** 65

Angel Y C and Bogy D B 1981 *J. Appl. Mech.* **48** 881
Bacon D R 1988 *IEEE Trans. Ultrasonics Ferroelectric Freq. Contr.* **UFFC-35** 152
Bentley P G, Dawson D G and Prine D W 1982 *NDT Int.* **15** 243
Bouchard G and Bogy D B 1985 *J. Acoust. Soc. Am.* **77** 1003
Bowers J E 1982 *Appl. Phys. Lett.* **41** 231
Butters J N and Leendertz J A 1971 *J. Measurement and Control* **4** 344
Dändliker R 1980 *Progress in Optics* vol 17, ed E Wolf (Amsterdam: North-Holland) pp 1–84
De La Rue R M, Humphryes R F, Mason I M and Ash E A 1972 *Proc. IEE* **119** 117
Dewhurst R J, Edwards C, McKie A D W and Palmer S B 1987 *Ultrasonics* **25** 315
Drain L E 1980 *The Laser Doppler Technique* (New York: Wiley) pp 85–118
Drain L E, Smith N and Dalzell W 1983 *Proc. Max Born Centenary Conf.*, SPIE vol 369 p 610
Francon M. 1979 *Laser Speckle and Applications in Optics* (New York: Academic)
Graff K F 1975 *Wave Motion in Elastic Solids* (Oxford: Clarendon) pp 312–23
Harihan P 1973 *Appl. Optics* **12** 143
Hiller W J and Meier G E A 1972 *Bericht 1/1972* Max Planck Institut fur Stromungsforschung, Gottingen
Hirao M and Fukuoka H 1982 *J. Acoust. Soc. Am.* **72** 602
Høgmoen K and Løkberg O J 1977 *Appl. Opt.* **16** 1869
Hutchins D A, Hu J and Lundgren K 1986a *Mater. Eval.* **44** 1244
Hutchins D A, Nadeau F and Cielo P 1986b *Can. J. Phys.* **64** 1334
Jablonowski D P 1978 *Appl. Optics* **17** 2064
Jones R and Wykes C 1983 *Holographic and Speckle Pattern Interferometry, Cambridge Studies in Modern Optics* vol 6 (Cambridge: Cambridge University Press)
Jungerman R L, Bowers J E, Green J B and Kino G S 1982 *Appl. Phys. Lett.* **40** 313
Jungerman R L, Khuri-Yakub B T and Kino G S 1984a *Mater. Eval.* **42** 444
—— 1984b *Appl. Phys. Lett.* **44** 392
Kline R A, Green R E Jr and Palmer C H 1978 *J. Acoust. Soc. Am.* **64** 1633
—— 1981 *J. Appl. Phys.* **52** 141
Kock W E 1975 *Engineering Applications of Lasers and Holography* (New York: Plenum)
Massey G A 1968 *Proc. IEEE* **56** 2157
Mezrich R, Etzold K and Vilkomerson D 1975 *Acoustic Holography VI* ed N Booth (New York: Plenum) p 165
Mezrich R, Vilkomerson D and Etzold K 1976 *Appl. Optics* **15** 1499
Monchalin J P 1985 *Rev. Sci. Instrum.* **56** 543
Monchalin J P, Héon R, Bouchard P and Padioleau C 1989 *Appl. Phys. Lett.* **55** 1612
Moss B C, Elwood, Drain L E and Scruby C B 1982 Unpublished
Moss B C and Milne J M 1980 *Community Bureau of Reference Final report* Contract 389/1/4/119/78/4BCRE
Moss B C and Scruby C B 1988 *Ultrasonics* **26** 179
Murray D and Ash E A 1977 *Proc. IEEE Ultrasonics Symp.* p 823
Palmer C H, Claus R O and Fick S E 1977 *Appl. Optics* **16** 1849
Palmer C H and Green R E Jr 1977 *Mater. Eval.* **35** no 10 107
Post D 1985 *Opt. Engng* **24** 663
Proctor T M Jr 1982 *J. Acoust. Soc. Am.* **71** 1163

Rudd E P, Mueller R K, Robbins W P, Skaar T, Soumekh B and Zhou Z Q 1987 *Rev. Sci. Instrum.* **58** 45

Reibold R 1980 *Acustica* **46** 149

Scruby C B 1985 *J. Acoust. Emission* **4** 9

—— 1987 *J. Phys. E: Sci. Instrum.* **20** 946

Scruby C B, Collingwood J C and Wadley H N G 1978 *J. Phys. D: Appl. Phys.* **11** 2359

Scruby C B and Moss B C 1985 *Rayleigh Wave Theory and Application* ed E A Ash and E G S Paige (Berlin: Springer) p 102

Scruby C B and Wadley H N G 1978 *J. Phys. D: Appl. Phys.* **11** 1487

Speake J H 1978 *Conf. on the Evaluation and Calibration of Ultrasonic Transducers, London, May 1977* (Guildford: IPC) p 106

Turner T M and Claus R O 1981 *Proc. IEEE Ultrasonics Symp., Chicago* p 384

Vest C M 1979 *Holographic Interferometry* (New York: Wiley)

Vilkomerson D, Mezrich R and Etzold K 1976 *Acoustic Holography VII* ed L W Kessler (New York: Plenum) p 87

Wadley H N G and Scruby C B 1983 *Int. J. Fracture* **19** 111

Wadley H N G, Scruby C B and Shrimpton G 1981 *Acta Metall.* **29** 399

Wagner J W 1985a *NDT Commun.* **2** 77

—— 1985b *Appl. Optics* **24** 2937

—— 1986 *Mater. Eval.* **44** 1238

Wehrmeister A E 1977 *Mater. Eval.* **35** 45

Wickramasinghe H K and Ash E A 1973 *Electron. Lett.* **9** 327

—— 1974 *Proc. Symp. on Optical and Acoustical Micro-electronics, Polytechnic Institute of New York* p 413

—— 1975 *Proc. IEEE Ultrasonics Symp.* p 496

Williams R V 1980 *Acoustic Emission* (Bristol: Adam Hilger)

Yoneda K, Tawata M and Hattori S 1981 *Japan. J Appl. Phys.* **20**, Suppl. 20–3, 61

5   Ultrasonic Generation by Laser

A number of different physical processes may take place when a solid surface is illuminated by a laser. At lower incident powers these include heating, the generation of thermal waves, elastic waves (ultrasound) and, in materials such as semiconductors, electric currents may be caused to flow. At higher powers, material may be ablated from the surface and a plasma formed, while in the sample there may be melting, plastic deformation and even the formation of cracks.

In this chapter we shall restrict discussion to laser power regimes that are suitable for non-destructive testing, and therefore focus most of our attention on the localized heating produced by the laser, which in turn generates the thermoelastic stresses and strains that act as an ultrasonic source (figure 5.1). We shall consider in turn the physical principles underlying the absorption of laser radiation, the generation and propagation of elastic waves within

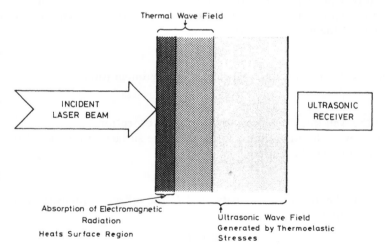

Figure 5.1   Electromagnetic radiation from the laser is absorbed in the surface region of a sample, causing heating. Thermal energy then propagates into the specimen as thermal waves. The heated region undergoes thermal expansion, and thermoelastic stresses generate elastic waves (ultrasound) which propagate deep within the sample. For typical Q-switched laser pulse durations, the thermal wave field only extends a few micrometres even in good conductors. Contrast the incidence of low-frequency modulated light where the thermal field extends millimetres or centimetres and is itself useful for materials characterization.

the sample, their characteristics and dependence upon experimental variables. Except for a very brief section (5.4.4) we shall not discuss generation in fluids, since much less work has been done here, and also because this process is only likely to be of limited relevance to non-destructive testing applications.

5.1 ABSORPTION OF ELECTROMAGNETIC RADIATION

A laser emits a beam of coherent radiation, whose wavelength may be in the infrared, visible, or ultraviolet part of the electromagnetic spectrum (see Chapter 1). When this is incident on a solid sample, in general some of the energy is absorbed by various mechanisms, depending upon the nature of the sample and the frequency of the radiation, while the remainder is reflected or scattered from the surface. We assume that the sample is too thick for any transmission to occur, and also that the intensity of the radiation is too low for ablation or damage processes to occur.

If the laser radiation is incident on a non-reflecting, absorbing material, then it is progressively attenuated as it penetrates into the sample. Assuming plane waves propagating in the z direction, if the fraction absorbed, δI, in any infinitesimal element is proportional both to the intensity I and the thickness of the element, δz, i.e.

$$\delta I = \gamma I \delta z \qquad (5.1)$$

where γ is the absorption coefficient, then by integration

$$I(z) = I(0)e^{-\gamma z} \qquad (5.2)$$

giving the well known exponential form of absorption.

If, however, the radiation is incident at the surface of a conductor, some of the energy will be reflected, so that

$$I = I' + I'' \qquad (5.3)$$

where I' is the absorbed intensity, and I'' the reflected intensity. Thus, if R is the reflectivity,

$$R = I''/I. \qquad (5.4)$$

Electromagnetic radiation interacts with the surface of a conductor causing electric currents to flow in the conduction-band electrons. Some of the energy is absorbed by resistive losses (e.g. electron–phonon scattering processes), while the remainder is reflected. The conduction electrons at the surface screen the interior of the metal from the radiation so that the absorption and reflection take place within a surface layer or 'skin'. A 'skin depth' can be defined (denoted by δ), such that the amplitude of the wave falls to $1/e$ of its initial value over a distance δ.

At longer wavelengths (e.g. in the infrared) classical physics can be used to calculate the skin depth. If σ and μ_r are the conductivity and relative permeability of the metal respectively, $\mu_0 = 4\pi \times 10^{-7}\,\mathrm{H\,m^{-1}}$ the permeability of free space, and v the frequency of the radiation, then, from Bleaney and Bleaney (1965),

$$\delta = (\pi\sigma\mu_r\mu_0 v)^{-1/2}. \tag{5.5}$$

This formula cannot be used at higher frequencies (e.g. in the visible and ultraviolet) where a quantum mechanical calculation must be made. Substituting $\sigma = 4 \times 10^7\,(\Omega\,\mathrm{m})^{-1}$ and $\mu_r \simeq 1$ for aluminium, the skin depth for Nd:YAG radiation at 1.06 μm is approximately 5 nm. We note that the skin depth is reduced for shorter wavelengths, as it is also for increased electrical conductivity or permeability.

Classical electromagnetic theory also permits the reflectivity of a clean metal surface to be expressed in terms of the skin depth (Bleaney and Bleaney 1965), i.e.

$$R = \frac{2 - 2\xi + \xi^2}{2 + 2\xi + \xi^2} \tag{5.6}$$

where $\xi = \mu_0\sigma c\delta$, c being the velocity of light. For most metals, and optical frequencies up to visible light, $\xi \gg 1$, so that approximately

$$R = 1 - 4/(\mu_0\sigma c\delta). \tag{5.7}$$

Substituting again the material constants for aluminium gives a theoretical reflectivity of 0.94 for 1.06 μm radiation. This is in reasonable agreement with the experimental value for polished aluminium, 0.93 (Kaye and Laby 1973). Increasing the optical wavelength further into the infrared increases the reflectivity, while in the ultraviolet more energy is absorbed, e.g. $R \sim 0.8$ for aluminium at 0.25 μm (Kaye and Laby 1973). Thus shorter wavelengths are likely to lead to more efficient generation of ultrasound by laser since more energy is absorbed into the metal. Reflectivities vary between different metals. Kaye and Laby (1973) give $R = 0.63$ for steel at 1.0 μm, and 0.38 at 0.25 μm. Generation of ultrasound by lasers should therefore be relatively efficient into steel, especially at shorter wavelengths. The reflectivity of copper changes very rapidly with wavelength ($R = 0.90$ at 1.0 μm, 0.26 at 0.25 μm, hence its reddish colour), so that visible or ultraviolet wavelengths are best for efficient generation. It should be noted, however, that these data are all for polished metal surfaces. In practice, the surface is likely to be roughened, oxidized or contaminated in some other way. These factors all have the effect of increasing the absorbed energy, and hence the efficiency of ultrasonic generation.

The absorbed energy mostly takes the form of heat so that, in the case of a metal, laser irradiation induces a temperature rise at the surface, while in the case of an insulator, depending upon the nature and magnitude of the absorption processes, it raises the temperature in the bulk of the sample also. In both cases, however, thermal conductivity ensures that the heat eventually becomes distributed throughout the sample. The temperature gradients induced set up stress and strain fields by thermal expansion. These in turn generate elastic waves in solids (figure 5.1), which may be of low (sonic) frequency or high (ultrasonic) frequency. The rise in temperature may also, under certain circumstances, cause melting, vaporization, or combustion of the surface material. The stresses set up may be sufficient to damage the material plastically or by crack formation.

Which of these processes occur, and which are significant, depend on the characteristics of the incident radiation (such as average incident power, peak power, whether continuous (with perhaps periodic modulation) or pulsed), and the properties of the material. For instance, for low continuous incident power at a solid surface of $\sim 1 \, \text{W cm}^{-2}$, the only effect is a steady rise in temperature. If, however, the optical power is modulated by, for instance, a mechanical chopper or acousto-optic modulator, thermal 'waves' can be generated (see Opsal and Rosencwaig 1982). These waves are the solution of the equation for thermal conduction for a periodic heat source. In the 1D case they decay rapidly with distance from the surface where the energy is absorbed, so that the effective depth of penetration, the thermal diffusion length, is given by

$$\Lambda_d = (2\kappa/\omega)^{1/2} \tag{5.8}$$

where the diffusivity

$$\kappa = K/\rho C. \tag{5.9}$$

K is the thermal conductivity, ρ the density, C the specific thermal capacity of the material, and ω is the angular frequency of the modulation.

For aluminium, the thermal diffusivity $\kappa \simeq 10^{-4} \, \text{m}^2 \, \text{s}^{-1}$, so that for frequencies of 1 and 100 Hz the diffusion length $\Lambda_d \simeq 5$ and 0.5 mm respectively. However, at 1 MHz, a more typical ultrasonic frequency, $\Lambda_d \simeq 5 \, \mu\text{m}$. Thus thermal waves are likely to be of major use for investigating relatively thin samples. The periodic heating of the surface of the sample by a modulated low-power laser acts as a source of acoustic waves, due to the periodic thermoelastic stresses generated. These are of low amplitude and generally need a phase-sensitive detection system to be observable. In practice their wavelength (60 m at 100 Hz in aluminium) is too large to probe the material, and they act solely as a method for coupling thermal effects in the sample to an acoustic detector. Detecting thermal waves acoustically can be a very convenient way of using the technique. Since they are not the subject of this book, we shall consider them no further. For further information the reader should refer to Rosencwaig and Gersho (1976), Opsal and Rosencwaig (1982), Tam (1986), and Murphy and Wetsel (1986).

The generation of ultrasonic waves suitable for the non-destructive examination of materials requires higher incident powers and frequencies than can be obtained using modulated continuous-waves lasers. It is instructive to estimate the order of magnitude of the incident optical power required. Let us assume that the absorption of energy δE in the material over a time δt causes thermal expansion in a small volume of material, V, so that the bulk strain is $\delta V/V$. It will later be shown (equation (5.45)) that

$$\delta V = \frac{3\alpha}{\rho C} \delta E \qquad (5.10)$$

where α, ρ and C are the coefficient of linear expansion, density and specific heat, respectively.

It will also be shown (equation (5.67)) that, neglecting the effects of boundaries, the displacement u_r at a distance r from the strained region is given by

$$u_r = \frac{B}{4\pi(\lambda + 2\mu)c_1 r} \delta V'(t - r/c_1). \qquad (5.11)$$

In this equation, c_1 is the compression velocity and $B = \lambda + \frac{2}{3}\mu$ is the bulk modulus, λ and μ being the Lamé elastic constants. Thus assuming $\lambda \simeq 2\mu$ for many common metals

$$u_r = \frac{1}{6\pi c_1 r} \frac{\delta V}{\delta t} \qquad (5.12)$$

$$= \frac{\alpha}{2\pi\rho C c_1 r} \frac{\delta E}{\delta t} \quad \text{from equation (5.10)} \qquad (5.13)$$

$$\simeq \frac{2.5 \times 10^{-16}}{r} \frac{\delta E}{\delta t} \qquad (5.14)$$

substituting values of α, ρ, C, c_1 for aluminium (see table 5.1). To generate a typical ultrasonic wave amplitude of 1 Å (10^{-10} m) at a distance of 100 mm,

Table 5.1 Values of parameters used in the text.

Symbol	Parameter	Aluminium	Mild steel
I_0	Absorbed power density	2×10^{10} W m^{-2}	2×10^{10} W m^{-2}
δE	Absorbed pulse energy	4×10^{-3} J	4×10^{-3} J
δt	Pulse length	20×10^{-9} s	20×10^{-9} s
ρ	Density	2.7×10^3 kg m^{-3}	7.9×10^3 kg m^{-3}
c_1	Compression velocity	6400 m s^{-1}	5960 m s^{-1}
α	Linear expansion	2.31×10^{-5} K^{-1}	1.07×10^{-5} K^{-1}
C	Specific heat capacity	880 J kg^{-1}K^{-1}	480 J kg^{-1}K^{-1}
K	Thermal conductivity	240 W m^{-1}K^{-1}	50 W m^{-1}K^{-1}
$\kappa = K/\rho C$	Thermal diffusivity	1.0×10^{-4} m^{2}s^{-1}	1.3×10^{-5} m^{2}s^{-1}

an instantaneous laser power approaching 10^5 W is required. This applies only to small or 'point' sources, but these are likely to be of most interest in ultrasonics. Such a high power from a cw laser is not a practical proposition. For instance, the sample would rapidly overheat. It is noted, however, that if a lower power cw laser were used and modulated at frequencies in the ultrasonic range, a lock-in amplifier would be needed to give narrow-band and hence high-sensitivity detection.

The need to generate ultrasound at frequencies in the range ~ 100 kHz to ~ 10 MHz with reasonable amplitude implies that the most effective method for ultrasonic generation is likely to be a short-pulse laser. Pulsed lasers can usually deliver instantaneous optical powers in excess of 1 MW, but only for the duration of the pulse. Unless the pulse repetition rate is very high, the average power incident on the sample remains low, typically of the order of 1 W or less, so that bulk heating is negligible. To generate ultrasonic frequencies in excess of 5 MHz, a very short optical pulse is required. The shorter the pulse, the more energy there is at higher frequencies, and the higher the peak power for a given pulse energy.

Let us assume that the ultrasonic pulse has the same profile as the optical pulse (i.e. there is no broadening due to thermal conductivity, etc). If the pulse is also assumed to be Gaussian, of the form $\exp(-t^2/2s^2)$, where s is the standard deviation, the Fourier transform of the pulse is proportional to $\exp(-2\pi^2 s^2 f^2)$. The Fourier transform falls to half its maximum value (i.e. -6 dB) at a frequency of

$$f(-6\,\mathrm{dB}) = 0.1874/s. \qquad (5.15)$$

Thus if $f(-6\,\mathrm{dB}) = 10$ MHz to ensure appreciable energy at ultrasonic frequencies, the pulse width (expressed as the standard deviation) must be approximately 20 ns. This confirms that a short-pulse laser is likely to be very suitable for ultrasonic generation, and the theory which follows will be restricted to such a source. The majority of published work has employed Q-switched laser pulses of duration 20–30 ns, although shorter pulses from nitrogen lasers (e.g. Tam and Leung (1984)), and picosecond mode-locked lasers (e.g. Dewhurst and Al'Rubai (1989)) have also been used. For work in the most common ultrasonic frequency range of 1–10 MHz, the exact choice of pulse length is not critical, provided it is < 50 ns. There will be more discussion of laser characteristics in §5.11.

5.2 TEMPERATURE DISTRIBUTIONS

5.2.1 Metals

The next step is to determine the form of the temperature distribution which in turn creates stresses and strains that generate elastic waves. Let us assume that the incident power is sufficiently low that the only effect of the incident

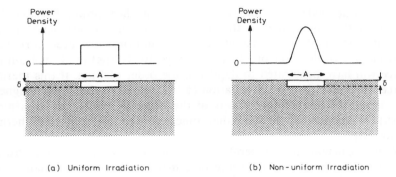

(a) Uniform Irradiation   (b) Non-uniform Irradiation

Figure 5.2 The variation of the incident power density across the irradiated region A.

laser pulse is to raise the surface temperature of the metallic sample. The laser is thus acting simply as a transient heat source. Let us further assume that the optical pulse is extremely short, so that thermal conductivity into the bulk of the sample can be neglected, that the laser uniformly irradiates an area A (figure 5.2(a)), and that absorption takes place uniformly throughout a depth equal to the skin depth δ. Then the temperature rise in the irradiated material is uniform and is given by

$$\delta T = \delta E / C\rho A\delta \tag{5.16}$$

i.e. the temperature rise δT is proportional to the absorbed energy density.

Let us now consider the effect of non-uniform irradiation (figure 5.2(b)), but still with absorption within a depth δ. If the incident power density and reflectivity are respectively $I(x, t)$ and R then, considering a small element of area dx, the absorbed energy is given by

$$(1 - R) \int_t I(x, t) \, dx \, dt = C\rho\delta \, dx \, \delta T(x). \tag{5.17}$$

Therefore

$$\delta T(x) = \frac{1 - R}{C\rho\delta} \int_t I(x, t) \, dt. \tag{5.18}$$

Thus the temperature distribution across the surface is the same as the energy density distribution in the optical pulse. The beam profile of a pure laser mode is generally Gaussian. Multi-mode beams, however, often give rise to hot spots, where $I(x, t)$ is a maximum.

Let us now suppose that a 100 mJ pulse uniformly irradiates a 10 mm² area of polished aluminium, with a reflectivity of 94%. Then equation (5.16) predicts a surface temperature rise of 5×10^4 K, i.e. well above the threshold for damage and vaporization. However, experimental data (e.g. Scruby *et al*

1980) show no evidence of damage or ablation at these power densities. The reason for the discrepancy is that this oversimplified model fails to take thermal conductivity into account. As will be shown below, even during the short lifetime of the laser pulse (in the range 10–100 ns) the heat from the source will have penetrated to depths of the order of 1 μm. If this thermal source 'thickness' is used in equation (5.16) instead of the electromagnetic skin depth, the temperature rise is of the order of 100 K, which is more realistic. It is concluded therefore that some account must be taken of thermal conductivity.

We now calculate the temperature distribution resulting from the absorption of a laser pulse at the surface of a metal without phase change. We assume that the thermal properties of the metal are independent of temperature, that a local thermal equilibrium is established during the pulse, and that negligible energy is lost from the surface by radiation. The differential equation for heat flow in a semi-infinite slab (half-space) with a boundary plane at $z = 0$ is given by

$$\nabla^2 T - \frac{1}{\kappa} \frac{\partial T}{\partial t} = -\frac{A}{K} \tag{5.19}$$

where $T(x, y, z, t)$ is the temperature distribution, $A(x, y, z, t)$ is the heat produced per unit volume per unit time, and K and κ are the thermal conductivity and diffusivity, respectively. The boundary conditions are that $T(x, y, z, 0) = 0$, and $T \rightarrow 0$ as $z \rightarrow \infty$, and no heat flux crosses the $z = 0$ plane.

For most practical situations in which a short-pulse laser irradiates a metal over an area of typically several mm^2, the depth to which heat is conducted during the duration of the pulse is much less than the area, so that a 1D treatment is appropriate. Thus

$$\frac{\partial^2 T(z, t)}{\partial z^2} - \frac{1}{\kappa} \frac{\partial T(z, t)}{\partial t} = -\frac{A(z, t)}{K}. \tag{5.20}$$

If the absorbed laser flux density is I_0, uniform across the irradiated area (figure 5.2(a)), and the laser is switched on instantaneously at zero time, then, from Carslaw and Jaeger (1959), for high absorption at the surface the solution of equation (5.20) is

$$T(z, t) = \frac{2I_0(\kappa t)^{1/2}}{K} \, \text{ierfc}\left(\frac{z}{2(\kappa t)^{1/2}}\right) \tag{5.21}$$

where

$$\text{ierfc}(\zeta) = \frac{1}{\sqrt{\pi}} e^{-\zeta^2} - \frac{2\zeta}{\sqrt{\pi}} \int_{\zeta}^{\infty} e^{-\xi^2} \, d\xi. \tag{5.22}$$

Thus

$$T(0, t) = \frac{2I_0(\kappa t)^{1/2}}{\sqrt{\pi}K}.$$ (5.23)

Suppose the laser generates a square pulse of duration t_0, i.e. is switched on at $t = 0$ and switched off at $t = t_0$. Equation (5.21) must still hold for $t < t_0$. However, for $t > t_0$ a second term must be added to represent switching off:

$$T(z,t) = \frac{2I_0(\kappa t)^{1/2}}{K} \text{ierfc}\left(\frac{z}{2(\kappa t)^{1/2}}\right)$$

$$- \frac{2I_0\sqrt{\kappa}(t - t_0)^{1/2}}{K} \text{ierfc}\left(\frac{z}{2\sqrt{\kappa}(t - t_0)^{1/2}}\right).$$ (5.24)

Figures 5.3 and 5.4 show calculations of temperature distribution from equations (5.21) and (5.24) for two metals, aluminium and mild steel, using the parameters given in table 5.1. Figure 5.3 shows that the temperature at

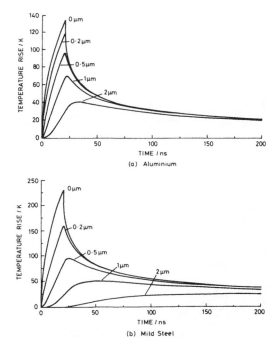

Figure 5.3   Rise in temperature as a function of time, for a range of depths below the surface of (a) aluminium and (b) mild steel, in response to an incident square pulse of laser energy, calculated from equations (5.21) and (5.24), using parameters in table 5.1.

the surface ($0\ \mu m$) rises for the first 20 ns (while energy is still being absorbed) to a maximum, which is higher in steel because steel has a lower thermal conductivity than aluminium. Once the laser pulse finishes, the temperature falls as the heat is conducted into the bulk. The maximum temperature rise at increasing depth below the surface is less and occurs later.

Figure 5.4 shows that rapid temperature changes (and hence transient thermoelastic stresses) occur within a relatively thin surface layer of the order of a few micrometres. Thus for most ultrasonic purposes the source can be considered to be at the surface.

The temperature distribution for more typical laser pulse shapes can be obtained by using Duhamel's theorem (Ready 1971, Carslaw and Jaeger 1959):

$$T(z, t) = \int_z^\infty \int_0^t \frac{I(t')}{I_0} \frac{\partial}{\partial t} \left(\frac{\partial}{\partial z} [T'(z', t - t')] \right) dz'\, dt' \qquad (5.25)$$

where $T'(z, t)$ is the solution for the case of a step increase in absorbed flux

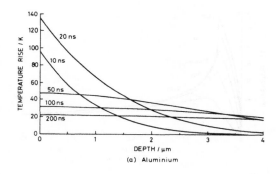

(a) Aluminium

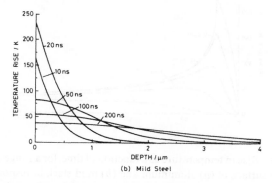

(b) Mild Steel

Figure 5.4 An alternative presentation of the data of figure 5.3, showing temperature distribution as a function of depth.

density, I_0. Substituting equation (5.21) in equation (5.25) yields

$$T(z, t) = \frac{\kappa^{1/2}}{K\pi^{1/2}} \int_0^t \frac{I(t - t')\exp(-z^2/4\kappa t')}{t'^{1/2}} \, dt'. \tag{5.26}$$

Assuming the pulse to be Gaussian in time, i.e. of the form $I(t) = I_1 \exp(-\xi t^2/\tau^2)$ where $\xi = 2.7726$ and τ is the pulse full width at half height, yields the results shown in figure 5.5. This figure is for aluminium, and still assumes $\delta E = 4 \times 10^{-3}$ J. For $\tau = 20 \times 10^{-9}$ s, I_1 the maximum power density is 1.9×10^{10} W m^{-2}. Comparison with figures 5.3 and 5.4 shows that the change to a Gaussian pulse produces modified temperature maxima and gradients at short times for the same total energy input. At longer times the pulse shape has little effect. Figure 5.6 shows the effect of a Gaussian pulse of the same energy but double the width, i.e. $\tau = 40 \times 10^{-9}$ s. The initial temperature rises are lower, but the long-term effect is the same. It is concluded that the time dependence of the laser pulse, and the peak power for a given energy, chiefly affect temperature distributions at short times and close to the incident surface. At longer times and larger distances, the pulse energy is the determining factor.

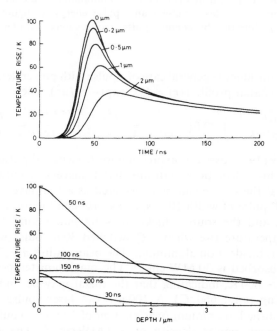

Figure 5.5 Temperature distribution in aluminium for a Gaussian time dependence of the laser pulse (maximum power density 1.9 MW cm^{-2}, $\tau = 20$ ns).

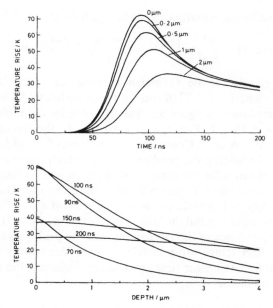

Figure 5.6 Temperature distribution in aluminium for a Gaussian time dependence of the laser pulse (same pulse energy as figure 5.5, but $\tau = 40\,\text{ns}$). Note that broadening pulses reduces maximum temperature rise.

Finally, we consider the general case of a pulse with power density $I_{\max}\,\Pi(t)$ and Gaussian spatial profile $I(r) = I_{\max}\exp(-r^2/d^2)$. Then

$$T(r,z,t) = \frac{I_{\max}\kappa^{1/2}d^2}{\pi^{1/2}K}\int_0^t \exp\left(\frac{-z^2}{4\kappa t'} - \frac{r^2}{4\kappa t' + d^2}\right)\frac{t'^{-1/2}\Pi(t-t')\,\mathrm{d}t'}{(4\kappa t' + d^2)}. \quad (5.27)$$

When Ready (1971) uses this expression for Q-switched pulses with widths $\sim 3 \times 10^{-8}\,\text{s}$, his calculations indicate that transverse conduction can be ignored, so that the source can be considered as a thin layer $\sim 1\,\mu\text{m}$ thick. For a 'normal' pulse of width $10^{-3}\,\text{s}$, transverse conduction is beginning to be significant, and the source thickness $\sim 300\,\mu\text{m}$. Figure 5.7 shows the calculated temperature rise when a Q-switched 30 ns pulse with Gaussian spatial profile is incident on aluminium. The beam radius $d = 7\,\text{mm}$, and the peak power absorbed is $30\,\text{MW cm}^{-2}$. The maximum temperature *rise* at the surface is $\sim 600\,\text{K}$, indicating that the metal may be undergoing local melting (melting point $660\,^\circ\text{C}$). A 100 mJ YAG laser (pulse length 20 ns) incident on polished aluminium over an area of $10\,\text{mm}^2$ would give rise to an absorbed peak power density of $\sim 2\,\text{MW cm}^{-2}$. The more modest temperature rise of $\sim 60\,\text{K}$ is unlikely to cause any discernible damage to the surface, which is consistent with observation.

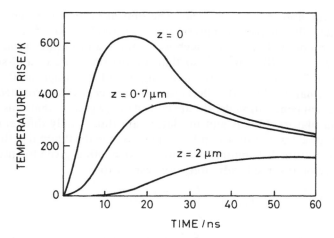

Figure 5.7 Calculated temperature rise for a 30 ns pulse with Gaussian spatial profile, radius 7 mm and peak power 30 MW cm^{-2} (Ready 1971).

5.2.2 Non-metals

We shall first assume that in a non-metal the absorption coefficient, γ, is relatively small, so that the radiation penetrates into the bulk of the material. This is in contrast to a metal, where γ is so large that all the radiation is absorbed within a few nanometres. For the absorption by a non-metal of laser radiation which is switched on at zero time to give a constant power density of I_0, equation (5.21) for the temperature rise is replaced by (Ready 1971)

$$T(z,t) = \frac{2I_0(\kappa t)^{1/2}}{K} \, \text{ierfc}\left(\frac{z}{2(\kappa t)^{1/2}}\right) - \frac{I_0}{\gamma K} \, e^{-\gamma z}$$

$$+ \frac{I_0}{2\gamma K} \exp(\gamma^2 \kappa t - \gamma z)\text{erfc}\{\gamma(\kappa t)^{1/2} - z/[2(\kappa t)^{1/2}]\}$$

$$+ \frac{I_0}{2\gamma K} \exp(\gamma^2 \kappa t + \gamma z)\text{erfc}\{\gamma(\kappa t)^{1/2} + z/[2(\kappa t)^{1/2}]\}. \quad (5.28)$$

Let us next consider the situation in which thermal conductivity is very small, i.e. when the flow of heat away from the region where energy is being absorbed is negligible during the timescale of the source. This is a very reasonable assumption for a short laser pulse irradiating a poor conductor, as we shall see. Then a 1D model is still applicable, so that if the intensity of the laser light at the surface is $I(0, t)$, the intensity $I(z, t)$ at a depth z is given by (equation (5.2))

$$I(z, t) = I(0, t)e^{-\gamma z} \quad (5.29)$$

where γ is the absorption coefficient. The optical penetration depth is $1/\gamma$. The thermal diffusion path in a given time t is given by $(4\kappa t)^{1/2}$. The thermal diffusivity in non-metallic solids such as ceramics and plastics typically lies in the range 10^{-7}–10^{-6} m^2 s^{-1}, so that for the timescale of a Q-switched laser pulse of 20×10^{-9} s, the thermal length is of the order of 10^{-7} m, i.e. $0.1 \, \mu$m. Thermal conductivity effects can thus be neglected provided $1/\gamma \gg 0.1 \, \mu$m. Optical penetration depths are considerably greater than this in most common insulators. Metals and insulators are thus totally different, for in metals the optical penetration depth is much less than the thermal path length.

The temperature distribution as a function of depth is given by integrating the intensity of absorbed radiation with respect to time:

$$T(z, t) = \int_0^t \frac{\gamma I(z, t')}{\rho C} \, \mathrm{d}t'$$
$$= \frac{\gamma e^{-\gamma z}}{C\rho} \int_0^t I(0, t') \, \mathrm{d}t'. \qquad (5.30)$$

Thus the temperature decays exponentially from a maximum at the surface $z = 0$ (figure 5.8), while the time dependence of the temperature is the same as the energy in the pulse, i.e. the integral of the power. Note that, as figure

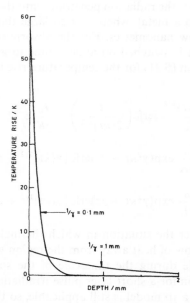

Figure 5.8 Temperature rise in an insulator calculated from equation (5.30) for incident energy density of 10^4 J m^{-2}, density 2500 kg m^{-3} and specific heat 700 J kg^{-1} K^{-1} as a function of depth, for two different absorption coefficients γ.

5.8 shows, the surface temperature rise becomes progressively higher with increasing absorption. Thus in a highly absorbing non-metal care must be taken not to damage the surface. Equation (5.30) shows that even higher temperatures would be reached in a material of low thermal capacity and density, such as a plastic.

5.3 THERMOELASTIC STRESSES

In the low-power regime in which the only effect of the incident laser pulse is a modest rise in temperature of the surface region of the sample, this must be accompanied by the thermal expansion of the hot material, which in turn generates stresses and strains within the sample. In this section we examine these thermoelastic stresses. We shall discuss the generation of elastic waves later. Some workers (e.g. White (1963), Gournay (1966), and Bushnell and McCloskey (1968)) have considered this is a 1D problem in which the pulse uniformly illuminates an infinitely large area of the specimen. Since for use as an ultrasonic source the laser pulse diameter is likely to be much smaller than the ultrasonic propagation path, a 3D solution is required. Thus more recent work has treated the generation of the elastic wave field by the thermoelastic laser source in three dimensions (Scruby *et al* 1980, Wadley *et al* 1984, Rose 1984, Doyle 1986, Aussel *et al* 1988). There are, however, situations where the 1D model is applicable, i.e. where the distance propagated into the solid is much smaller than the diameter of the laser beam, and we shall thus discuss this model first.

5.3.1 1D model

For the 1D model as first discussed by White (1963) the radiation uniformly heats the surface of a semi-infinite body (or half space), defined by $z = 0$ (figure 5.9(a)). As a result of the temperature rise $\delta T(z, t)$, there is, in the absence of any stresses or constraints, a strain ε_{zz} given by

$$\varepsilon_{zz} = \frac{\partial u(z, t)}{\partial z} = \alpha \delta T(z, t) \tag{5.31}$$

where u is the z component of the particle displacement and α is the coefficient of linear thermal expansion.

We note that in this 1D model, there are no strains in the x or y directions, i.e. $\varepsilon_{xi} = \varepsilon_{yi} = 0$. This source will thus generate only compressive waves perpendicular to the surface with planar wavefronts. No shear waves can be generated.

When only part of the surface is heated (figure 5.9(b)) there are lateral constraints from the rest of the body. Under these conditions the stress–strain

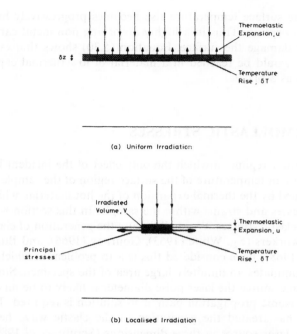

(a) Uniform Irradiation

(b) Localised Irradiation

Figure 5.9 Comparison of (a) uniform with (b) localized irradiation.

relationship becomes

$$\sigma_{zz} = (\lambda + 2\mu)\varepsilon_{zz} - 3B\alpha\delta T \tag{5.32}$$

where $B = \lambda + \frac{2}{3}\mu$ is the bulk modulus of elasticity. The equation of motion is thus

$$\rho \frac{\partial^2 u}{\partial t^2} = (\lambda + 2\mu) \frac{\partial^2 u}{\partial z^2} - 3B\alpha \frac{\partial T}{\partial z}. \tag{5.33}$$

In the absence of external constraints normal to the surface, the stress across the surface must be zero, i.e. $\sigma_{zz}(z = 0) = 0$. However, there are non-zero thermoelastic stresses immediately below the surface, and these can be obtained by solving the equation of motion. There is a non-zero strain, ε_{zz}, at the surface, given by

$$\varepsilon_{zz}(z = 0) = \frac{3B\alpha\delta T}{\lambda + 2\mu}. \tag{5.34}$$

This manifests itself in a rise of the surface given by

$$u = \frac{3B\alpha}{\lambda + 2\mu} \delta z \delta T \tag{5.35}$$

where δz is the depth over which there is a temperature rise, assumed constant (figure 5.9(a)). Since the temperature rise is related to the deposited energy density $\delta e = \delta E / A$ by equation (5.16), i.e.

$$\delta T = \frac{\delta e}{\rho C \delta z} \tag{5.36}$$

therefore, from equation (5.35), substituting $3B = 3\lambda + 2\mu$, we obtain

$$u = \left(\frac{3\lambda + 2\mu}{\lambda + 2\mu}\right)\left(\frac{\alpha}{\rho C}\right)\delta e \tag{5.37}$$

$$= 2(\alpha/\rho C)\delta e \tag{5.38}$$

for Poisson's ratio of $\frac{1}{3}$, when $\lambda = 2\mu$. This is double the value which would be obtained by integration of equation (5.31). The reason is that this equation has no constraints so that the body expands in all directions, whereas equation (5.38) has the constraint of the rest of the material, so that all the expansion has to take place in the z direction. This latter situation is close to physical reality for a laser pulse irradiating a finite area of a large specimen (figure 5.9(b)).

Let us briefly estimate the magnitude of the displacement. Suppose the laser pulse is 100 mJ, uniformly irradiating an area of 100 mm^2, with a reflectivity of 0.9. Thus the absorbed energy density is $\delta e = 100$ J m^{-2}. For aluminium, $\alpha/\rho C \simeq 10^{-11}$ m^3 J^{-1}, so that $u \simeq 2 \times 10^{-9}$ m (i.e. 2 nm).

5.3.2 3D model

For a 3D model, which more closely represents physical reality when a laser pulse is incident at a metal surface, stresses and strains with components parallel to the surface must also be taken into account, i.e. $\varepsilon_{1i} \neq 0$, $\varepsilon_{2i} \neq 0$, $\sigma_{1i} \neq 0$, $\sigma_{2i} \neq 0$ where $i = 1, 2$ or 3. Thus shear as well as compressive stresses may be generated within the solid, although at the surface the only non-zero stresses are the compressive σ_{11} and σ_{22}.

For the purposes of calculating the elastic wave field generated by the laser source, it is convenient to consider the source as a centre of expansion or dilatation that takes place at a point (or at least within a very small volume, V). If the centre of expansion is buried within the bulk of the material, assumed isotropic, then it is equivalent to the insertion of a small extra volume of material at that point, δV. This produces the following strains locally:

$$\varepsilon_{11} = \varepsilon_{22} = \varepsilon_{33} = \frac{1}{3}\frac{\delta V}{V} \tag{5.39}$$

$$= \alpha \delta T. \tag{5.40}$$

This produces the same stress field locally as a hydrostatic pressure, thus:

$$\sigma_{11} = \sigma_{22} = \sigma_{33} \tag{5.41}$$

$$= B \frac{\delta V}{V} \tag{5.42}$$

where $B = \lambda + \frac{2}{3}\mu$ as before.

Thus the point expansion is equivalent to the sudden appearance of three equal and orthogonal pairs of forces (called force dipoles) (figure 5.10(a)). The dipole strength is given by the product of magnitude of each force and the distance separating them, i.e. the product of the stress and the volume, thus:

$$D_{11} = D_{22} = D_{33} = B\delta V. \tag{5.43}$$

(a) Centre of expansion buried in body

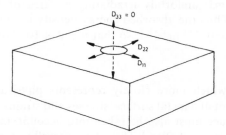

(b) Centre of expansion in plane of surface

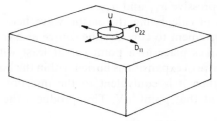

(c) Centre of expansion of non-zero thickness

Figure 5.10 Schematic diagram showing centre of expansion (dilatation) and associated dipolar stress fields, when (a) buried in a body, (b) in plane of surface of body, and (c) at surface, but with finite thickness.

The thermoelastic laser source is not buried within the sample, but at the surface. Let us therefore allow the source to approach vanishingly close to the surface (figure 5.10(b)). The plane stress boundary conditions at the surface imply that there can be no net stress with a component perpendicular to the surface, i.e. $\sigma_{33} = \sigma_{31} = \sigma_{32} = 0$. If the source has zero thickness in the z direction, this boundary condition implies that $D_{33} = 0$, leaving only the horizontal dipoles D_{22} and D_{33}, with stresses parallel to the surface only (figure 5.10(b)). If, as in reality, the source has a small, non-zero thickness, and the material is able to expand upwards perpendicular to the surface (figures 5.9(b) and 5.10(c)), there is a measurable displacement, calculated above.

It is instructive to estimate the magnitude of the thermoelastic stresses as a function of laser power. If energy $\delta E = (1 - R)E$ from the pulse is absorbed within a volume V to give a temperature rise δT then, from equations (5.40) and (5.16), neglecting thermal conductivity, the bulk strain is given by

$$\frac{\delta V}{V} = 3\alpha\delta T = \frac{3\alpha\delta E}{\rho C V}. \tag{5.44}$$

Thus the source is equivalent to the insertion of a small extra volume of material δV given by

$$\delta V = \frac{3\alpha}{\rho C}\,\delta E. \tag{5.45}$$

This is the same as equation (5.10) which was used to estimate typical ultrasonic displacements. Thus for 1 mJ of energy absorbed uniformly over an area of 10 mm^2, substituting values of α, ρ and C (see table 5.1 or 5.2 (p 273)) for aluminium, yields a value of $\delta V \simeq 3 \times 10^{-14}$ m^3. In iron (or mild steel) δV is smaller, i.e. $\simeq 10^{-14}$ m^3, chiefly because of the higher density. Thus from equation (5.42) the dipole strengths are $D_{11} = D_{22} \simeq 2.2 \times 10^{-3}$ N m per mJ of absorbed energy in aluminium and 1.4×10^{-3} N m per mJ in mild steel.

From equation (5.41) the volume V must also be known in order to calculate the stress. This is more difficult because V increases with time due to thermal diffusion. We assume that the source has a thickness of 2 μm, which from figures 5.3–5.6 is the depth within which most of the temperature rise occurs in aluminium. Then, for the above example, $V \simeq 2 \times 10^{-11}$ m^3, so that $\delta V/V \simeq 2 \times 10^{-3}$. The thermoelastic stresses are thus of the order of $B\delta V/V \simeq 10^8$ Pa mJ^{-1} absorbed energy in aluminium.

An alternative method is to estimate the mean maximum temperature within a depth of 2 μm from figures 5.3 and 5.5, and this is $\sim$60 K for 4 mJ energy. Thus from equations (5.44) and (5.41) $\sigma \simeq 10^8$ Pa per mJ. In aluminium the yield stress is of the same order. It is thus relatively easy to exceed the yield stress locally and cause permanent damage if too high a Q-switched laser pulse energy is used. In aluminium alloys yield is somewhat higher ($\sim 5 \times 10^8$ Pa). The situation is less serious in steels where the thermoelastic stresses are of the same order as in aluminium, i.e. 2×10^8 Pa per mJ, but where yield stresses are closer to 10^9 Pa.

These stresses are of the same order of magnitude as those due to ablation at higher laser powers, to be discussed in the next section. It is at first surprising that the thermoelastic stresses are as high. The reason is that the Q-switched laser pulse deposits heat energy so quickly that very little is conducted away, so that a substantial temperature rise is contained within a small well defined volume, generating high compressive stresses due to expansion of the material.

5.4   OTHER EFFECTS

At higher incident power densities and pulse energies, effects other than simple heating may occur. The calculations in the previous section indicate that high surface temperatures can readily be generated in metals with Q-switched lasers. In this section we shall discuss the main consequences of high temperatures, namely melting, vaporization and plasma formation. First, however, we must briefly discuss radiation pressure.

5.4.1   Radiation pressure

Electromagnetic radiation incident on a solid surface generates a radiation pressure due to the change in momentum of the photons as they are reflected from or absorbed by the surface. The magnitude of the pressure, p, can be calculated from the power density of the incident wave I and the reflectivity R as follows:

$$p = \frac{(1 + R)I}{c} \tag{5.46}$$

where c is the velocity of light. For a laser pulse of incident energy δE, duration δt, uniformly illuminating an area A, the average radiation pressure during the pulse is given by

$$p = \frac{(1 + R)\delta E}{cA\delta t}. \tag{5.47}$$

By way of example, a pulse of energy 100 mJ and duration 20 ns incident over an area of 10 mm^2 will generate a radiation pressure of 3000 N m^{-2} for a reflectivity of 0.8. The mean force experienced normal to the surface during the pulse is thus 0.03 N. Radiation pressure and the resulting forces and stresses on the solid sample are generally several orders of magnitude smaller than those due to thermoelastic effects and ablation, and can therefore be largely ignored for the purposes of ultrasonic generation by laser.

5.4.2 Surface melting

As the incident optical power density is increased, the surface temperature increases until the melting point of the solid, here assumed to be a metal, is reached. Laser-induced melting is very important to welding technology, and is discussed in some detail in, for instance, Ready (1971). For the purposes of this book we shall only deal briefly with this topic insofar as it affects the generation of ultrasound. For laser welding, one desires effective melting of the surface without excessive vaporization. For laser ultrasound generation there is little value in moving from the purely thermoelastic regime to the melting regime, without proceeding to even higher power densities when vaporization occurs. In addition, with Q-switched short-pulse lasers the power regime in which melting occurs without vaporization is too narrow to be of very much practical interest. Melting of the surface is likely to be accompanied by damage or permanent changes to material properties on cooling, and is best avoided unless accompanied by vaporization and ablation, when the ultrasonic wave field is intensified.

The conditions for which melting occurs depend on the nature of the material and the laser pulse. For metals such as aluminium and steel, melting is likely to occur when the incident power density is of the order 10^6–10^7 W cm^{-2} for a Q-switched pulse.

5.4.3 Vaporization, ablation and plasma formation

Vaporization of the specimen surface is relatively easy to produce with pulsed lasers. For common metals and typical Q-switched laser pulses it occurs for power densities above $\sim 10^7$ W cm^{-2}. The average power in a typical 100 mJ, 20 ns duration pulse is 5 MW. This is probably below the threshold for vaporization if the pulse is unfocused (i.e. area ~ 1 cm^2). However, if the beam is focused, e.g. to an area of ~ 1 mm^2 when the incident power density is $\sim 5 \times 10^8$ W cm^{-2}, appreciable vaporization occurs and a plasma is formed (figure 5.11).

A full discussion of the physical processes that take place in this regime is beyond the scope of this book, and the reader is referred for instance to Ready (1971) and Krehl *et al* (1975). In simple terms, as the incident optical power increases, the surface temperature rises until the boiling point of the material is reached, and some material is vaporized, ionized and a spark or plasma formed. As Ready (1971) discusses, at lower powers using normal (i.e. not Q-switched) pulse lasers, thermal conductivity into the bulk of the metal is important, controlling the amount of metal that is vaporized.

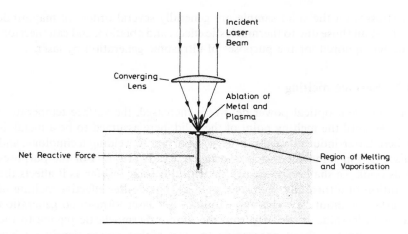

Figure 5.11 Schematic diagram to show ablation of surface material and net reactive force on sample.

However, at the much higher powers obtainable by Q-switching, the *rate* at which heat is supplied is too fast for appreciable conduction to occur, so that the amount of metal vaporized depends almost entirely on its latent heat of vaporization. Ready gives the following approximate expression for the power density I_c above which thermal conductivity can be neglected:

$$I_c \geqslant 2L\rho\kappa^{1/2}\tau^{-1/2} \tag{5.48}$$

where τ is the pulse duration, L the latent heat required to vaporize the solid, κ the thermal diffusivity and ρ the density of the metal. Thus for normal pulse lasers ($\tau \sim 10^{-3}$ s), $I_c \sim 10^7$ W cm^{-2} for common metals, while for Q-switched lasers ($\tau \sim 2 \times 10^{-9}$ s), $I_c \sim 10^9$ W cm^{-2}.

For normal (i.e. long) pulse lasers it is probably adequate to assume that an equilibrium is reached in which the surface remains at the melting point, so that the rate of material removed (ζ) is given by (Ready 1971)

$$\zeta = \frac{I}{\rho[L + C(T_v - T_0)]} \tag{5.49}$$

where T_0 and T_v are the initial and vaporization temperatures, I is the incident power density, ρ is the material density and C the specific thermal capacity. Thus for $I = 10^8$ W cm^{-2} and substituting values from table 5.2 (p 273), the rate of material removal from an aluminium sample would be $\zeta \sim 130$ m s^{-1}, and from steel (i.e. iron) 70 m s^{-1}.

This ablation of material from the surface (figure 5.11) produces a net stress σ in reaction against the sample. This is simply calculated from Newton's second law of motion as the rate of change of momentum, i.e.

$$\sigma = \frac{I\zeta}{L + C(T_v - T_0)} \tag{5.50}$$

$$= \frac{I^2}{\rho[L + C(T_v - T_0)]^2}. \tag{5.51}$$

For an incident power density of 10^8 W cm^{-2} the stress in aluminium is thus estimated to be ~ 50 MPa, and in steel ~ 40 MPa, i.e. of similar magnitude for the two metals, somewhat surprisingly. Some estimate of the threshold to ablation can be made with similar assumptions, i.e. that the absorbed laser energy must exceed the thermal energy required to raise the surface material up to its boiling point. The data of table 5.2, assuming a heated region of 1 μm thickness in aluminium and 0.5 μm in steel (estimated from the 20 ns pulse data of figure 5.4), yield similar ablation thresholds in the two materials of $\sim 4 \times 10^7$ W cm^{-2} absorbed power density, which agrees surprisingly well with experiment (see §5.9.2). If the irradiation area is known, then the net force can be calculated, e.g. for an area of 1 mm^2, the force is 50 N, and for 0.1 mm^2 it is 5 N.

With Q-switched lasers, power densities in excess of 10^9 W cm^{-2} are easy to attain. Because of the extremely high rate of change of power input during the nanosecond pulse, different phenomena must be considered than for longer pulses. We must consider the effects caused by the plasma generated in association with ablation. First the plasma exerts a high pressure on the surface which in turn suppresses vaporization of the material by raising the boiling point of the material well above its normal value. Secondly, it absorbs light from the laser pulse, acting as a shield, but also becoming extremely hot. Thirdly, as it expands it produces an impulse reaction on the surface, and fourthly it radiates some of its heat back on to the surface, maintaining its high temperature for some time after the incident laser pulse power has started to fall.

Ready (1971) notes also that Q-switched laser pulses are relatively ineffective in removing material from the target, compared with normal pulses. At very high power densities a small mass of metal absorbs the energy, vaporizes at a very high temperature almost explosively, and the resulting plasma screens the remaining metal from the pulse. Furthermore, the high pressure of the plasma on the surface tends to suppress metal loss. While this may be a disadvantage for laser drilling, for example, it is a positive benefit for NDE, where the minimum damage to the sample is required.

Theoretical and experimental data on plasma densities, ablation velocities, surface temperatures and reactive stresses on the sample can be found in, for example, Krehl *et al* (1975). Figure 5.12 shows their computed surface and subsurface temperature for an aluminium sample irradiated by a Q-switched pulse of 25 ns FWHH and power $\sim 10^9$ W cm^{-2}. Although this

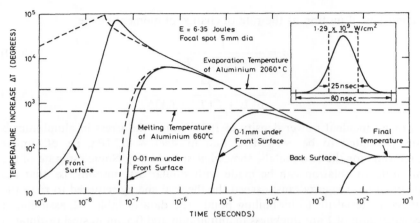

Figure 5.12 Computed surface and subsurface temperatures for an aluminium sample irradiated by a 25 ns pulse of power 10^9 W cm^{-2}. Full curves Gaussian pulse, broken curves rectangular pulse. (Krehl *et al* 1975.)

calculation does not take into account phase changes at the surface or the complexities of the plasma/light interactions, it illustrates well the fact that the surface and immediate subsurface of the metal remain at high temperatures (i.e. above the boiling point) for considerably longer ($\sim 10\ \mu s$) than the laser pulse duration (25 ns).

The same authors report measurements of the normal stress on the sample surface produced by the ablation and plasma (the form of which is shown

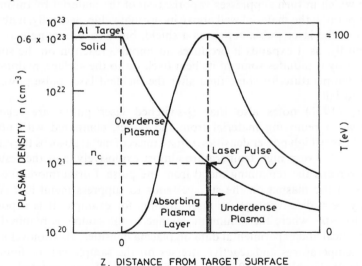

Figure 5.13 Plasma density and temperature (T) as functions of distance from sample surface (Krehl *et al* 1975).

in figure 5.13). Maximum stresses for the same laser power were 800 bar (8×10^7 Pa) in vacuum and half this value in air. The total impulse was ~ 6 g cm s^{-1} in vacuum, and somewhat larger in air. In both cases the stress pulse lasts considerably longer than the laser pulse. The focal spot size was 6 mm, so that the peak reactive force experienced by the sample was approximately 80 N in air. Their calculations of ion densities, etc, predict a momentum transfer in broad agreement with the measured impulse, and also show that $\sim 90\%$ of the impulse is due to the ions in the plasma, and only $\sim 10\%$ due to ablation of neutral particles from the surface.

From the point of view of ultrasonic generation by laser, the ablation/plasma regime is best considered as a source in which a normal stress pulse is applied to the surface. While the leading edge of the stress pulse has a similar risetime (~ 20 ns) to the laser pulse and is therefore capable of generating high-frequency ultrasound, the trailing edge tends to be considerably lengthened.

5.4.4 Generation in liquids and gases

Ultrasound can be generated in a liquid as a consequence of irradiation by a high-power laser pulse. Below a certain threshold the main effects are thermoelastic. A complication is that absorption of the light will take place along the whole length of the beam which passes through the liquid, so that a linearly extended ultrasonic source with exponentially decreasing amplitude (equation (5.29)) may result. At higher powers (e.g. as a result of focusing the incident laser beam) the liquid is vaporized within a small region, and a high-amplitude stress wave (initially a shock wave) radiates out from the vaporized region. Von Gutfeld and Melcher (1977) have investigated generation in a constrained liquid, which is discussed in §5.5.4, and von Gutfeld (1980) has carried out a preliminary study relating to medical applications, which is referred to in §6.6. Otherwise laser generation in liquids will not be considered further in this book. For further information the reader is referred to Sigrist and Kneubuhl (1978), Emmony (1985) and Berthelot (1989).

A recent discovery has been the air-breakdown source (Edwards *et al* 1989). A TEA CO$_2$ laser pulse is focused to a point in the air just above the surface of the specimen. At very high power densities, the air breaks down to form a plasma. This generates a detonation wave in the air which strikes the surface of the specimen and causes a normal reactive force there, which in turn acts as a source of ultrasonic waves in the solid. Although the force exerted on the surface is impulsive and resembles that generated by ablation of material from the surface, there is no damage to the specimen surface, making this air-breakdown source truly non-destructive.

5.4.5 Effect of raised specimen temperature

High-temperature NDT is likely to be one of the most important commercial applications for laser-generated ultrasound, so that we must ask what effect a raised specimen temperature might have on the generation process. With the exception of a few high-temperature studies (e.g. Calder *et al* 1980, Dewhurst *et al* 1988) most of the work on laser-generated ultrasound has been carried out at room temperature.

Several effects can be predicted as the temperature is raised. In the first place, the threshold for a transition from the thermoelastic regime to ablation is likely to occur at a somewhat lower power density, for the simple reason that less absorbed heat is required in order to raise the temperature of the metal to its boiling point. Thus more care is needed to restrict the incident laser power density if it is desired to keep within the non-destructive thermoelastic regime.

Turning now to the efficiency of ultrasonic generation in the thermoelastic regime, two main effects must be considered. The first is the change in surface reflectivity (and hence absorption), and the second is the change in the factor $3\alpha/\rho C$ (equation (5.45)) that links absorbed energy to elastic strain. Dealing first with the reflectivity, the authors were unable to find suitable data in an explicit form, and so only a rough estimate will be made of the magnitude of the effects here. From equations (5.5) and (5.7) the absorption $(1 - R)$ is proportional to $(\mu_r/\sigma)^{1/2}$, assuming a constant electromagnetic frequency which is low enough for a classical treatment to be valid. Now let us assume that the specimen is aluminium and that the temperature is raised by approximately 500° above ambient in an inert atmosphere so that the surface does not oxidize. The magnetic permeability μ can be assumed to be unity throughout, so that $(1 - R) \propto (\text{resistivity})^{1/2}$. A rise in temperature of $\sim 500°$ raises the resistivity of aluminium by a factor of ~ 3.5 (deduced from Kaye and Laby 1973), so that the absorption of laser radiation in the infrared can be estimated to increase by $\sim 80\%$. Thus raising the temperature of the specimen to just below its melting point causes considerably more electromagnetic energy to be absorbed from the pulse, and a potentially stronger ultrasonic source in consequence.

In iron or steel, there is the added complication of variations in permeability. From Spooner (1927) the initial permeability of electrolytic iron increases more than two-fold for a temperature rise from 0 to 500 °C and even more dramatically just before the Curie temperature (~ 770 °C). Above the Curie temperature a non-magnetic austenitic structure is formed, so that there is a drop to unity in permeability, which is also accompanied by a significant change in electrical resistivity. From this it is speculated that the proportion of laser energy absorbed should initially rise with increasing temperature in iron and steel, but then should fall above the Curie temperature. Such a variation in reflectivity is likely to have a marked effect on the efficiency of ultrasonic generation in ferromagnetic materials as the temperature is raised.

Secondly, the thermoelastic coupling factor $3\alpha/\rho C$ varies with temperature. It can be shown by substitution of values from table 5.2 to be $\sim 3 \times 10^{-11}$

for aluminium at 0 °C. It increases by ~13% at 500 °C, also making the thermoelastic source more efficient at elevated temperatures. Note, however, that the predominant effect is the decrease in reflectivity as the temperature is raised.

It is more difficult to estimate changes of efficiency in the ablation regime. However, the decrease in reflectivity should still have the effect of giving increased efficiency. In addition, less specific heat needs to be extracted from the heat pulse to raise the surface temperature to its boiling point to cause ablation. If equation (5.49) can be assumed to hold, and if the surface temperature of aluminium is raised from 0 to 500 °C, then only ~80% of the absorbed power is needed to cause the same rate of material removal, i.e. the ablation process becomes ~20% more efficient. Making the same estimate for steel, we find that very approximately 35% less absorbed optical power is needed to cause the same rate of ablation at 1000 °C as at 0 °C.

These admittedly somewhat elementary calculations indicate that raising the specimen temperature should lead to more efficient laser generation of ultrasound in both thermoelastic and ablation regimes.

5.5 CONSTRAINED SURFACES

The chief benefit of laser-generated ultrasound is that it can be used remotely, i.e. without contact with the specimen, and without couplant or modification to the surface. Thus the major part of this chapter deals with ultrasonic generation at a free surface, either through thermoelastic or ablative stresses. There are nevertheless situations where it is required to use laser generation when the surface is constrained in some way; either by a transparent solid layer, or by a liquid. We shall therefore consider briefly some of the effects that accompany generation of ultrasound by laser at a constrained surface. The effects of surface modification or constraint can be very complex, since optical absorption, conversion to ultrasonic energy and ultrasonic propagation are all affected by changes at surfaces. We restrict discussion, for the most part qualitative, to the following: coating the surface with a thin solid layer of, for instance, paint or rust and roughness, covering the surface with a transparent solid such as glass, covering the surface with a transparent liquid, and finally constraining a thin layer of liquid between a transparent solid and the sample.

5.5.1 Surface coating and roughness

The first effect of a surface coating such as paint or rust is to change the reflectivity of the surface. The maximum reflectivity and minimum absorption is obtained with a polished surface. Applying a layer of paint or rust increases the absorbed energy. Thus at low power the local temperature rises, and the thermoelastic stresses and intensity of the ultrasonic source also increase. Increasing the roughness of the surface has a similar effect, reducing reflectivity. A coating of matt black paint increases the absorption almost to 100%.

However, there is sometimes a second effect which changes not only the strength of the source but also its nature. Because of the high absorptivity of, for instance, black paint, there is enough energy to vaporize the paint, which is blown off, causing a normal reaction on the surface similar to metal ablation. Subsequent pulses are incident on plain metal, so that the effect is generally limited to the first few shots. Consistent data are difficult to obtain. The combination of stresses in the surface for the first shot is likely to be similar to figure 5.14(a).

In the high-power ablation regime the effect of paint, etc, is less significant, since there is already a large 'blow-off' force. The presence of paint, rust or dirt is still however likely to intensify the first shot as in the thermoelastic regime.

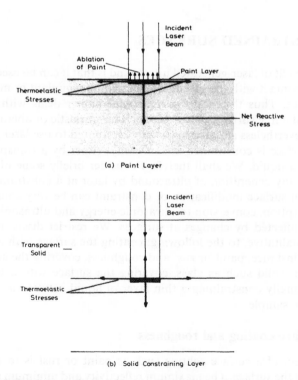

Figure 5.14 Schematic diagram showing stresses induced when a laser pulse is incident on a sample surface covered by (a) a thin layer of paint, and (b) a thick transparent solid.

5.5.2 Transparent solid covering

Neglecting absorption in the transparent covering, e.g. glass, assumed also to be well bonded to the sample, the major effect is likely to be the change in elastic boundary conditions at the surface of the sample. The surface is no longer free, and can support tractions with normal components, i.e. $\sigma_{i3} \neq 0$. Thus, in addition to the dipolar stresses parallel to the surface (D_{11}, D_{22}), there must also be dipolar stresses normal to the surface (i.e. $D_{33} \neq 0$). This is illustrated in figure 5.14(b). In general $D_{11} = D_{22} \neq D_{33}$, unless the two media have identical elastic properties (density and moduli) in which case the source appears as if it is buried in the sample $(D_{11} = D_{22} = D_{33})$. Thus constraint by a solid significantly modifies the thermoelastic source. The effect at high power is largely of academic rather than practical interest. Ablation of metal and flame formation will be reduced, and damage to the interface in the form of a disbonded region is likely to ensue.

5.5.3 Transparent liquid covering

There are expected to be some similarities with the transparent solid, although the elastic properties of the two media will always be disparate because liquids cannot support shear stresses. Thus the first effect, other than a change in absorptivity, is to add a normal dipolar stress to the solid surface (cf figure 5.14(b)).

There are however additional effects due to the thermal properties of the liquid. Liquids generally have higher expansion coefficients than solids, and it would be expected therefore that thermoelastic stresses in the liquid, heated from the solid, would exert an appreciable normal force on the solid. At higher powers the liquid would vaporize (at a lower temperature and lower incident power than the metal), exerting a strong normal force on the solid. Thin layers of liquid tend to evaporate, exerting a momentum-transfer reactive force resembling ablation, even at lower power densities (figure 5.15(a)). In a thin liquid (or solid) layer there are likely also to be ultrasonic reflections interfering with the main sound field generated by the source.

5.5.4 Constrained liquid

The source in which a thin liquid layer is constrained between the sample and a transparent solid (figure 5.15(b)) has been investigated in detail by von Gutfeld and Melcher (1977). This technique combines a number of effects, all of which modify the form of the ultrasonic source. Firstly, the sample surface is constrained, secondly there is thermal expansion in the liquid, while thirdly the liquid is itself constrained, enhancing stresses due to expansion in the liquid. All these effects tend to generate a strong stress normal to the sample surface which may predominate over the stresses parallel to the surface.

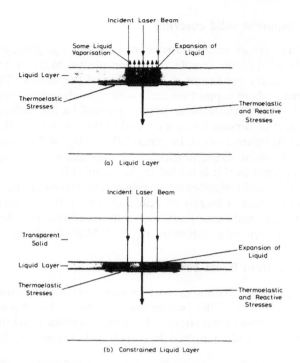

Figure 5.15 Schematic diagram showing stresses induced when a laser pulse is incident on a sample surface covered by (*a*) a layer of liquid, and (*b*) a layer of liquid constrained by a transparent solid.

To summarize, all four techniques for modifying the surface of the sample introduce large stresses normal to the surface, which are otherwise absent for the thermoelastic source at a free surface. The form of the ultrasonic source is thus substantially modified. As will be seen in §5.8, the different combination of stresses generates a different ultrasonic sound field, with considerable enhancement in compression-wave amplitude, especially propagating normal to the surface. The constrained surface source generates a sound field more akin to that of ablation, but at lower power densities and without any damage. However, the need to constrain with for instance a liquid, reduces the usefulness of the laser-ultrasonic source for many non-destructive testing applications.

5.6   ULTRASONIC WAVE PROPAGATION IN UNBOUNDED SOLIDS

A sudden change in stress in a region of a body, such as caused by laser irradiation, must act as a source of elastic (stress) waves which then redistribute the stresses throughout the body. The characteristics of the ultrasonic (i.e. elastic) wave field such as amplitude, frequency and angular variation, depend upon the size and nature of the source, i.e. the changing stress field. For the purposes of this chapter we shall restrict discussion to isotropic elastic solids. These are able to support the propagation of both longitudinal (also known as compression) waves and transverse (shear) waves in the bulk, and various surface waves along their boundaries. The equations of motion are thus more complex than for the propagation of sound through a fluid.

Following Achenbach (1973), the particle displacement, u, at any point in an elastic solid can be expressed in terms of a scalar potential Φ, and a vector potential Ψ, thus:

$$u = \nabla\Phi + \nabla \times \Psi. \tag{5.52}$$

The two potentials satisfy the following equations:

$$\nabla^2\Phi - \frac{1}{c_1^2}\frac{\partial^2\Phi}{\partial t^2} = 0 \tag{5.53}$$

$$\nabla^2\Psi - \frac{1}{c_2^2}\frac{\partial^2\Psi}{\partial t^2} = 0 \tag{5.54}$$

where $c_1 = [(\lambda + 2\mu)/\rho]^{1/2}$ and $c_2 = (\lambda/\rho)^{1/2}$ are the compression- and shear-wave velocities respectively. Thus the compression and shear waves in an elastic continuum are uncoupled, except at boundaries where boundary conditions couple the longitudinal and transverse displacements. The longitudinal wave equation (5.53) has the following one-dimensional form for the propagation of plane waves parallel to the x direction:

$$\frac{\partial^2\Phi}{\partial x^2} - \frac{1}{c_1^2}\frac{\partial^2\Phi}{\partial t^2} = 0. \tag{5.55}$$

It can readily be shown by substitution that the general solution is of the form:

$$\Phi = f(t - x/c_1) + g(t + x/c_1) \tag{5.56}$$

where f and g are any functions, representing positive and negative going waves respectively of velocity c_1. We very often need only consider the first term in equation (5.56). Of equal importance to plane waves in the study of laser ultrasonics are spherical waves propagating out from a point source. Assuming spherical symmetry, the general solution of equation (5.53) in polar coordinates is

$$\Phi = \frac{1}{r}f(t - r/c_1) + \frac{1}{r}g(t + r/c_1). \tag{5.57}$$

For a point source of ultrasound, only outgoing spherical waves have physical significance, so that we can neglect the second term in equation (5.57) to obtain

$$\Phi = \frac{1}{r} f(t - r/c_1).$$ (5.58)

Analogous equations can be derived for the transverse displacements due to shear waves. Thus equations (5.57) and (5.58) are replaced by

$$\Psi = \frac{1}{r} f(t - r/c_2) + \frac{1}{r} g(t + r/c_2)$$ (5.59)

$$= \frac{1}{r} f(t - r/c_2) \qquad \text{for outgoing waves.}$$ (5.60)

It is convenient now to introduce two special mathematical functions, the Dirac delta function, $\delta(z)$, and the Heaviside function $H(z)$. The delta function is zero everywhere except at $z = 0$, where it has infinite amplitude. The area under the function is however unity, i.e.

$$\int_{-\infty}^{\infty} \delta(z) \, dz = 1.$$

The heaviside function is zero for $z < 0$, and unity for $z \geqslant 0$. It follows that the function $H(z)$ is the integral of $\delta(z)$. In ultrasonics an ideal short pulse can be represented by $\delta(t)$, while a step ('edge') pulse can be represented by $H(t)$.

If a point disturbance, whose time dependence is $\delta(t)$, acts as a source of elastic waves at the origin of an infinite solid, then from equation (5.58) the compressional component can be represented mathematically by

$$\Phi = \frac{1}{4\pi r} \delta(t - r/c_1).$$ (5.61)

For a source with $H(t)$ time dependence, equation (5.61) is replaced (again at large r) by

$$\Phi = \frac{1}{4\pi r} H(t - r/c_1).$$ (5.62)

The solution of the wave equation for a point excitation in time (i.e. $\delta(t)$ time dependence) and space is called the Green's function. In this case it is the Green's function for an infinite elastic body, i.e. an unbounded domain.

We are now in a position to calculate the displacement wave field which is generated by the two most important types of source in the study of laser ultrasonics, i.e. the centre of expansion (dilatation), and the point force excitation, representing the thermoelastic and ablation sources respectively.

5.6.1   Centre of expansion in an unbounded solid

It was shown in §5.3.2 that the absorption of radiation at a point causes a centre of expansion with a localized dilatation, δV, and that this is equivalent to the sudden appearance of three equal and orthogonal force dipoles, D_{11}, D_{22}, D_{33}, such that

$$D_{11} = D_{22} = D_{33} = B\delta V. \qquad (5.63)$$

Achenbach (1973) shows that the elastic wave field due to a centre of compression is given by

$$u_i = \frac{1}{4\pi c_1^2} \frac{\partial}{\partial x_i} \left(\frac{1}{r} f(t - r/c_1) \right). \qquad (5.64)$$

This equation only has a term for waves propagating at velocity c_1. Thus the centre of compression does not generate shear waves, only compression waves. In Achenbach (1973) the dipoles are composed of forces separated by a small distance h, each dipole having strength $h^{-1}\rho f(t)$. Translating this into the above notation we obtain that

$$\rho f(t) = D_{ij} = B\delta V(t). \qquad (5.65)$$

Substituting for $f(t)$ in equation (5.64) and transferring to spherical coordinates because the wave field must have spherical symmetry about the source, we therefore obtain

$$u_r = \frac{1}{4\pi c_1^2} \frac{\partial}{\partial r} \left(\frac{B}{\rho r} \delta V(t - r/c_1) \right) \qquad (5.66)$$

$$= \frac{B}{4\pi(\lambda + 2\mu)} \left(\frac{\delta V'(t - r/c_1)}{c_1 r} + \frac{\delta V(t - r/c_1)}{r^2} \right)$$

$$\simeq \frac{B}{4\pi(\lambda + 2\mu)c_1 r} \delta V'(t - r/c_1) \qquad (5.67)$$

where the second term, which varies as $1/r^2$, has been neglected because it will make a negligibly small contribution at large r. Figure 5.16 gives the definition of u_r along with the perpendicular displacement u_θ. In the above and following equations the following shorthand notation is used for derivatives of functions:

$$\delta V'(\xi) = \frac{\partial}{\partial \xi} (\delta V). \qquad (5.68)$$

If thermal conductivity effects can be neglected then, from equation (5.45), we can substitute for δV to obtain

$$u_r = \frac{B}{4\pi(\lambda + 2\mu)c_1 r} \frac{3\alpha}{\rho C} \delta E'(t - r/c_1). \qquad (5.69)$$

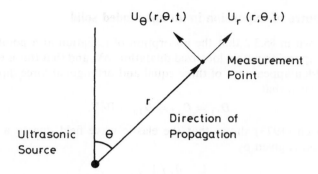

Figure 5.16 Diagram to define displacements u_r and u_θ at a point r, θ, due to a source at the origin.

Thus, from equation (5.67), the displacements in the wave field are proportional to the rate of change of volumetric strain, and hence from equation (5.69) to the power (dE/dt) in the optical pulse under conditions when thermal conductivity effects can be neglected. The energy in the optical pulse controls the area under the displacement pulse waveform. We note in passing that the ultrasonic pulse area is less affected by reduced frequency response in any measurement system than the pulse height. Figure 5.17(a)

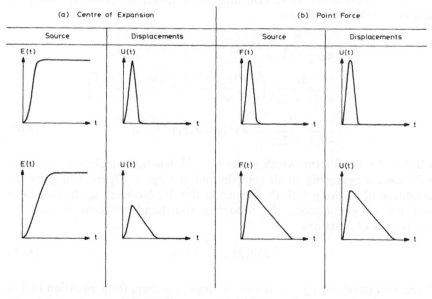

Figure 5.17 Far-field displacement waveform due to a compression wave for source functions of different time dependence. Note that in the thermoelastic regime (a) the displacement is the derivative of the absorbed energy function, whereas in the ablation regime (b) it is the same as the applied force function.

illustrates that two optical pulses of identical energy, but different time dependence, may generate ultrasonic displacements of different magnitude. The first pulse is shorter and generates a higher amplitude ultrasonic pulse than the second, although the areas under the pulses should be equal. The bulk compression waveform takes the form of a unipolar pulse, which makes it ideal for many ultrasonic applications.

5.6.2 Point force in an unbounded solid

It was shown in §5.4.3 that at higher incident laser power, ablation of metal from the surface of the specimen applies an impulsive force normal to the surface. Before discussing the effects of surfaces, however, we first calculate the wave field from a time-varying point load in an infinite solid.

Following Aki and Richards (1980) or Achenbach (1973), if a point load of time dependence $F_j(t)$ ($j = 1, 2, 3$) acts at the origin, the displacement wave field u_i ($i = 1, 2, 3$) at a point $r = (x_1, x_2, x_3)$ is given by

$$u_i(t) = \frac{3\gamma_i\gamma_j - \delta_{ij}}{4\pi\rho r^3} \int_{r/c_1}^{r/c_2} t' F_j(t' - t)\, dt'$$

$$+ \frac{\gamma_i\gamma_j}{4\pi\rho c_1^2 r} F_j(t - r/c_1)$$

$$+ \frac{\delta_{ij} - \gamma_i\gamma_j}{4\pi\rho c_2^2 r} F_j(t - r/c_2) \qquad (5.70)$$

where γ_i and γ_j are the sets of direction cosines of the vectors u_i and F_j respectively, $\delta_{ij} = 1$ if $i = j$, $\delta_{ij} = 0$ if $i \neq j$.

The first term in equation (5.70) behaves like r^{-2} for sources whose lifetime is much less than the difference between the compression- and shear-wave arrival times, e.g. very short pulses such as generated by Q-switched lasers. The second and third terms both behave like r^{-1}, and therefore dominate as $r \to \infty$. They therefore represent displacements in the far field. The first term dominates for $r \to 0$, and is therefore a near-field term.

The second term in equation (5.70) describes a wave which propagates at the compression-wave speed c_1. If θ is the angle between F and r, the radial displacement can be simplified to

$$u_r(t) = \frac{\cos\theta}{4\pi\rho c_1^2 r} F(t - r/c_1)$$

$$= \frac{\cos\theta}{4\pi(\lambda + 2\mu)r} F(t - r/c_1). \qquad (5.71)$$

The third term in equation (5.70) describes a transverse (shear) wave which propagates with velocity c_2. The displacement is perpendicular to the direction of propagation (figure 5.16), and is given by

$$u_\theta(t) = \frac{\sin\theta}{4\pi\rho c_2^2 r} F(t - r/c_2)$$

$$= \frac{\sin\theta}{4\pi\mu r} F(t - r/c_2). \tag{5.72}$$

We note that the time dependence of both components of displacement is the same as the time dependence of the force $F(t)$, rather than a derivative as in the case of the centre of expansion. Figure 5.17(b) gives examples of the ultrasonic pulse shape for two force functions.

5.7 PROPAGATION IN BOUNDED SOLIDS

The infinite solid used for the calculations of the previous section is of academic interest only. The effects of boundaries in all practical solids are very important, and indeed are often the dominating factor. Many of the most important characteristics of laser-generated ultrasound arise through elastic interactions with boundaries and other discontinuities. There are certain conditions imposed upon the stresses and strains at boundaries, and these can only be satisfied by modification of the elastic wave field. The major effects of such boundaries can be summarized as follows.

(a) Modification of the stress field associated with a source that is situated at the boundary. This is particularly important in the case of the centre of expansion generated by a laser pulse in the thermoelastic regime.
(b) Transmission through the boundary into a second medium, where refraction will occur.
(c) Reflection from the boundary, with change of amplitude and phase. Two or more boundaries may give rise to an infinite series of reflected wavefronts.
(d) Conversion of elastic wave energy at the boundary into a different bulk wave propagation mode.
(e) Generation of surface or interface waves that will propagate along the boundary.
(f) Attenuative losses due to inhomogeneities at the boundary.

In all cases, the complexity of the problem of calculating the elastic response of a solid to some stimulation increases dramatically with the number of boundaries and discontinuities. Analytical solutions are readily available for a semi-infinite body, usually known as a half space, which has just one boundary. Analytical and numerical solutions also exist for infinite parallel-sided plates, which have just two boundaries. For more complicated solids, with more than two boundaries, approximations must generally be made, e.g. to a half space or infinite plate. Such approximations are usually adequate for the first part of the received signal and/or for the far-field components of the elastic response. The reader is directed to one of the many excellent books (e.g. Achenbach 1973, Graff 1975) on elastic wave propagation for a fuller discussion of the effect of boundaries. In this chapter we shall restrict discussion to topics of direct relevance to laser-generated ultrasound.

The first effect to consider is reflection from a plane rigid boundary. Reflection of elastic waves in solids is more complex than the reflection of, for instance, sound waves in a fluid, or light in a transparent solid. This is because the boundary conditions can mostly be satisfied only if there is some mode conversion. Thus at a plane steel–air boundary, where transmission into the air is negligible, an incident compression wave generates not one, but two 'reflected' waves, the second being a shear wave.

If a plane compression wave of amplitude A_0 is incident at angle θ_0 on the boundary, there will be a reflected compression wave of amplitude A_1 at angle θ_1 and a shear wave of amplitude A_2 at angle θ_2 (figure 5.18(a)). The shear wave is designated an SV wave because its plane of polarization is vertical, i.e. in a plane perpendicular to the boundary. The relationships that follow have been developed for continuous plane waves of single frequency, but since there is no frequency dependence in the equations, they are equally applicable to broadband pulses of ultrasound. The equations are, however, only applicable in the far field, when the wavefronts from a point source can be approximated by plane waves. Thus it can be shown (Achenbach 1973) that

$$\theta_1 = \theta_0 \tag{5.73}$$

$$\frac{\sin \theta_2}{\sin \theta_0} = \frac{c_2}{c_1} = \frac{1}{k} \tag{5.74}$$

$$A_1 = A_0 \frac{\sin 2\theta_0 \sin 2\theta_2 - k^2 \cos^2(2\theta_2)}{\sin 2\theta_0 \sin 2\theta_2 + k^2 \cos^2(2\theta_2)} \tag{5.75}$$

$$A_2 = A_0 \frac{2k \sin 2\theta_0 \cos 2\theta_2}{\sin 2\theta_0 \sin 2\theta_2 + k^2 \cos^2(2\theta_2)}. \tag{5.76}$$

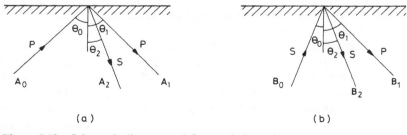

(a)   (b)

Figure 5.18 Schematic diagram to define symbols used in equations (5.73)–(5.80), for (a) incident compression, and (b) incident shear waves at a plane boundary.

In a similar manner a shear wave of amplitude B_0, incident at angle θ_0, is reflected into a compression wave and a shear wave (figure 5.18(b)), of amplitudes B_1 and B_2, and at angles θ_1 and θ_2 respectively. Then

$$\theta_2 = \theta_0 \tag{5.77}$$

$$\frac{\sin \theta_1}{\sin \theta_0} = \frac{c_1}{c_2} = k \tag{5.78}$$

$$B_1 = B_0 \frac{-k \sin 4\theta_0}{\sin 2\theta_0 \sin 2\theta_1 + k^2 \cos^2(2\theta_0)} \tag{5.79}$$

$$B_2 = B_0 \frac{\sin 2\theta_0 \sin 2\theta_1 - k^2 \cos^2(2\theta_0)}{\sin 2\theta_0 \sin 2\theta_1 + k^2 \cos^2(2\theta_0)}. \tag{5.80}$$

For $\theta_0 = 0$, $B_1 = 0$ and $B_2 = -B_0$. Thus a shear wave at normal incidence is reflected as a shear wave of equal amplitude. The phase change of π means in this notation that the direction of displacement is continuous at the boundary. We note that $B_1 = 0$ and $B_2 = B_0$ for $\theta_0 = \frac{1}{4}\pi$, so that an SV wave incident at 45° is reflected purely as an SV wave at the same angle. We also note that θ_0 is only real for $\theta_0 < \theta(\text{critical}) = \sin^{-1}(1/k)$.

For $\theta_0 > \theta(\text{critical})$ the 'reflected' compression (P) wave becomes a surface-skimming compression wave propagating along the boundary, whose amplitude decreases exponentially with depth. This surface P wave also radiates shear-wave energy into the bulk at the critical angle as it propagates. For $\theta_0 > \theta(\text{critical})$ the reflected shear-wave amplitude also becomes complex. The real and imaginary parts are best expressed as the product of a modulus (giving its magnitude) and a phase factor. As θ_0 varies about the critical angle, there are changes in phase of the signal, in a similar way to the changes in phase of the shear-wave radiation from the centre of compression and point force sources which are discussed in §§5.8.1 and 5.8.2 (see especially equations (5.97) and (5.98)).

Finally we note that an SH wave (shear with displacement parallel to the plane of the boundary) is reflected without mode conversion, and that therefore the reflected amplitude is equal to the incident amplitude. There is a phase change of π on reflection.

The free boundary is capable of supporting not only surface waves of a compressional type as discussed above, but also surface-skimming shear waves and, most important of all, Rayleigh surface waves. In a Rayleigh wave, the motion of the particles at the surface is elliptical. Their normal displacement is about 1.5 times their tangential displacement. The displacements decay exponentially with depth into the material. Thus the normal displacement at a depth of one Rayleigh wavelength is reduced to 0.2–0.25 of its value at the surface depending upon Poisson's ratio.

Rayleigh waves propagate parallel to the surface with a velocity given by c_R, where c_R is the root of the following equation:

$$\left(2 - \frac{c_R^2}{c_2^2}\right)^2 - 4\left(1 - \frac{c_R^2}{c_1^2}\right)^{1/2}\left(1 - \frac{c_R^2}{c_2^2}\right)^{1/2} = 0. \tag{5.81}$$

This equation has to be solved numerically, but approximate values for the Rayleigh wave velocity can be obtained from the following expression (Sinclair 1982):

$$c_R = \frac{c_2}{1.14418 - 0.25771v + 0.12661v^2} \tag{5.82}$$

where v is Poisson's ratio. For aluminium ($v = 0.34$), and steel ($v = 0.29$), the Rayleigh-wave velocity is calculated to be 0.9335 and 0.9258 times the shear velocity respectively. There is a special case for $v = 0.25$, when this factor is exactly $2[\sqrt{(3 + \sqrt{3})}]^{-1}$.

Because the Rayleigh waves only propagate in two dimensions across the surface, their geometrical attenuation with distance is less than for bulk waves. Thus the amplitude of circular waves from a point surface source decays as $r^{-1/2}$. Thus they dominate at large distance from the source over all bulk P and S waves (which attenuate as r^{-1}), and also over surface-skimming P and S waves (which attenuate as r^{-2}).

The Rayleigh wave is associated with a singularity in the solution of the wave equation. As will be shown in §5.10.2, the simplest form is for a normal point force, when the time dependence of the Rayleigh arrival takes the form of a square-root singularity. It is thus a more difficult wave arrival to analyse quantitatively than the bulk-wave δ function arrivals. Rayleigh waves have been studied in great detail; for further details see Viktorov (1967).

If transmission through the boundary and/or coupling into a second medium is permitted, the situation becomes even more complicated. In general there will be compression and shear (if the second medium is elastic) waves on both sides of the boundary. If the medium above the boundary is liquid, then there is still wave propagation along the interface, but it now takes the form of what is sometimes called a 'leaky Rayleigh wave'. This is somewhat similar to a Rayleigh wave if the liquid is of a much lower acoustic impedance than the solid, but the surface displacements cause energy to radiate away as compression waves into the liquid at the critical angle for the Rayleigh–compression wave conversion (approximately 30° for aluminium or steel with water). There will also be interface waves when the second medium is a solid, depending upon the elastic properties. Under certain conditions of the densities and velocities in the two media, Stoneley waves are propagated.

When two boundaries are in close proximity, as in a thin parallel-sided plate, the displacements in one surface (due for instance to the propagation of a Rayleigh wave) interact with the other surface. Thus the two surfaces become coupled together to form a new type of wave motion, known in this case as Lamb (or plate) waves. Their propagation characteristics, including velocity, are now partially governed by the plate thickness. Lamb waves are discussed in more detail in §5.10.3.

5.8 RADIATION PATTERNS FOR LASER ULTRASONIC SOURCES

The laser generation of ultrasound takes place at, or close to, the surface of most solids. Boundary conditions and other effects must therefore be taken into account when calculating the radiation each type of source produces. We consider the thermoelastic, ablation and constrained surface sources in turn.

5.8.1 Thermoelastic source

We first neglect any finite extent of the source and any effects due to thermal conductivity, so that the thermoelastic source can be represented as a point dilatation (centre of expansion) at the surface. The far-field radiation pattern has been derived in three separate ways.

The first method, which was used by Hutchins *et al* (1981b), is to represent the source as a combination of surface stresses. As discussed in §5.3, the only non-zero components of stress take the form of two orthogonal dipoles, $D_{11} = D_{22} = B\delta V$, parallel to the surface. In the xz plane, the contribution from D_{22} is effectively zero at large distances from the source (this condition corresponds to $\theta = \frac{1}{2}\pi$ in equation (5.71). The first stage towards calculating the radiation from the dipole D_{11} is to consider the radiation from a single force which is lying tangential to the surface. The radiation pattern in this case is derived both by Miller and Pursey (1954) and Lord (1966) for an infinitely long strip subjected to a sinusoidal force. The angular dependence of the amplitude of the compression wave radiated by this strip is

$$u_r \propto \frac{\sin 2\theta_0 (k^2 - \sin^2 \theta_0)^{1/2}}{(k^2 - 2\sin^2 \theta_0)^2 + 4\sin^2 \theta_0 (1 - \sin^2 \theta_0)^{1/2}(k^2 - \sin^2 \theta_0)^{1/2}}. \quad (5.83)$$

Hutchins *et al* (1981b) use this expression in the first published comparison of theory with experimental directivity pattern data. They point out that, although it is for an unbalanced force, rather than a dipole, there is good correlation with experiment.

Equation (5.83) can be used for ultrasonic waves or pulses of any temporal profile because there is no explicit or implicit dependence upon frequency. Furthermore, this equation can also be extended to a force dipole, by summing the wave field, u_r, due to a positive force at $x_1 + dx_1$ and a negative force at x_1:

$$u_r(x_1 + dx_1) - u_r(x_1) = \frac{du_r}{dx_1} dx_1$$

$$= \left(\frac{du_r}{dr}\right)\left(\frac{dr}{dx}\right)dx_1 \qquad (5.84)$$

where the direction cosine $dr/dx_1 = \sin\theta_0$. Thus the angular directivity of the radial displacement field from the force dipole D_{11} is given by

$$\frac{du_r}{dr}\sin\theta_0 \propto \frac{\sin\theta_0 \sin 2\theta_0 (k^2 - \sin^2\theta_0)^{1/2}}{(k^2 - 2\sin^2\theta_0)^2 + 4\sin^2\theta_0(1 - \sin^2\theta_0)^{1/2}(k^2 - \sin^2\theta_0)^{1/2}}. \qquad (5.85)$$

The expression for the shear-wave directivity for an unbalanced force (see Miller and Pursey (1954) or Lord (1966)) can be used in a similar manner to deduce the equivalent expression for the dipole source as follows:

$$u_\theta \propto \frac{k\sin(4\theta_0)}{k(1 - 2\sin^2\theta_0)^2 + 4\sin^2\theta_0(1 - \sin^2\theta_0)^{1/2}(1 - k^2\sin^2\theta_0)^{1/2}}. \qquad (5.86)$$

Equations (5.85) and (5.86) are shown plotted in figure 5.19.

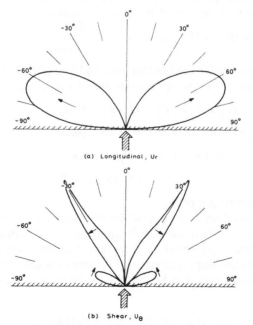

Figure 5.19 Calculated angular dependence (directivity) of (*a*) longitudinal, and (*b*) shear waves generated in aluminium by a pulsed laser source in the thermoelastic regime (Scruby *et al* 1982).

The second derivation of these expressions can be made by representing the thermoelastic source as a centre of expansion which is located at a vanishingly small distance below the surface. The source itself only generates compression waves. However, interactions with the boundary generate compression- and shear-wave fields, both of which we wish to calculate. The interaction of spherical wavefronts radiated by the source with the nearby surface is a relatively difficult problem, but one which has been solved directly by Rose (1984), using an integral transform method.

In the third method, the seismic reciprocity principle (Knopoff and Gangi 1959) is invoked. The reciprocal problem is somewhat easier to solve if only the wave field at infinity is required. Using this principle, the source and receiver positions can be interchanged. However, considerable care has to be taken, especially when spatial derivatives are involved. The mathematical formulation uses the Green's function notation. In the case of a point force F_j acting at the point x' at time t', the displacement vector u_i at point x at time t is given by (assuming the summation convention)

$$u_i = G_{ij}(x, t; x', t')F_j. \tag{5.87}$$

In this formulation of the Green's function it is assumed that the source has δ-function time dependence. By way of example, one Green's function for an unbounded solid was given in equation (5.70). Using the above notation, and excluding the first, near-field term in this equation:

$$G_{ij}(r, t; 0, 0) = \frac{\gamma_i \gamma_j}{4\pi\rho c_1^2 r} \delta(t - r/c_1) + \frac{\delta_{ij} - \gamma_i \gamma_j}{4\pi\rho c_2^2 r} \delta(t - r/c_2). \tag{5.88}$$

Applying reciprocity (equation 7 of Burridge and Knopoff (1964)), the Green's function in equation (5.87) is equal to its reciprocal, thus:

$$G_{ij}(x, t; x', t') = G_{ji}(x', -t'; x, -t). \tag{5.89}$$

Thus if the force operates at time $t' = 0$, this becomes

$$G_{ij}(x, t; x', 0) = G_{ji}(x', 0; x, -t)$$

$$= G_{ji}(x', t; x, 0) \tag{5.90}$$

by shifting the time origin by t.

The force is now at the point x and the displacement at the point x'. The Green's function (defined for instance in equation (5.88)) is the same, meaning that the displacement measured at x' at time t in response to a force at x will be the same as the displacement at x at time t for the same force source, but at x'. The Green's function for the dilatation source is expressed in this notation by

$$G^{\mathrm{H}}_{ij,j'}(x, t; x', 0) = G^{\mathrm{H}}_{ji,j'}(x', t; x, 0) \tag{5.91}$$

i.e. differentiating both sides of equation (5.90) with respect to x'_j, which is represented by the subscript j' following the comma (for a further explanation see Scruby (1985)). The source is now assumed to have Heaviside (step) function time dependence in the calculation of the Green's function, which is represented by the superscript H.

Note that, when the source and receiver positions are reversed, differentiation still takes place at the same point in space (in this case at x'). Differentiation in the left-hand side of equation (5.91) implies that the source is comprised of a set of three spatial derivatives of forces, i.e. three force dipoles at point x'. Differentiation in the right-hand side (the reciprocal case) is now of the displacement u_i, to yield its derivative which is represented by a strain. In general the strain tensor is given by

$$\varepsilon_{ij} = \tfrac{1}{2}(\partial u_i/\partial x_j + \partial u_j/\partial x_i). \tag{5.92}$$

In the present case, differentiation yields the dilatational strain:

$$\partial u_j/\partial x_j = \varepsilon_{jj}. \tag{5.93}$$

In the reciprocal problem the source, which was a sum of dipoles, is no longer differential in the reciprocal, and is simply a force F_i. Thus the recpirocal problem of finding the displacement at infinity of a set of force dipoles (i.e. a centre of expansion) at a point vanishingly close to the surface of a half space, is to calculate the dilatational strain at the same point in response to plane waves impinging upon the surface from a force source at infinity.

First consider P waves generated by this force. At a point close to the surface, the dilatational strain is given by the sum of the direct compression wave (assumed to have unit amplitude), and a P wave reflected from the nearby surface. We need not consider the mode-converted SV wave because it does not generate dilatational strain. We can use the plane-wave reflection coefficients because the boundary conditions are satisfied exactly for the sum of the incoming wave plus the reflected and mode-converted waves. Substituting the reflection coefficient from equation (5.75) and setting $A_0 = 1$, we obtain

$$u_r(\theta_0) \propto 1 + \frac{\sin 2\theta_0 \sin 2\theta_2 - k^2 \cos^2(2\theta_2)}{\sin 2\theta_0 \sin 2\theta_2 + k^2 \cos^2(2\theta_2)} \tag{5.94}$$

$$= \frac{2 \sin 2\theta_0 \sin 2\theta_2}{\sin 2\theta_0 \sin 2\theta_2 + k^2 \cos^2(2\theta_2)}. \tag{5.95}$$

Substituting $\cos\theta_0 = (1 - \sin^2\theta_0)^{1/2}$, $\cos\theta_2 = (1 - \sin^2\theta_0)^{1/2}$, $\sin\theta_2 = \sin\theta_0/k$ yields the same angular dependence as equation (5.85). A rigorous derivation of the P-wave directivity using reciprocity, but based on Gauss's theorem, is made by Kino and Stearns (1985). The various methods outlined above all produce the same analytical expressions for the directivity.

Reciprocity can be used similarly to deduce the shear-wave directivity for the thermoelastic source. Since in the reciprocal case we need to calculate a purely dilatational strain immediately below the surface, we must only include the mode-converted P wave generated by an SV wave incident on the surface. We omit both the SV wave reflected from the surface and the direct SV wave because they generate shear strain. Thus the shear-wave directivity is given directly by the expression for the reflection of shear into mode-converted compression waves (equation (5.79)):

$$u_\theta(\theta_0) \propto \frac{\sin 4\theta_0}{\sin 2\theta_0 \sin 2\theta_2 + k^2 \cos^2(2\theta_2)}. \tag{5.96}$$

Substituting $\cos\theta_0 = (1 - \sin^2\theta_0)^{1/2}$, $\cos\theta_1 = (1 - \sin^2\theta_1)^{1/2}$, $\sin\theta_1 = k\sin\theta_0$ yields the same angular dependence as equation (5.86).

Equation (5.85) has $u_r(\theta, t)$ real for $|\theta_0| \leqslant 90°$, indicating that the phase of the compression-wave field remains zero throughout. Thus the time dependence of a short ultrasonic pulse remains the same for all angles, e.g. $u_r(\theta, t) \sim \delta(t - r/c_1)$ for all θ. By way of contrast, however, $u_\theta(\theta)$ is only real for $|\theta_0| \leqslant \sin^{-1}(1/k)$ (equation (5.86)). This is the critical angle, which is approximately $30°$ in aluminium. Thus it has zero phase in this region, and $u_\theta(\theta, t) \sim \delta(t - r/c_2)$. However, $u_\theta(\theta, t)$ becomes complex for $|\theta_0| > \sin^{-1}(1/k)$. The imaginary (i.e. phase shifted by $\frac{1}{2}\pi$) counterpart of the function $\delta(t)$ is the function $\Pi(t) = (\pi t)^{-1}$. Thus, if $|\theta_0| > \sin^{-1}(1/k)$,

$$u_\theta(\theta, t) = \mathrm{Re}[u_A \delta(t - r/c_1) + iu_B \Pi(t - r/c_2)] \tag{5.97}$$

where

$$u_A + iu_B \propto \frac{k\sin 4\theta_0}{k(1 - 2\sin^2\theta_0)^2 + 4\sin^2\theta_0(1 - \sin^2\theta_0)^{1/2}(1 - k^2\sin^2\theta_0)^{1/2}}. \tag{5.98}$$

This equation shows that the phase varies continuously when θ_0 exceeds the critical angle. This in turn implies that the shape of the ultrasonic waveform also varies. For a short-pulse source, the waveform will approximate to $\delta(t)$ for θ_0 just greater than the critical angle. However, for θ_0 close to $45°$, the phase approximates to $\frac{1}{2}\pi$, so that the waveform in turn approximates to $\Pi(t)$. This is a bipolar function, which switches between $-\infty$ and $+\infty$ at $t = 0$. Thus the waveform appears as a bipolar pulse. In intermediate regimes between $\sin^{-1}(1/k)$ and $90°$ the waveform will be an unsymmetrical bipolar pulse.

Experimental data (Hutchins et al 1981b) for the angular directivity of the compression-wave field of the thermoelastic source are shown in figure 5.20. These authors also obtained data for the shear directivity using a compression-wave transducer (presumably identifying the signal as shear by its arrival time), but it is more difficult to compare this with the theory since the measurement was of the perpendicular displacement, and the above equations are for the tangential displacement. Cooper (1985) measured the shear directivity from the thermoelastic source using a piezoelectric shear-wave transducer, and obtained reasonable agreement with the theory.

Hutchins et al (1981b) showed that the agreement between experiment and theory was best when the diameter of the source was less than the acoustic wavelength corresponding to the characteristic frequency of the transducer.

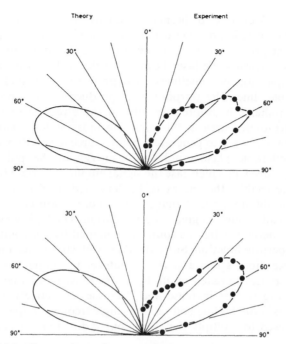

Figure 5.20 Comparison of experimental data (Hutchins *et al* 1981b) with calculation for a thermoelastic laser source, with (*a*) a 1 mm aperture beam, and (*b*) a 3 mm aperture beam (1 MHz piezoelectric receiver on 50 mm radius hemisphere).

They investigated the effect of increasing the source diameter, and found that additional lobes began to appear, due to interference between waves originating from different parts of the source. Theoretically, finite source size could be taken into account for a single frequency (a reasonable approximation for the piezoelectric receivers used) by multiplying equations (5.85) and (5.86) by an aperture function. For a source with cylindrical symmetry, this function is (Miller and Pursey 1954)

$$\Upsilon(\theta_0) = \frac{J_1(2\pi f a \sin \theta_0/c_1)}{2\pi f a \sin \theta_0/c_1} \tag{5.99}$$

where a is the source radius, f is the frequency and $J_1(x)$ is a first-order Bessel function. This function consists of a broad, large-amplitude central maximum, surrounded by smaller amplitude rings, which get progressively smaller as θ increases. These rings manifest themselves as sidelobes in the ultrasonic data.

The experimental data (figure 5.20) do not show the zero predicted for $\theta_0 = 0$ in the compression-wave radiation pattern, even for the smaller source diameter. The non-zero signal could be due to a combination of factors, including a small component still due to source size, near-field terms in the wave field, and effects due to thermal conductivity, which have so far been neglected. The relative importance of these factors will be discussed in §5.9.1.

An important practical thermoelastic source is the line source, produced for example by focusing the laser beam onto the surface with a cylindrical lens. Beam expansion is sometimes used prior to the cylindrical lens to produce a longer line. The purpose is to produce a directional radiation pattern with a sharp, high-intensity lobe in the direction of interest. The width of the irradiated area in this direction must be less than 1 mm. The strong focusing necessary to generate a point source of this dimension would generate considerable plasma unless the energy in the pulse were also considerably reduced, in which case the signal amplitude would be very small. The line source enables the energy to be deposited over a sufficiently large area to prevent ablation, but so that the width of the source in the direction of interest is small enough to give good resolution. The bulk-wave directivity pattern of the thermoelastic line source is obtained by convolution with the appropriate aperture function. Since the thermoelastic source must still have two equal force dipoles D_{11} and D_{22}, whatever its shape, the areas under the ultrasonic pulses radiated in all directions must remain constant, while their heights and widths increase and decrease respectively as the source width decreases. For the line source, the single-frequency aperture function that must be used to multiply the point source directivity is (Lord 1966)

$$\Upsilon(\theta_0) = \frac{\sin(2\pi f b \sin \theta_0 / c_1)}{2\pi f b \sin \theta_0 / c_1} \tag{5.100}$$

where $2b$ is the breadth of the line source. Hutchins et al (1981) present directivity data for the thermoelastic line source. Normal to the line, it has a similar directivity to the point source.

5.8.2 Ablation source

Let us first assume that when ablation occurs, the thermoelastic stresses (which would always accompany ablation in practice) are so small that they can be neglected. Thus, as discussed in §5.4.3, ablation can be represented simply as a time-varying force acting normal to the surface. The approximation to a point source is likely to be a good one in most situations, since ablation is generally produced by focusing the laser pulse to very small dimensions.

The radiation pattern from a normal force excitation is derived in Miller and Pursey (1954) and Lord (1966) for sinusoidal time dependence. It was first applied to the ablation source by Hutchins et al (1981b). Because there is again no frequency dependence in the expressions for the directivity, they can be applied to short ultrasonic pulses. The angular dependences of the compression and shear waves for the ablation source are shown in figure

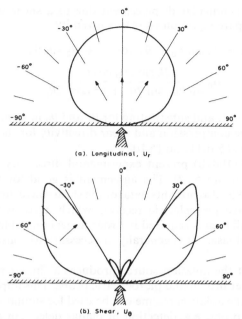

(a). Longitudinal, U_r

(b). Shear, U_θ

Figure 5.21 Calculated angular dependence (directivity) of (a) longitudinal, and (b) shear waves generated in aluminium by a pulsed laser source in the ablation regime (Scruby *et al* 1982).

5.21, and are given respectively by

$$u_r(\theta_0) \sim \frac{2k^2 \cos \theta_0 (k^2 - 2\sin^2 \theta_0)}{(k^2 - 2\sin^2 \theta_0)^2 + 4\sin^2 \theta_0 (1 - \sin^2 \theta_0)^{1/2}(k^2 - \sin^2 \theta_0)^{1/2}} \quad (5.101)$$

$$u_\theta(\theta) \sim \frac{\sin 2\theta_0 (1 - k^2 \sin^2 \theta_0)^{1/2}}{k(1 - 2\sin^2 \theta_0)^2 + 4\sin^2 \theta_0 (1 - \sin^2 \theta_0)^{1/2}(1 - k^2 \sin^2 \theta_0)^{1/2}}. \quad (5.102)$$

These expressions can also be derived using the reciprocity approach used in §5.8.1. The reciprocal problem to calculating the compression and shear radiation patterns is to derive the normal displacements at a point on the surface generated in turn by plane compression and shear waves incident at that surface. This normal displacement is readily calculated by summing the resolved displacements in the incident wave, the reflected wave and the mode-converted wave. Thus, using the notation of equations (5.75) and (5.76), the perpendicular displacement due to a compression wave incident at θ_0 (compare equation (4.12)) is given by

$$u = A_0 \cos \theta_0 - A_1 \cos \theta_0 + A_2 \sin \theta_2$$

$$= A_0 \frac{2k^2 \cos 2\theta_2 \cos \theta_0}{\sin 2\theta_0 \sin 2\theta_2 + k^2 \cos^2 2\theta_2}. \quad (5.103)$$

Similarly, the perpendicular displacement due to a shear wave incident at an angle θ_0 (compare equation (4.14)) is given by

$$u = B_0 \sin \theta_0 - B_1 \cos \theta_1 + B_2 \sin \theta_0$$

$$= B_0 \frac{2k \sin 2\theta_0 \cos \theta_1}{\sin 2\theta_0 \sin 2\theta_1 + k^2 \cos^2 2\theta_0}. \tag{5.104}$$

Putting $A_0 = B_0 = 1$, and substituting for θ_1 and θ_2 in terms of θ_0 yields the expressions for the compression and shear directivity for the ablation source given in equations (5.101) and (5.102).

Hutchins *et al* (1981b) present experimental directivity patterns for the ablation source (figure 5.22). The agreement is good for the compression wave. Cooper (1985) also presents data for the shear-wave directivity obtained using a shear-wave piezoelectric receiver which agree with theory. Since focusing to a small area is required in order to generate ablation except with the most powerful lasers, it is generally unnecessary to consider finite source size effects.

It is noted that the ablation source produces a similar radiation pattern to a small-diameter broadband piezoelectric probe. This implies that a pulsed laser working in the ablation regime can be used for similar measurements to the piezoelectric probe, e.g. detection of planar defects in a solid specimen that are parallel to the surface. This is in contrast to the thermoelastic source, which produces a quite different directivity pattern, which cannot readily be reproduced by a piezoelectric probe. The thermoelastic source is likely to be of most value when an angled ultrasonic beam is required. These points will be discussed further in Chapter 6.

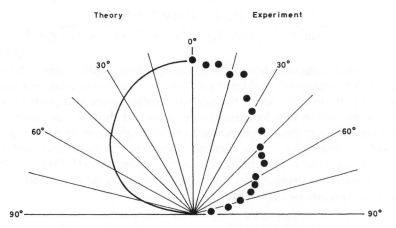

Figure 5.22 Comparison of experimental data (Hutchins *et al* 1981b) with calculation for a laser-induced ablation source.

5.8.3 Constrained surfaces

The compression-wave directivity for a laser pulse incident on a metal surface constrained, firstly with a glass slide (figure 5.23), and secondly with a thin coating of oil, has been measured by Hutchins *et al* (1981b) and Cooper (1985). The radiation pattern in each case resembles the ablation source, rather than the thermoelastic source. This is to be expected, because every type of surface constraint must add a perpendicular stress to the source, which is otherwise absent in a pure thermoelastic source.

If the surface is coated with a material such as oil, which is vaporized by the laser, then there will be an additional unbalanced force perpendicular to the surface (figure 5.15(*a*)) and thus the same directivity as from ablation of metal might be expected. For surface constraint alone, i.e. a change of boundary conditions without any additional thermal effects such as vaporization or fluid expansion, the addition of a force dipole perpendicular to the surface might be expected (figure 5.14(*b*)).

If a metal specimen is covered by a very thick layer which is totally transparent to the laser pulse, but which has identical elastic properties to the metal, then the source should resemble a buried centre of expansion. There should be three equal orthogonal dipoles as discussed in §§5.5.2 and 5.6.1, no shear-wave field, and a spherically symmetrical compression-wave field. If the transparent layer has different elastic properties from the metal, the force dipole perpendicular to the surface of the metal will be unequal to the other two dipoles. Furthermore, the source will be subject to boundary conditions caused by the interface. If we can assume that the perpendicular dipole is much larger than the other two dipoles, then the directivity pattern for this buried single dipole should be as shown in figure 5.23. This figure

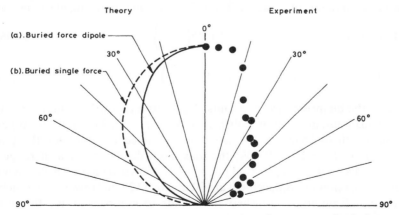

Figure 5.23 Comparison of experimental data for a constrained source (Hutchins *et al* 1981b) with directivity of buried force dipole and buried single force.

shows that agreement with the experimental data for surface constraint by a layer of glass is better for the dipole than for an unbalanced force (also shown).

Comparison of figure 5.23 with figures 5.19(a) and 5.20 explains the large increase in the compression-wave amplitude at the epicentre ($\theta_0 = 0$) when the surface is constrained in the thermoelastic regime (Von Gutfeld and Melcher 1977, Hutchins *et al* 1981a). The effect of surface constraint is to redistribute the ultrasonic energy in the directivity pattern and concentrate it into the perpendicular direction. The theory predicts an infinitely large increase in the far-field compression-wave amplitude, because the value for the thermoelastic source is zero at $\theta_0 = 0$. In practice the increase is large, but not infinite, because the thermoelastic amplitude at $\theta_0 = 0$ is not zero due to small near-field terms.

5.8.4   Comparison of far-field wave amplitudes

We shall now compare the amplitudes of the ultrasonic wave fields generated by each of the types of source situated on the surface of a half space, as considered above. First, for the thermoelastic source, the strength of the far-field compression waves is given by combining equation (5.69), which predicts the wave field in an unbounded solid, with equation (5.95), which takes into account the effect of the boundary on the far-field radiation:

$$u_r^{\text{Th}}(\theta) = \left(\frac{2 \sin 2\theta_0 \sin 2\theta_2}{\sin 2\theta_0 \sin 2\theta_2 + k^2 \cos^2(2\theta_2)} \right) \left(\frac{3\alpha B}{4\pi(\lambda + 2\mu)\rho C c_1 r} \right) \delta E' \quad (5.105)$$

where $\delta E' = \mathrm{d}(\delta E)/\mathrm{d}t$ is the absorbed optical power, and $\sin \theta_2 = \sin \theta_0 / k$. The maximum value of the trigonometrical function is 0.6 and occurs at $\theta_0 = 64°$ for Poisson's ratio $\nu = \frac{1}{3}$ (e.g. aluminium). Thus, substituting values of the materials constants (table 5.2) we obtain the maximum displacement as

for aluminium $\qquad u_{\text{max}}^{\text{Th}} \simeq 1.8 \times 10^{-16} \left(\frac{\delta E'}{r} \right)$ $\qquad\qquad\qquad$ (5.106a)

for steel $\qquad u_{\text{max}}^{\text{Th}} \simeq 5 \times 10^{-17} \left(\frac{\delta E'}{r} \right)$ $\qquad\qquad\qquad$ (5.106b)

where all the quantities are in SI units. Thus, for an absorbed energy of 6 mJ (e.g. 100 mJ pulse incident on polished aluminium with 94 % reflectivity) and a pulse length of 25 ns, to give a power (assumed constant for the pulse duration) of $\delta E' = 2.4 \times 10^5$ W, the maximum displacement at a range of 100 mm is 4.4×10^{-10} m, i.e. 0.44 nm. 6 mJ of absorbed energy in steel would generate a maximum displacement at 100 mm of 0.12 nm. However, in practice the lower reflectivity of polished steel tends to compensate for the lower value of $3\alpha/\rho C$.

If, as commonly happens, the measurements are made at a surface (e.g. a large hemisphere) which is locally normal to the incident P waves, then the calculated amplitudes must be multiplied by two to take into account

Table 5.2 Materials constants for aluminium and mild steel from Kaye and Laby (1986), estimated from values for iron if steel data not available, or calculated from other constants in the table.

Symbol	Parameter	Aluminium	Mild steel
ρ	Density	2700 kg m^{-3}	7900 kg m^{-3}
μ	Shear elastic modulus	26.1 GPa	82.2 GPa
B	Bulk elastic modulus	75.5 GPa	169.2 GPa
λ	Lamé elastic constant	58.1 GPa	114.4 GPa
c_1	Compression velocity	6400 m s^{-1}	5960 m s^{-1}
c_2	Shear-wave velocity	3150 m s^{-1}	3240 m s^{-1}
c_R	Rayleigh velocity	2940 m s^{-1} .	3000 m s^{-1}
v	Poisson's ratio	0.345	0.291
k	Velocity ratio, c_1/c_2	2.03	1.84
α	Linear expansion	$2.31 \times 10^{-5} \text{ K}^{-1}$	$1.07 \times 10^{-5} \text{ K}^{-1}$
C	Thermal capacity (273 K)	$880 \text{ J kg}^{-1}\text{K}^{-1}$	$480 \text{ J kg}^{-1}\text{K}^{-1}$
C	Thermal capacity (773 K)	$1130 \text{ J kg}^{-1}\text{K}^{-1}$	$670 \text{ J kg}^{-1}\text{K}^{-1}$
C	Thermal capacity liquid	$1175 \text{ J kg}^{-1}\text{K}^{-1}$	$824 \text{ J kg}^{-1}\text{K}^{-1}$
T_f	Melting point	934 K	1810 K
T_v	Boiling point	2790 K	3030 K
L_f	Latent heat of fusion	397 J kg^{-1}	247 J kg^{-1}
L_v	Latent heat of vaporization	$10\,778 \text{ J kg}^{-1}$	6258 J kg^{-1}
$3\alpha/\rho C$	Thermoelastic coupling	$2.9 \times 10^{-11} \text{ m}^2\text{s}^{-1}$	$8.5 \times 10^{-12} \text{ m}^2\text{s}^{-1}$

boundary conditions (i.e. equation (4.12) with $\theta = 0$). Thus the maximum displacement at 100 mm from 6 mJ absorbed energy is ~ 0.9 nm in aluminium.

Next, we calculate the P-wave field due to the ablation source at the surface of a half space. This time we must combine equation (5.71) for the unbounded solid with equation (5.103) for the boundary effects:

$$u_r^{Ab}(\theta) = \left(\frac{2k^2 \cos 2\theta_2 \cos \theta_0}{\sin 2\theta_0 \sin 2\theta_2 + k^2 \cos^2 2\theta_2} \right)\left(\frac{F}{4\pi\rho c_1^2 r} \right) \qquad (5.107)$$

where θ_2 is defined as before and $F(t)$ is the ablative force. The maximum value of the angular dependence is 2 and occurs at $\theta_0 = 0$. Thus

for aluminium $\qquad u_{max}^{Ab} \simeq 1.5 \times 10^{-12}\left(\dfrac{F}{r}\right) \qquad (5.108a)$

for steel $\qquad u_{max}^{Ab} \simeq 5.7 \times 10^{-13}\left(\dfrac{F}{r}\right). \qquad (5.108b)$

In §5.4, ablative forces of ~ 5 N to ~ 50 N were estimated in both materials depending upon source dimensions. In the latter case the P-wave displacement at 100 mm from the source in a half space would be 0.75 nm in aluminium or 0.3 nm in steel. If the measurement is made at a second surface which is perpendicular to the direction of maximum amplitude (e.g. a parallel-sided plate), then a further factor of 2 must be applied, to give a maximum displacement in aluminium of 1.5 nm. For a peak force of 8 N, as deduced by Dewhurst *et al* (1982) from their experimental data, the maximum displacement should thus be 0.24 nm.

We note therefore that under typical experimental conditions, the *maximum* compression amplitude in the radiation pattern is of a similar order of magnitude for both types of source. The reason that the thermoelastic source frequently appears weaker than ablation is either because it is measured well away from its maximum (e.g. at the epicentre), or because source dimension effects broaden the P-wave pulse, thereby reducing its height. This latter effect is unlikely to be as important for ablation since a greater degree of focusing is usually employed.

Wave-field amplitudes are more difficult to estimate for constrained or otherwise modified surfaces, due to the complexity of the physical processes that may take place. Assuming that surface constraints only change the boundary conditions at the source, the chief effect is to change the directivity pattern, and redirect the energy away from its previous maximum at $\sim 60°$ towards the normal to the surface. There is only likely to be a small change in *maximum* amplitude in the radiation pattern under these conditions.

When other processes occur in the source, such as expansion of fluid in contact with the surface, or ablation of fluid, there is likely to be a considerable enhancement of compression amplitude in the wave field in addition to boundary-change effects. In the case of fluid ablation, the maximum amplitude could be considerably greater than for metal ablation, depending upon factors such as the volatility of the fluid.

5.9   ULTRASONIC BULK WAVEFORMS IN PLATES

The parallel-sided plate is the simplest body in which to study ultrasonic waveforms generated by a pulsed laser. The majority of reported measurements have been carried out with the source and receiver directly opposite one another, so that the direction of wave propagation of interest is normal to the plate surfaces. In this position, the receiver is said to be at the epicentre of the source. The reasons for the choice of orientation are fairly evident. Firstly, because the wavefronts are approximately parallel to the surface (neglecting wavefront curvature), broadband measurements of the compression waveforms can be made with a source and receiver of finite dimensions, since pulse broadening due to source or receiver will be minimized. Secondly, because of the high symmetry in this orientation, it is easier to calculate the form of the wave arrivals, and hence interpret the experimental data. Thirdly, many ultrasonic applications make use of measurements made perpendicular to the surface of a plate, either in transmission or pulse-echo.

The epicentre is, however, not a good position in which to investigate the shear-wave field from a cylindrically symmetrical source, even if a broadband shear-wave transducer were readily available. It also transpires that the compression-wave field is different in form at the epicentre than at other orientations because the far-field thermoelastic term is zero here.

5.9.1 Thermoelastic waveforms at epicentre

As part of a comprehensive study of epicentral waveforms in the thermoelastic and ablation regimes (Scruby *et al* 1980, Dewhurst *et al* 1982), a calibrated capacitance transducer was used to measure the waveforms, which were then recorded either digitally (figure 5.24) or else photographically. Although a wide range of lasers could be used (see Aindow *et al* 1982), the results presented here are all for a laboratory pulsed laser, a Q-switched neodymium–YAG system, capable of delivering single pulses $\leqslant 100$ mJ at 1.06 μm wavelength. The pulse duration (i.e. full width at half height, FWHH) was in the range 20–30 ns, and typically 24 ns. The specimen was generally a metallic disc (e.g. aluminium alloy or mild steel) of variable thickness.

Figure 5.25(a) shows the waveform generated by an incident pulse of 31 mJ, using a 4 mm diameter aperture, and irradiating a 25 mm aluminium disc. Apart from a small positive initial pulse, the whole waveform is negative, implying that the surface at the receiver is depressed inwards. Hutchins and Nadeau (1983) have confirmed this waveform using a laser interferometer receiver. Intuitively, a rise of the surface would be expected from a centre of expansion, and this is indeed the case for all orientations except the epicentre. The epicentre constitutes a special case, which arises because the plane stress boundary conditions ($\sigma_{13} = \sigma_{23} = \sigma_{33} = 0$) cause a cancellation of the far-field term in the compression-wave field, as demonstrated by the directivity pattern of figure 5.19(a). Thus the waveform of figure 5.25(a) must be due to near-field terms, as will be shown below.

The normal displacement to the surface (u_3) at the epicentre can be determined exactly if the thermoelastic source is represented as a centre of

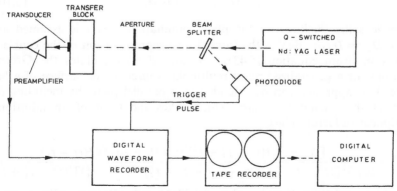

Figure 5.24 Apparatus to investigate laser-generated ultrasonic waveforms at the epicentre. Piezoelectric or capacitance transducers may be used, and signals may be recorded either digitally (with a tape recorder link or direct) or by using an oscilloscope (Scruby *et al* 1980, Dewhurst *et al* 1982).

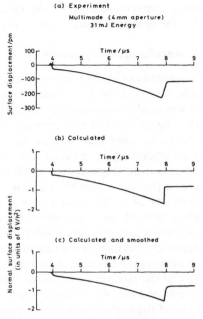

Figure 5.25 Comparison of experimental epicentre waveform (Scruby *et al* 1980) acquired in the thermoelastic regime, using the apparatus of figure 5.24, with theory, based on equation (5.110).

expansion located at a point vanishingly close to the surface of a plate. The approach is to calculate the Green's function for a parallel-sided plate, defined by the following equation (using the same notation as equation (5.91)):

$$u_3 = G_{3j,j}^{\mathrm{H}}(h, t; 0, 0)D_{jj} \tag{5.109}$$

where h is the thickness of the plate, summation over j is assumed and $D_{11} = D_{22} = D_{33} = B\delta V$ is the force dipole representation of the centre of expansion (from equation (5.43)). Sinclair (1979) calculates the Green's functions for a generalized point multipolar source at the surface of a half space. This approach can be extended to a parallel plate, by including the effect of the second boundary which takes the form of an additional multiplicative factor. Thus

$$u_3(t) = \frac{[X_1 + Y_1 t(t^2 - t_1^2 + t_2^2)^{1/2}](3\lambda + 2\mu)\delta V H(t - t_1)}{6\pi\mu h^2[(t^2 - t_1^2 + \frac{1}{2}t_2^2)^2 - t(t^2 - t_1^2 + t_2^2)^{1/2}(t^2 - t_1^2)]^3(t^2 - t_1^2 + t_2^2)}$$

$$+ \frac{[X_2 + Y_2 t(t^2 - t_2^2 + t_1^2)^{1/2}](3\lambda + 2\mu)\delta V H(t - t_2)}{6\pi\mu h^2[(t^2 - \frac{1}{2}t_2^2)^2 - t(t^2 - t_2^2 + t_1^2)^{1/2}(t^2 - t_2^2)]^3(t^2 - t_2^2 + t_1^2)}$$

$$\tag{5.110}$$

where $t_1 = h/c_1$ and $t_2 = h/c_2$ are the compression- and shear-wave arrival times respectively, and X_1 and Y_1 are defined as follows:

$$X_1 = y(y + w)t_1^2[5y^4 + (10u + 9v)y^3 + (5u^2 + 18uv + 4v^2)y^2$$
$$+ uv(9u + 8v)y + 4u^2v^2] - (2y + w)t_1^2[2y^5 + 4(u + v)y^4$$
$$+ 2(u^2 + 4uv + v^2)y^3 + 4uv(u + v)y^2 + 2u^2v^2y]$$

$$Y_1 = -y(y + w)t_1^2[5y^3 + (7u + 6v + 2w)y^2 + (4wv + 8uv + 6uw - 3w^2)y$$
$$- 2vw^2 + 8uvw - uw^2] + (2y + w)t_1^2[2y^4 + 2(u + v + 2w)y^3$$
$$+ 2(uv + 2uw + 2vw + w^2)y^2 + 2w(2uv + uw + vw)y + 2uvw^2]$$

and where $y = t^2 - t_1^2$, $u = t_1^2$, $v = t_2^2$ and $w = \frac{1}{2}t_2^2$.

The coefficients X_2 and Y_2 are defined by $X_2 = -X_1$, $Y_2 = -Y_1$, where this time $y = t^2 - t_2^2$, $u = t_2^2$, $v = t_1^2$, and $w = \frac{1}{2}t_1^2$.

Figure 5.25(b) shows the result of using equation (5.110) to calculate the displacement in a 25 mm aluminium block, as in the experiment described earlier in this section, which provided the data of figure 5.25(a). The theoretical waveform given by equation (5.110) was also convolved with a Gaussian broadening function (figure 5.25(c)), to simulate the effect of both laser pulse width and limited bandwidth in the recording system: hence the fact that the P and S arrivals are smoothed, rather than the sharp steps predicted by the theory for a source with $H(t)$ time dependence. The agreement with experiment is indeed very good. There is just one exception: there is a small positive pulse at the commencement to the waveform which is not predicted by the theory. Leaving this point, which will be discussed later in this section, theory supports the experimental observation that the compression-wave arrival is a negative step, rather than the positive signal that might have been be expected for a centre of expansion.

The explanation of this theoretical result lies in the boundary conditions at the source surface. The centre of expansion can be supposed to be buried a small depth, δz, below the surface, in which case the waveform would have the form of figure 5.26(b). The negative pulse arises because the positive pulse term of the half-space solution (figure 5.26(a)) has added to it a reflection in the nearby surface which reverses the polarity of the pulse. As the source depth δz is decreased, the positive pulse (the far-field term) interferes with the reflected pulse, and is eventually cancelled out by it, to leave only the step-function term (figure 5.26(c)).

The magnitude of this compression-wave step can be calculated from equation (5.110) by substituting $t = t_1$, thus:

$$u_{3p} = \frac{8B\delta V}{\pi\mu h^2 k^5} \qquad \text{where } k = c_1/c_2. \qquad (5.111)$$

If $v = \frac{1}{3}$, then $k = 2$ and $B = \frac{8}{3}\mu$, so that

$$u_{3p} = \frac{2\delta V}{3\pi h^2}. \qquad (5.112)$$

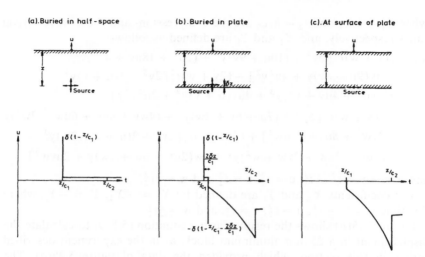

Figure 5.26 Calculated epicentre waveform for centre of expansion (point dilatation) situated (*a*) in a half space, (*b*) just below one surface of a plate, and (*c*) at one surface of a plate.

The strength of the shear arrival, which appears as positive step, is given by substituting $t = t_2$ in equation (5.110) thus:

$$u_{3s} = k^2 u_{3p} = \frac{8\delta V}{3\pi h^2} \qquad \text{for } \nu = \tfrac{1}{3}. \qquad (5.113)$$

We observe from these two expressions that the amplitude of the whole epicentre waveform falls with the square of the distance from the thermoelastic source to the receiver. It must therefore be comprised of near-field terms, since the far-field pulse (as calculated e.g. in equation (5.106)) is inversely proportional to the distance. It is understandable that the perpendicular displacement at the epicentre due to the far-field shear arrival should only be a second-order term, but it is more surprising that this is also true of the compression-wave arrival.

We now discuss the small positive compression pulse observed on the experimental data, but absent from the theoretical data (figure 5.25). It could in principle be due to a number of distinct mechanisms, as follows.

(1) Radiation pressure

This would be expected to generate a positive pulse. Taking a mean force of 0.01 N for 31 mJ incident energy estimated as in §5.4.1, and substituting it in the formula for the epicentre displacement due to a point force (see §5.9.2, equation (5.116)) yields a displacement of $\sim 10^{-12}$ m. This is almost an order of magnitude smaller than the pulse observed in figure 5.25(*a*).

(2) Residual ablation

If the surface is covered with dirt, grease or oxide, there could be a small amount of ablation, generating a positive pulse (as discussed in the next section). However, the small positive pulse in question is observed to occur consistently after many laser shots, by which time the surface should have become completely clean. If the incident laser power density is high, so that it is operating only just within the thermoelastic regime, then local 'hot spots' might cause small amounts of metal ablation. This, however, would be expected to vary somewhat from shot to shot (because of variations in the multimode beam profile) and also with laser energy, whereas the observation is that the pulse amplitude does not vary at constant energy, and is also approximately proportional to incident energy.

(3) Finite source and receiver dimensions

The compression wave only takes the form of a step at the epicentre: at other orientations it is a positive pulse, so that the effect of finite source and/or receiver dimensions might be expected to incorporate an off-epicentre pulse term into the compression-wave arrival. To investigate this effect, calculations and experimental measurements were made, in which a 1 mm diameter point-contact receiver was systematically moved across the surface of a 50 mm aluminium plate away from the epicentre. The ultrasonic source was a 1 mm diameter Nd:YAG laser pulse in the thermoelastic regime. Figure 5.27(a) shows calculations for the far-field P- and SV-wave pulses away from the epicentre. The values for P waves are obtained by taking the source function (equation (5.105)), substituting $r = h \sec \theta$ for the propagation path, and multiplying this by a receiver function for P waves at oblique incidence to the surface (equation (4.12) or (5.103)). A corresponding expression can be derived for SV waves. For small angles θ_0, the geometrical parts of both expressions tend to vary with θ_0 as $\sin^2 \theta_0 \to x^2/h^2$. Figure 5.27($a$) confirms that, for small x, the pulse magnitude should increase with the square of the distance from the epicentre.

In the experimental data (figures 5.28 and 5.27(b)) the shear arrival pulse decreases to zero as the epicentre is approached, as predicted by the calculations. However, the compression-wave arrival initially decreases as predicted, but then reaches a constant small value. This suggests that the initial P-wave pulse which is the subject of discussion is a separate effect from the inclusion of off-epicentre terms.

In a further experiment, the effect of changing the diameter of the irradiated area in the thermoelastic source was investigated, both for mild steel and aluminium. It can be seen (figure 5.29) that in neither case does increasing the diameter from 1 to 9 mm cause an increase in the initial pulse, as might be expected if off-epicentre terms were the cause; rather, its amplitude tends to decrease slightly due to pulse broadening.

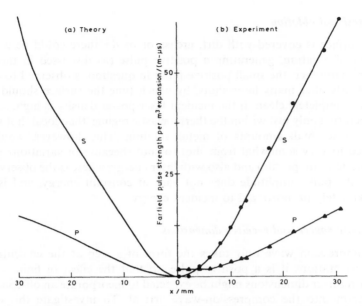

Figure 5.27 Comparison of (*a*) calculated far-field P and S arrival strengths in a plate for a thermoelastic source as a function of distance (*x*) off epicentre with the experimental data of figure 5.28.

(4) Thermal conductivity

This factor has so far been neglected during the calculation of the waveforms. Its effect is expected to make the source progressively extend deeper and deeper into the sample with increasing time. Thus from figure 5.4(*b*) the thermal wavefront has to a first approximation only reached a depth of $\sim 1.5\ \mu$m after 10 ns, $\sim 2.1\ \mu$m after 20 ns, and $\sim 3.4\ \mu$m after 50 ns in mild steel. The speed of propagation of the thermal wavefront through aluminium is much higher (figure 5.4(*a*)). Thus, as time progresses, the thermoelastic stress boundary (which is related to the rate of change of temperature) also moves into the bulk of the material away from the surface, so that elastic waves from it reach the receiver a short time before they would if the source remained at the surface. There is thus some degree of incomplete cancellation (compare figure 5.26(*b*)) of the direct P wave received at the transducer by the P wave which propagates initially in the opposite direction and is then reflected back to the transducer by the nearby surface.

Because the stress boundary is propagating into the solid with a speed dependent upon the thermal diffusivity, a convolution of the wave propagation function with the thermal conduction is required to compute the final waveform. In such a model it would be expected (by analogy with the Doppler effect) that the direct far-field pulse component would be shifted to higher

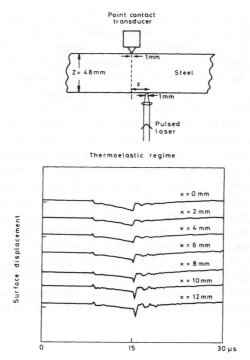

Figure 5.28 Experimental data obtained for a thermoelastic source at small distances off epicentre in steel.

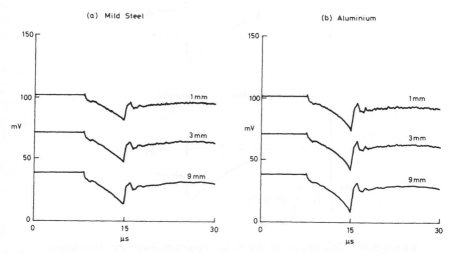

Figure 5.29 Effect of changing diameter of irradiated area on epicentre waveforms in thermoelastic regime in (*a*) mild steel, (*b*) aluminium.

frequency and thereby sharpened up. Similarly, the second pulse, reflected in the surface, and therefore of opposite polarity, would be expected to arrive progressively later and also be shifted to lower frequency, i.e. broadened out.

When a one-dimensional model of thermal conductivity is incorporated into the theory for the epicentre waveform (equation (5.110)), the result shown in figure 5.30(b) is obtained for aluminium (Sinclair 1982). The calculated waveform shows clearly the initial P-wave pulse, which is absent from figure 5.30(a) where thermal effects were excluded. The pulse is of the correct order of magnitude to correlate with experiment (figure 5.25(a)), suggesting that thermal conductivity effects are sufficient to explain the appearance of the pulse. Note from the inset to figure 5.30(b) that the second, negative, pulse is both markedly delayed and broadened compared with the positive pulse. This is as expected for a material such as aluminium where, as figure 5.4(a) shows, the thermal wavefront has penetrated to a depth in excess of $3\,\mu m$ in $10\,ns$, which is a substantial fraction of the speed of compression waves in this material. Although not shown, the model was also used to calculate the waveform for mild steel, substituting the appropriate materials parameters. It correctly predicted a positive pulse of reduced amplitude. More recently, Doyle (1986) has addressed this problem rigorously, and has confirmed that the initial spike is indeed due to the effects of thermal diffusion.

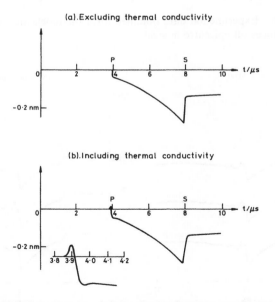

Figure 5.30 Comparison of theoretical epicentre waveform (a) excluding, (b) including thermal conductivity effects, for aluminium (Dewhurst et al 1982).

Thermal conductivity effects are only likely to increase the lateral dimensions of a typical source of a few millimetres by less than 1%, which is negligible compared with the depth effect. In any case, increasing the source diameter should broaden the pulses and reduce their height, but without any other major effect on the contribution from the thermal conductivity.

It is concluded that, while both the off-epicentre and thermal conductivity terms may make some contribution in general to this small positive pulse, the latter predominates for the experimental data considered above. Off-epicentre terms may dominate in, for instance, materials with much lower thermal conductivity, or for other source and receiver geometries.

The above theory only calculates the waveform for the direct P- and SV-wave arrivals. In practice, there will be additional arrivals in a parallel plate, which are due to the reflection of waves backwards and forwards across the plate. The first additional arrival will be a compression wave that has been reflected once at each surface, and hence has traversed the plate three times. It will therefore arrive at time $3h/c_1$, where h is the thickness of the plate.

The experimental waveforms described above were mostly obtained using a capacitive receiver. This is a broadband device, which requires special surface treatment in the form of flatness and polish. If the electrode of the capacitive transducer is a plate, then some correction may be necessary to allow for its diameter. Provided the diameter is much less than the plate thickness, then the only effect is to broaden the wave arrivals slightly, due to small phase shifts between the signal at the centre and edges of the plate. If the electrode is a sphere ('ball'), as demonstrated by Cooper (1985), then its diameter is negligible, except for measurements on very thin plates. Measurements by laser interferometer do not require any such correction since the receiver diameter can be focused to be many orders of magnitude less than the thickness of the plate, and can therefore be made at locations other than the epicentre without loss of fidelity.

If instead of a broadband receiver, a resonant piezoelectric transducer is employed, then the effect is to convolve the theoretical waveforms derived earlier with the impulse response of the transducer, which is likely to be a sinusoid with approximately exponential decay. In figure 5.31(a), which shows the output from a 5 MHz piezoelectric transducer sensitive to longitudinal waves, the step function of the compression-wave arrival of figure 5.25 is replaced by a decaying sinusoid. We note that this longitudinal transducer is still sensitive to the shear arrival, although there is much less reverberation than for the P arrival.

In addition to including off-epicentre terms as discussed above, changing the dimensions of the source may broaden the wave arrivals in a similar manner to enlarging the receiver. The spatial profile of the laser beam, e.g. whether it is Gaussian (as produced for instance by single-mode operation) or more complex (as produced by many multimode lasers), is unlikely to be

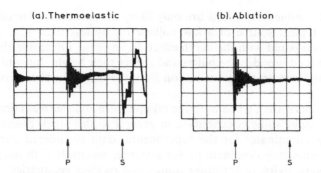

Figure 5.31 Epicentre waveforms ($1\,\mu s\ div^{-1}$) recorded by a 5 MHz piezoelectric transducer on a 25 mm aluminium disc, in (a) the thermoelastic and (b) the ablation regime with gain -20 dB (Scruby *et al* 1981).

a critical factor unless the beam dimensions are large enough for near-field effects to be important. Provided the receiver is far enough away for the source to be approximated by a point, the important factor is the total energy deposited at the surface, and its time dependence, assuming that the power density across the whole irradiated area remains below the threshold for other effects such as melting or ablation. Experimental data confirm this to be the case, showing that the amplitude of the epicentre waveform is directly proportional to the laser pulse energy in the thermoelastic regime. Extended sources are discussed in §5.9.5.

The laser pulse can alternatively be transmitted down a multimode fibre before impinging on the specimen. When the irradiated area of the specimen was ~ 2.5 mm diameter, the source was found to be thermoelastic (Dewhurst *et al* 1988), and very similar waveforms were generated to that shown in figure 5.25. Sessler *et al* (1985), Dewhurst and Al'Rubai (1989) and Harada *et al* (1989) have all generated ultrasound by means of picosecond lasers. Both a thermoelastic regime at low power density and an ablation regime at higher power density were observed, and the waveforms were consistent with those obtained using 20 ns pulse lasers. Faster ultrasonic risetimes were, however, obtained from the picosecond laser source when a very broadband detector was employed.

5.9.2 Ablation waveforms at epicentre

Experimental data at the epicentre confirm that, at higher power densities, when ablation takes place and a plasma is formed, the ultrasonic waveforms and hence the types of acoustic source are different from the thermoelastic regime considered in the previous section. In order to study these changes, it is convenient to increase the power density while keeping the incident energy constant, by using a lens to vary the diameter of the incident laser beam (figure 5.32). Below about 20 MW cm^{-2} the waveform does not change

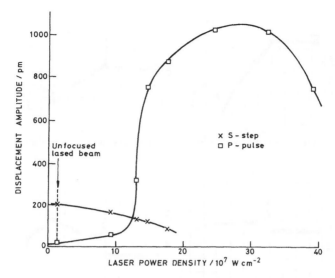

Figure 5.32 Compression and shear arrival amplitudes at epicentre as functions of laser power density at a constant incident energy of 33 mJ (Dewhurst *et al* 1982).

significantly, which is as expected for a source entirely within the thermoelastic regime. Between 20 and 100 MW cm^{-2} the small initial P-wave pulse increases in amplitude, while the shear-wave step decreases in amplitude, suggesting that some change of state may be taking place. Above 100 MW cm^{-2} incident power density, there is a huge increase in P-wave amplitude, as strong ablation takes place. Figure 5.33 shows the changes that take place in the epicentre waveforms. As the power density is increased, the compression-wave arrival first becomes dominated by a pulse rather than a step, of increasing amplitude ((a)–(c)), and then later ((d)–(f)) the pulse is modified into a step. The positive pulse (and step) is consistent with ablation causing a recoil force normal to the surface. At these higher powers, a visible plasma is observed, and some damage takes place in the form of shallow craters on the surface.

The energy in the ablation regime can also be varied while keeping the power density constant, by using a combination of lenses and neutral density filters. Under these conditions, Dewhurst *et al* (1982) found that the compression pulse height was proportional to the incident laser energy. They also found that the shape of the waveform was independent of energy, provided the power density was kept constant. Thus, in this ablation regime, the power density controls the type of source, in particular the relative strengths of the normal and tangential stresses set up, because it controls the local temperature rise. At constant power density, an increase in energy must mean a corresponding increase in source area, causing more metal to be ablated, and hence an acoustic source of greater strength.

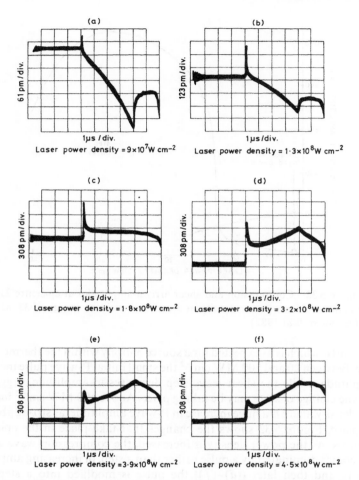

Figure 5.33 Epicentre waveforms recorded on a 25 mm thick aluminium sample, as a 33 mJ laser pulse is focused to generate progressively higher power density (Dewhurst *et al* 1982).

In order to calculate the epicentre waveforms in this regime, ablation can be modelled as a point force acting normal to the surface of the specimen (cf §5.8.2). Using a similar notation to §5.9.1, it can be shown that the normal surface displacement, u_3 (corresponding to the first two wave arrivals) generated at the epicentre of a parallel plate of thickness h, in response to a normal point excitation, $F_3 H(t)$, on the other surface is given by

$$u_3 = G_{33}(h, t; 0, 0)F_3. \tag{5.114}$$

Using the formalism of Sinclair (1979), the Green's function (defined here for a Heaviside function rather than δ-function source) can be calculated analytically for this configuration of source and receiver, so that the displacement is given by

$$u_3(t) = \frac{\frac{1}{2}t^2t_2^2(t^2 - t_1^2 + \frac{1}{2}t_2^2)^2 F_3 H(t - t_1)}{2\pi\mu h[(t^2 - t_1^2 + \frac{1}{2}t_2^2)^2 - t(t^2 - t_1^2)(t^2 - t_1^2 + t_2^2)^{1/2}]^2}$$

$$- \frac{\frac{1}{2}t^2t_2^2(t^2 - t_2^2)(t^2 + t_1^2 - t_2^2) F_3 H(t - t_2)}{2\pi\mu h[(t^2 - \frac{1}{2}t_2^2)^2 - t(t^2 - t_2^2)(t^2 + t_1^2 - t_2^2)^{1/2}]^2} \tag{5.115}$$

where as before $t_1 = h/c_1$ and $t_2 = h/c_2$. These are essentially the same equations as in Knopoff (1958), who also includes an expression for the next arrival, which is the twice-reflected P wave.

Equation (5.115) predicts a waveform (figure 5.34(a)) with a P arrival, given by the first term, that 'switches on' as a step at time t_1, and then slowly rises until time t_2, when the second term starts contributing. This is the shear arrival. Since the second term has a factor $(t^2 - t_2^2)$, it does not 'switch on'

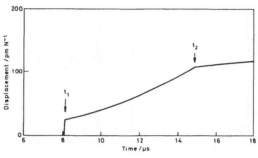

(a) Force step of height IN as source

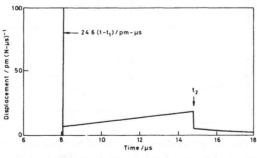

(b) Force pulse of area IN-µs as source

Figure 5.34 Epicentre waveforms for a point force acting normal to a surface (and simulating ablation), calculated from equation (5.115). Source time dependence modelled as (a) a Heaviside step function, and (b) a delta-function pulse.

as a step, but rather as a parabolic ramp. By substituting $t = t_1$, the magnitude of the P-wave step is found to be

$$u = \frac{c_2^2 F_3}{\pi \mu c_1^2 h}.$$ (5.116)

Note that, since equation (5.115) is for a Heaviside-function source, it is suitable for modelling the later stages of ablation. The expression for a *δ-function* source, required for instance to model the earlier stages of ablation, must be obtained by *differentiation* with respect to time. As an example, figure 5.34(*b*) shows the waveform for a source whose time dependence is a pulse. This was calculated by differentiation of equation (5.115).

The acoustic source generated in the study of ablation referred to above (figure 5.33) can be modelled as a combination of thermoelastic and ablative stresses, by adding together the epicentre displacements which would result from each source type taken separately (figure 5.35). Good agreement with

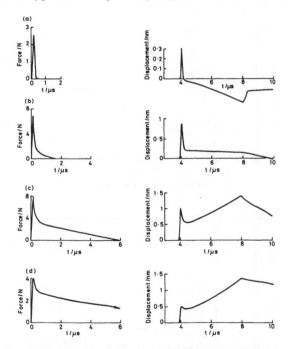

Figure 5.35 Epicentre waveforms calculated by assuming the source to be a combination of normal force F_3 (of time dependence shown) and centre of expansion δV (Dewhurst *et al* 1982). Values of F_3 and δV are estimated from figure 5.33: (*a*) $F_3 = 0.4$ N, $\delta V = 12 \times 10^4 \, \mu m^3$, compare figure 5.33(*b*); (*b*) $F_3 = 6.9$ N, $\delta V = 6 \times 10^4 \, \mu m^3$, compare figure 5.33(*c*); (*c*) $F_3 = 8.0$ N, $\delta V = 0$, compare figure 5.33(*c*); (*d*) $F_3 = 4.0$ N, $\delta V = 0$, compare figure 5.33(*f*).

experiment can be obtained, provided the time dependence of the ablative force is allowed to vary. It was necessary to use a short force pulse at lower power densities to obtain a good fit with the experimental data, implying that the plasma did not persist much after the optical pulse had finished. At higher power densities ($\geqslant 10^8$ W cm^{-2}), however, the 'tail' of the force pulse in the model had to be progressively lengthened to fit experiment, implying that ablation persists for a long time after the end of the incident optical pulse. This observation was consistent with other work on plasma formation, discussed in §5.4. At the highest powers studied, the peak force begins to decrease, which is at first somewhat surprising. However, the impulse (i.e. time integral of the force) transmitted to the specimen continues to increase. These effects were thought to be due to shielding of the surface by the plasma, and to long recombination times in very hot plasmas.

Dewhurst *et al* (1982) also noted a marked decrease in the amplitude of the shear arrival at high power densities. Although care is always needed in interpreting the second-order shear arrival at the epicentre, this observation is seen to be consistent with a redistribution of energy from shear to compression modes of propagation as the power density is increased and ablative stresses replace thermoelastic stresses.

The waveforms generated by picosecond lasers in the ablation regime (Dewhurst and Al'Rubai 1989, Harada *et al* 1989) are generally similar to those obtained with ~ 20 ns pulse lasers. Furthermore, if a multimode fibre is used to transmit the light from a Q-switched laser to a very small area of the specimen surface to give a high incident power density (Dewhurst *et al* 1988), then ablation waveforms are generated whose characteristics are very similar to those generated in the absence of the fibre.

The air-breakdown source, which is generated when a pulsed CO_2 TEA laser is focused just above the surface of the specimen (Edwards *et al* 1989), generates waveforms that are similar to those generated by ablation, at high powers, with a step arrival for the compression wave, although the source mechanism is different. There is no damage to the surface using this source, even though high-amplitude short ultrasonic pulses are generated.

In concluding this subsection, we note that three distinctly different types of waveform can be produced, simply by varying the optical power density. At low powers the P wave takes the form of a negative step (figure 5.25(*a*)); at intermediate powers a positive pulse (figure 5.33(*b*)), and at high powers a positive step (figure 5.33(*f*)). All of these are broadband arrivals with rise times generally only slightly longer than the optical pulse length. This makes them eminently suitable for NDT applications employing short-pulse ultrasonics.

5.9.3 Constrained surface waveforms at epicentre

Two examples from Dewhurst *et al* (1982) of epicentre waveforms for an unfocused beam of 1.06 μm laser radiation incident on constrained surfaces are shown in figure 5.36. Comparison of these waveforms with, for instance,

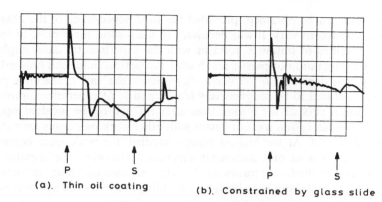

(a). Thin oil coating

(b). Constrained by glass slide

Figure 5.36 Epicentre waveforms recorded on an aluminium sample by a capacitive receiver for a laser source constrained by (Dewhurst *et al* 1982) (*a*) a thin coating of light oil ($0.5\ \mu s\ div^{-1}$); (*b*) a 1 mm glass slide cemented to surface ($1.0\ \mu s\ div^{-1}$).

figure 5.25, confirms that the cause of the dramatic increase in compression-wave amplitude reported in work using piezoelectric transducers (Fox 1974, von Gutfeld and Melcher 1977, Hutchins *et al* 1981a) is the replacement of the weak step of the thermoelastic waveform with a strong pulse arrival.

Because of the complex combination of effects discussed in §5.5, especially when a liquid is involved, it is not possible to produce theoretical waveforms in the same manner as previously. The source might initially be expected to have some of the character of a buried thermoelastic source, to give two approximately equal and opposite compression-wave pulses (cf figure 5.26(*b*)). For a reasonably thin constraining layer, a train of echoes from the top surface and interface is more probable, the periodicity, t', being given by

$$t' = 2h'/c'_1 \qquad\qquad (5.117)$$

where h' and c'_1 are the thickness of and longitudinal acoustic velocity in the layer. The data of figure 5.36 generally confirm this effect. The first echo in each case has the opposite polarity from the direct arrival.

As discussed in §5.5.3, constraint by liquids or other materials (including adhesives) with higher coefficient of expansion and higher volatility than the sample, is likely also to generate a transient force on the same due to expansion or vaporization. This would produce a positive pulse which would probably be qualitatively indistinguishable from the effect of constraint. The enhancements reported by Hutchins *et al* (1981b) do not vary greatly from one constraining medium to another (table 5.3), which tends to imply that the dominant mechanism here is the change in surface boundary conditions.

We now propose to try and estimate the enhancement caused by this change in boundary conditions as follows. Suppose that the constraining

Table 5.3 Enhancement (dB) compared with unmodified surface (Hutchins *et al* 1981a).

Surface treatment	Sample Materials		
	Aluminium	Brass	Mild steel
Constraining layers	30	28	33
Light oil	25	27	26
Silicone resin	21	24	26
Water	21	23	24
Acetone	25	24	32
Matt black paint	22	36	38
Matt white paint	25	29	22

layer is transparent, but has the same elastic properties as the sample, so that there is no elastic boundary. Then the thermoelastic source should act as if it is buried in a homogeneous medium. If we further assume that the pulse lasts a time τ, and that the rate of change of volume $\delta V'$ can be approximated by $\delta V' = \delta V / \tau$, then equation (5.67), which predicts the displacement due to centre of expansion buried in a whole space, can be written as

$$u_r = \frac{B\delta V}{4\pi(\lambda + 2\mu)c_1 r\tau}. \tag{5.118}$$

In order to calculate the normal displacement at the surface of a half space, we must multiply by the receiver function of equation (5.103):

$$u(\text{buried}) = \frac{2k^2 \cos 2\theta_2 \cos \theta_0}{\sin 2\theta_0 \sin 2\theta_2 + k^2 \cos^2 2\theta_2} \frac{B\delta V}{4\pi(\lambda + 2\mu)c_1 r\tau} \tag{5.119}$$

$$= \frac{B\delta V}{2\pi(\lambda + 2\mu)c_1 h\tau} \tag{5.120}$$

at the epicentre, where $\theta_0 = \theta_2 = 0$ and $r = h$. From equation (5.111), the magnitude of the compression arrival (a step) from the thermoelastic source at a free surface is

$$u(\text{surface}) = \frac{8B\delta V}{\pi\mu h^2 k^5}. \tag{5.121}$$

The enhancement due to perfect constraint is given by the ratio of equation (5.120) to equation (5.121):

$$\frac{u(\text{buried})}{u(\text{surface})} = \frac{hk^3}{16c_1\tau} \tag{5.122}$$

where the substitution $\lambda + 2\mu = k^2\mu$ has been made. Thus for $v = \frac{1}{3}(k = 2)$:

$$\frac{u(\text{buried})}{u(\text{surface})} = \frac{h}{2c_1\tau}. \tag{5.123}$$

Substituting typical experimental values of $h = 22\,\text{mm}$, $\tau = 24\,\text{ns}$ and $c_1 = 6300\,\text{m s}^{-1}$ for aluminium yields a P-wave enhancement of 37 dB (i.e. a factor of ~ 70) for a perfect constraining layer.

Suppose now that the same sample is covered by a transparent layer of for instance, glass. For a wave at normal incidence from the metal, the reflection coefficient for the P wave can be written as (Achenbach 1973)

$$\frac{A_1}{A_0} = \frac{(\rho c_1)_{\text{glass}} - (\rho c_1)_{\text{Al}}}{(\rho c_1)_{\text{glass}} + (\rho c_1)_{\text{Al}}} \tag{5.124}$$

where the same notation for the amplitudes A_1 and A_0 is used as in equation (5.75). Thus the far-field amplitude from a source at the interface can be calculated by then using reciprocity, in an analogous manner to equation (5.94), i.e.

$$u \sim 1 + \frac{A_1}{A_0} = \frac{2(\rho c_1)_{\text{glass}}}{(\rho c_1)_{\text{glass}} + (\rho c_1)_{\text{Al}}} \tag{5.125}$$

$$= 0.84.$$

where typical values of density and velocity have been substituted.

Thus the enhancement afforded by a glass constraining layer is likely to be approximately $70 \times 0.84 \simeq 60$, i.e. 35 dB. If the constraining layer is water instead of glass, and all other effects can be neglected, then the enhancement can similarly be estimated as 21 dB. These values are in reasonably good agreement with the experimental data of Hutchins *et al* (1981a) in table 5.3. However, as they discuss, for a thin layer of volatile liquid (e.g. light oil, water, acetone) rapid vaporization might be expected to contribute a large normal force. In the case of the paint layers, the enhancement is due both to vaporization and to a drastic reduction in the reflectivity of the specimen.

5.9.4 Off-epicentre waveforms

At the beginning of §5.9 it was noted that most laboratory investigations of laser-generated bulk waves have been carried out at the epicentre for various mainly practical reasons. However, although a very large number of ultrasonic applications do involve the propagation of elastic waves perpendicular to the source surface, such as thickness measurement and planar flaw detection, there are others where an angled beam may be more appropriate. In the case of ablation and surface constraints, the maximum in the compression-wave directivity is normal to the surface, so that studies of epicentre waveforms are highly relevant to the majority of possible applications. However, in the thermoelastic regime, the form of the directivity (figure 5.20), with lobes centred on an orientation of $\sim 60°$, suggests angled compression-wave beams. Thus, although the epicentre is useful in studying the physics of the thermoelastic source, other orientations are likely to be more useful in the context of applications of the technique.

At an orientation of 45° for instance, the P wave is a short symmetrical pulse (figure 5.37(a)), and totally different from the epicentral response. When the capacitor is replaced by a damped resonant piezoelectric transducer, the P arrival becomes the impulse response of that transducer (figure 5.37(b)).

This common practical ultrasonic orientation of 45° was selected for the measurement of compression waveform amplitudes in a 48 mm steel plate, as a function of incident laser energy and power density. Using a broadband, point-contact receiver, it was shown that the amplitude of the P arrival at this orientation was 4.3 times greater than the arrival at the epicentre in the thermoelastic regime. It was confirmed that the P arrival at 45° was a positive (i.e. compressive) pulse (figure 5.38(a)) rather than a negative step. These observations can be shown to be consistent with the theory outlined previously. Thus, the amplitude at angle θ_0 is given, either by combining equation (5.105) with the receiver function (equation (5.103)), or by combining

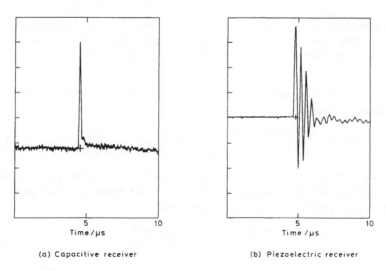

(a) Capacitive receiver   (b) Piezoelectric receiver

Figure 5.37 Comparison of off-epicentre (45°) waveforms due to a thermoelastic source in steel, recorded by (a) a capacitive transducer, and (b) a piezoelectric transducer with a damped crystal of nominal frequency 2.5 MHz (source–receiver distance 50 mm).

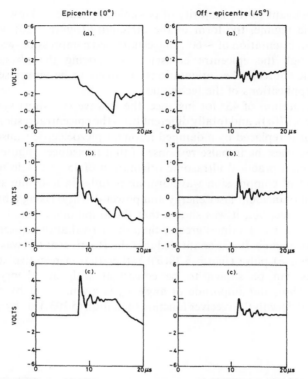

Figure 5.38 Waveforms recorded using a point-contact piezoelectric transducer at $0°$ and $45°$, as a function of incident laser energy, focused to a 1 mm diameter spot: (a) 10.8 mJ, (b) 28 mJ, (c) 54 mJ.

equation (5.119) with the source directivity function (equation (5.95)). In both cases we obtain

$$u(\theta) = \frac{4k^2 \sin 2\theta_0 \sin 2\theta_2 \cos 2\theta_2 \cos^2 \theta_0}{(\sin 2\theta_0 \sin 2\theta_2 + k^2 \cos^2 2\theta_2)^2} \frac{B\delta V}{4\pi(\lambda + 2\mu)c_1 h\tau} \quad (5.126)$$

where the sample thickness is h, and $h = r \cos \theta_0$, r being the path length.

Substituting $u(0)$ from equation (5.121) for the epicentre case, and also $\lambda + 2\mu = k^2\mu$, yields for the ratio of amplitudes:

$$\frac{u(\theta)}{u(0)} = \frac{\sin 2\theta_0 \sin 2\theta_2 \cos 2\theta_2 \cos^2 \theta_0}{(\sin 2\theta_0 \sin 2\theta_2 + k^2 \cos^2 2\theta_2)^2} \frac{hk^5}{8c_1\tau}. \quad (5.127)$$

Substituting $\theta_0 = 45°$, $k = 1.84$ and hence $\theta_2 = 22.6°$ for steel yields

$$\frac{u(\theta)}{u(0)} = 0.115 \frac{h}{c_1\tau}. \quad (5.128)$$

Substituting the experimental values of $h = 48$ mm, $c_1 = 5.96$ mm μs^{-1} and $\tau = 0.2\,\mu s$, gives a value of 4.6 for the ratio of the amplitude at 45° to 0°, which is in reasonably good agreement with experiment.

When the laser energy was increased so that a transition was made from the thermoelastic to ablation regime, it was observed that the behaviour off epicentre was different from the epicentre in two important respects. Somewhat surprisingly, the shape of the compression pulse at 45° hardly changed, although there was some indication of pulse broadening at the highest powers (figure 5.38). Furthermore, there was no dramatic increase in amplitude at the onset of ablation (figure 5.39). The approximately constant slope of the data at 45° indicates that, *per unit energy*, the ablation amplitudes were not appreciably larger than those in the thermoelastic regime. This is in strong contrast to the epicentre data.

It is concluded that, except close to the epicentre or at very high powers, the thermoelastic and ablation regimes are of broadly similar efficiency as ultrasonic sources. This observation is consistent with the conclusions about typical ultrasonic amplitudes in the two regimes made in §5.8.4, and also with the similarity in the magnitude of the stresses believed to be induced by the two effects (§5.3.2). In addition, thermoelastic stresses still make a

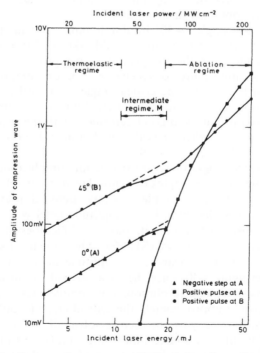

Figure 5.39 Compression-wave amplitudes as functions of incident laser energy and power at orientations of 0° and 45° (Scruby 1986).

significant contribution to the observed ultrasonic amplitudes at moderate power densities in the ablation regime. Finally it is noted that, prior to the formation of a full plasma, there is a dip corresponding to an intermediate regime (figure 5.39). This is possibly the result of energy being absorbed as latent heat to cause changes of state in the metal, i.e. melting combined later with vaporization.

Off epicentre should be better suited to the study of shear waves than the epicentre, where both the thermoelastic and ablation sources exhibit cylindrical symmetry. Figures 5.19(b) and 5.21(b) indicate that at orientations other than the epicentre, the shear radiation patterns are complex, with the possibility of zeroes, singularities and changes of phase.

There has been less interest in applications of laser-generated shear waves than compression waves, and hence fewer experimental data are available. Shear waves can be received directly by purpose-built transducers whose piezoelectric elements are sensitive to displacements tangential to the surface. Alternatively, at oblique incidence away from the epicentre, they can be detected by measuring the resolved normal surface displacement they induce, as discussed in §4.1.3. Because broadband perpendicular surface displacement is more readily measured than tangential, the following data on off-epicentre shear arrivals were obtained in this manner.

Within the critical angle ($\sim 30°$ in aluminium, $\sim 33°$ in steel) the shear arrival from the thermoelastic source should take the form of a pulse with similar time dependence to the compression wave, provided no appreciable broadening due to finite source or receiver size occurs. Figure 5.40 includes data for 15°, which is within the critical angle, in a steel plate. The shear arrival in this waveform is approximately four times greater than the compressional arrival. This is in good agreement with the theory which was used to derive figure 5.27, where the calculated ratio of shear- to compression-wave amplitude at 15° is 3.9 for steel.

Outside the critical angle there are two main complications. Firstly, the shear amplitude becomes a complex function. The resulting phase shift means that the arrival is no longer a simple pulse, as discussed in §5.8.1. Secondly, the shear wave is preceded by a large-amplitude head wave (H) that may under some circumstances obscure it, and give difficulties in interpretation. Figure 5.40 shows the shear arrival (S) at 65°, where the head wave is evident. Theory also predicts zero shear-wave amplitude at 45°. It is difficult to confirm this from figure 5.40, since the shear arrival would be too close to the head wave for adequate resolution. For completeness, figure 5.40 also gives the waveform at approximately the critical angle ($\sim 30°$).

Wave propagation theory can be used to solve fully the plate problem for a wide range of multipolar sources, including simulations of both thermoelastic and ablation, and predict complete waveforms off epicentre, including near-field terms. Since only the far-field terms are needed for the majority

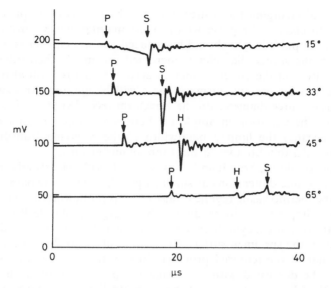

Figure 5.40 Waveforms recorded by a point-contact piezoelectric transducer in the thermoelastic regime at different orientations relative to the source normal on a 48 mm steel plate.

of ultrasonic applications, we shall pursue this no further and recommend the interested reader to Pao *et al* (1979), Wadley *et al* (1984) or Hsu (1985).

5.9.5 Extended sources

If the diameter of the thermoelastic source is increased, the effect on the epicentre is relatively small, provided the diameter remains significantly less than the specimen thickness, and takes the form of a broadening of the step arrivals of the P and S waves. A rigorous treatment of source diameter has been made by Bresse and Hutchins (1989), who compare the results of theoretical modelling of the source, in terms of spatially varying distributions of forces, with experimental data. From their study, they were able to conclude that the main contributions to the displacement waveform arise from the edge of a source with constant amplitude across the beam. Thus this particular source acts as an edge-wave-only transducer.

The effects of increasing source diameter on off-epicentre waveforms are more marked. We now discuss the result of varying the source dimensions by means of lenses, while keeping the incident energy constant, so that there is no change from the thermoelastic to ablation regime or vice versa. Within the former regime, the strengths of the force dipoles in the source, and hence the strengths (defined here as the pulse areas rather than heights) of the far-field wave arrivals are proportional to the total deposited energy. The main effect of changing the source diameter by means of a spherical lens should be to change the pulse width, while maintaining a constant pulse area. Thus the pulse height should be reduced as the diameter is increased. If, however, the source diameter is increased too much, then other effects due to variation of the Green's function across the source will also become significant, but for present purposes these will be ignored. An example of changing the source diameter on the waveform recorded at 45° is shown in figure 5.41. The variation in amplitude here is less than might have been expected, because the limited bandwidth of the receiving transducer has already considerably broadened the 1 mm diameter pulse.

For large-amplitude broadband ultrasonic pulses it is clearly advantageous to keep the source dimensions as small as possible. The drawback with this is that in the thermoelastic regime it raises the power density so that ablation occurs. The simplest way to produce a sharp large-amplitude P-wave pulse off epicentre is to use a cylindrical converging lens to focus the incident light into a line on the specimen surface (Aindow *et al* 1980). The line should be perpendicular to the required propagation direction. In this way sufficient energy can be deposited without raising the power density so high as to cause ablation. For a line source of approximate dimensions 1 mm × 10 mm, waveforms can be recorded in the 45° orientation and either perpendicular or parallel to the direction of the line (figure 5.42). Comparison with figure 5.41 shows how effective this method is at raising the amplitude of both the

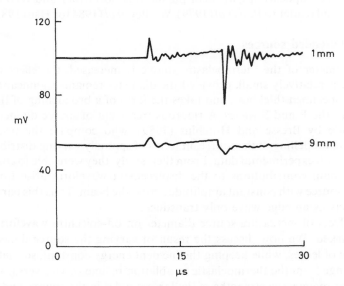

Figure 5.41 Showing the effect of varying the thermoelastic source diameter on a bulk waveform recorded by a point-contact piezoelectric transducer at an orientation of 45° on a 48 mm steel plate.

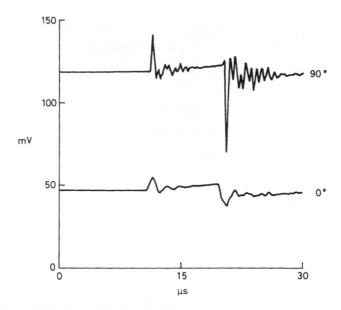

Figure 5.42 Showing bulk waveforms recorded at an orientation of 45° on a 48 mm steel plate, in response to a 1 mm × 10 mm thermoelastic line source, in directions (a) perpendicular, and (b) parallel to the line.

P and SV arrivals in the 90° orientation. The enhancement could be even greater using a broader band receiver.

For most ultrasonic applications it is a considerable advantage to have a directional beam, so that the line source is likely to have many uses, especially in the thermoelastic regime and at oblique angles to the surface. An even more directional beam has been demonstrated at TNO (Vogel and Bruinsma 1988) using fibre-optic technology. The light from the pulsed laser is divided along a number of optical fibres of different length (figure 5.43), so that the optical pulses impinge upon different points on the surface of the sample at progressively delayed times. The effect is to produce an ultrasonic phased array, which launches a highly directional beam at a pre-determined angle into the bulk of the material. By varying the delay times in the optical fibres, the beam can be effectively steered.

Focusing by means of a spherical or cylindrical lens can also be used in the ablation regime to change to pulse width and/or beam direction. It tends to have a smaller effect because generally speaking the pulse is already well focused in order to produce a strong plasma. The line source is therefore likely to be useful where sharp pulses from a weak plasma are required, and where the extent of surface damage from a more powerful laser is to be minimized.

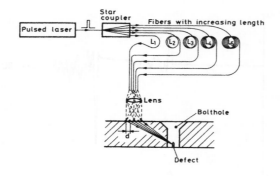

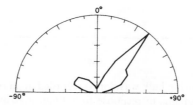

Figure 5.43 (*a*) Experimental arrangement for beam-steering of laser-generated ultrasound, using an optical-fibre phased array (Vogel and Bruinsma 1988). (*b*) Measured longitudinal directivity pattern using a phased array with five elements spaced at 2.5 mm, and with a 0.25 μs delay between elements.

5.10   ULTRASONIC SURFACE AND GUIDED WAVES

In addition to the bulk waves considered in the previous section, the pulsed laser can be used to generate surface acoustic waves as reported by Ledbetter and Moulder (1979), Lee and White (1968) and Aindow *et al* (1980). It is possible to generate all types of surface and guided wave by laser, but generally speaking Rayleigh waves have the largest amplitude on thick specimens, and are therefore of most interest. Lamb waves dominate on thin plates, and various extensional and torsional modes in rods. The laser offers the possibility of remote, non-contact generation of surface acoustic wave pulses of similar high bandwidth to laser-generated bulk waves. For optimum experimental results, a non-contact broadband receiver, e.g. capacitive transducer or laser interferometer, should be used, in an experimental arrangement similar to that illustrated in figure 5.44. Because of their importance, most of this section will be concerned with Rayleigh waves, but we shall briefly consider other types of guided wave in §§5.10.3 and 5.10.4.

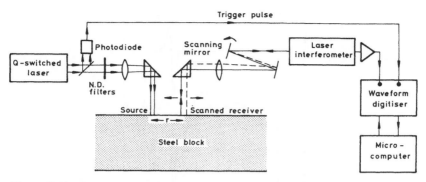

Figure 5.44 Apparatus used to study the laser generation of ultrasonic Rayleigh waves (Scruby and Moss 1985).

5.10.1   Surface waves from a thermoelastic source

In an early study, referred to in Scruby *et al* (1982), the receiver was a calibrated capacitor, with a linear rather than circular active plate. The laser source was focused into a line of similar dimensions to the capacitor plate, and then oriented parallel to it. The specimen was of steel and the source–receiver separation was 40 mm. The waveform (figure 5.45(*a*)) is dominated by the Rayleigh arrival, which takes the form of an asymmetrical bipolar pulse. Ahead of this is a weak arrival at a time corresponding to propagation at the compression-wave speed: this must be a surface-skimming P wave (known sometimes as the lateral wave).

The waveform is of too low amplitude for accurate measurement by the less sensitive interferometer at this range without extensive signal averaging. Thus figure 5.45(*b*) shows data obtained by optical detection at a range of 10 mm, and with averaging over 25 shots to reduce noise. The source was again a line (20 mm long × 0.5 mm wide), and the incident energy per pulse 80 mJ. Note that, at this shorter range, the Rayleigh pulse is more asymmetrical, and the P wave, although still weak, is relatively more intense than in (*a*). In Scruby and Moss (1985) the amplitudes of the compression and Rayleigh arrivals were measured as a function of range. It was found that the Rayleigh wave decayed with distance as $r^{-1/2}$, consistent with geometrical spreading in two dimensions, whereas the P-wave amplitude decayed as r^{-1}.

A theoretical understanding of the origin of this waveform can be obtained by making reference either to the work of Chao (1960) or Pekeris (1955). The former calculates the dynamical response of an elastic half space (including the perpendicular surface displacement) to a tangential surface force with Heaviside time dependence, i.e. the Green's function, $G^{H}_{31}(r, t; 0, 0)$. Pekeris (1955) includes in his paper the calculation of the tangential displacement in response to a perpendicular point loading with Heaviside time dependence, i.e. $G^{H}_{13}(0, t; r, 0)$. Equation (5.90) can be applied to show

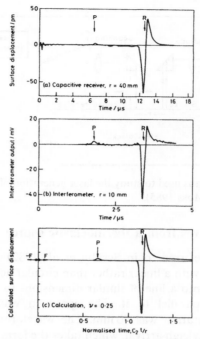

Figure 5.45 Comparison of experimental ((a), (b))) surface waveforms with calculation (c) for a laser ultrasonic source in the thermoelastic regime.

that this problem is the inverse of the problem solved by Chao (1960). Both workers derive the following equations to describe the required surface motion for Poisson's ratio of 0.25 (i.e. $c_1/c_2 = k = \sqrt{3}$):

$$u_3(\zeta) = 0 \qquad \text{for } \zeta < 1/k \qquad\qquad (5.129)$$

$$u_3(\zeta) = \frac{\sqrt{(\frac{3}{2})}\zeta F}{16\pi^2 \mu r} \{6K(\ell) - 18\Pi(8\ell^2, \ell) + (6 - 4k)\Pi[-(12k - 20)\ell^2, \ell]$$
$$+ (6 + 4k)\Pi[(12k + 20)\ell^2, \ell]\} \qquad \text{for } 1/k < \zeta < 1 \qquad (5.130)$$

$$u_3(\zeta) = \frac{\sqrt{(\frac{3}{2})}\zeta\kappa F}{16\pi^2 \mu r} \{6K(\kappa) - 18\Pi(8, \kappa) + (6 - 4k)\Pi[-(12k - 20), \kappa]$$
$$+ (6 + 4k)\Pi[(12k + 20), \kappa]\} \qquad \text{for } 1 < \zeta < \gamma \qquad (5.131)$$

$$u_3(\zeta) = \frac{\sqrt{(\frac{3}{2})}\zeta\kappa F}{16\pi^2 \mu r} \{6K(\kappa) - 18\Pi(8, \kappa) + (6 - 4k)\Pi[-(12k - 20), \kappa]$$

$$+ (6 + 4k)\Pi[(12k + 20), \kappa]\} + \frac{F\zeta}{8\pi\mu r(\tau^2 - \gamma^2)^{1/2}} \qquad \text{for } \gamma < \zeta$$
$$\qquad\qquad (5.132)$$

where F is the magnitude of the force, normalized time, $\zeta = c_2 t/r$, $\ell^2 = \frac{1}{2}(3\zeta^2 - 1)$, $\kappa = 1/\ell$ and the Rayleigh arrival time, $\gamma = \frac{1}{2}[(3 + k)^{1/2}]$. The functions $K(x)$ and $\Pi(y, x)$ are elliptic integrals defined by

$$K(x) = \int_0^{\pi/2} \frac{d\xi}{(1 - x^2 \sin^2 \xi)^{1/2}} \qquad \Pi(y, x) = \int_0^{\pi/2} \frac{d\xi}{(1 + y \sin^2 \xi)(1 - x^2 \sin^2 \xi)^{1/2}}.$$

The normalized displacement $U(\zeta) = -u(\zeta)\pi\mu r/F$ is plotted in figure 5.46(a). Note that the Rayleigh arrival (R) takes the form of a square root singularity (equation (5.131)) of negative polarity, followed by a slowly rising 'wash'. The compression arrival (P) is a very broad positive pulse, and the shear (S) a barely discernible change in higher order derivative.

As discussed earlier in this chapter, the thermoelastic source can be represented by a combination of force dipoles. In particular, when calculating displacements along a line perpendicular to a line source, it is a good approximation to represent the source by a single dipole parallel to the direction of propagation. The contribution of the other dipole is very small.

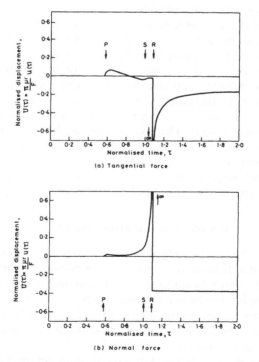

Figure 5.46 Vertical displacement generated by (a) tangential and (b) normal forces, calculated from equations (5.129)–(5.132) and (5.134)–(5.137), based on Pekeris (1955). The ordinate and abscissa are defined in the text.

We require therefore to calculate the Green's function for a source which is a dipole D_{11}, i.e.

$$G^H_{31,1'}(r, t; 0, 0) = - \frac{\partial}{\partial r} (G^H_{31}(r, t; 0, 0)) \qquad (5.133)$$

since in this case differentiation with respect to x'_1 at the source is equivalent to differentiation with respect to the source–receiver distance, r. This is the same as the spatial differentiation procedure that was used to derive equation (5.85) from equation (5.83).

The derivative of figure 5.46(a) has singularities at $-\infty$ followed immediately by $+\infty$. If the derivative is convolved with a Gaussian pulse, to simulate the effects of (a) limited bandwidth in the receiver circuitry, (b) a finite duration laser pulse, and (c) the finite spatial extent of the source and receiver, then the waveform of figure 5.45(c) is produced. We note that considerable care must be taken in performing the convolution, since the singularities occur at an irrational value of ζ. Unless the strength of the singularity is correctly allowed for, the resulting amplitude will depend upon the digitization interval chosen for the calculation.

There is good overall agreement with the experimental curves in figure 5.45. There are slight differences in the relative sizes of the positive and negative portions of the bipolar Rayleigh-wave pulse. These are believed to be due firstly to errors in assigning a width to the Gaussian smoothing function, and secondly to assuming that finite source size can be represented by such a function. In the latter case some form of spatial integration over the area of the source would be more correct; this is closer to the approach followed by Cooper (1985). The second source of error is likely to be the assumption that $\nu = 0.25$, whereas for steel $\nu \simeq 0.29$. The authors are not aware of an explicit analytical calculation of this half-space Green's function for general ν, although it can, for instance, be obtained as a special case using the formalism of Pao et al (1979) or Hsu (1985) for a parallel plate.

5.10.2 Surface waves from an ablation source

Figure 5.47(a) shows the displacement waveform detected by a laser interferometer, at a distance of 10 mm from an ablation source generated by focusing the pulsed laser beam to a ~ 0.6 mm diameter spot, on a steel plate. When instead the laser pulse was focused sharply onto a line, other conditions remaining the same, the surface waveform of figure 5.47(b) was generated. Two main differences can be observed: first the Rayleigh-wave arrival (R) is of shorter duration and higher bandwidth in figure 5.47(b); secondly the 'wash' following R extends for a longer time in figure 5.47(a). The first difference is due partly to the larger diameter of the point-focus spot compared with the width of the line, and partly to the longer duration of the plasma at the higher power densities that obtain in figure 5.47(a). The longer wash

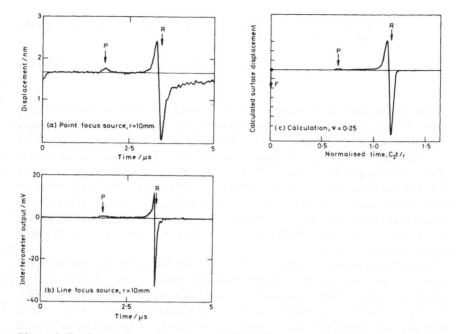

Figure 5.47 Comparison of experimental ((a), (b)) surface waveforms with calculation (c) for a laser ultrasonic source in the ablation regime.

in figure 5.47(a) is also believed to be due to the extended duration of the ablation source function, and will be discussed later in this section.

If the data of figure 5.47 are compared with figure 5.45 it is seen that, while the P wave is of the same polarity in both cases, the polarity of the bipolar Rayleigh-wave arrival is reversed. However, experimental data confirm that P and R decay with increasing distance in the same manner for the ablation as for the thermoelastic source, i.e. as r^{-1} and $(\sqrt{r})^{-1}$ respectively.

A theoretical understanding of the ablation waveform can again be obtained by reference to the work of Pekeris (1955), who calculates the normal displacement due to a normal point force loading (known as Lamb's problem). Again he restricts his derivation to $v = 0.25$ in order to simplify computation. Assuming a point force with Heaviside time dependence, and defining the variables as before where possible,

$$u_3(\zeta) = 0 \qquad \text{for } \zeta < 1/k \qquad (5.134)$$

$$u_3(\zeta) = -\frac{F}{32\pi\mu r}\left(6 - \frac{k}{(\zeta^2 - 0.25)^{1/2}} - \frac{(3k + 5)^{1/2}}{(0.75 + 0.25k - \zeta^2)^{1/2}}\right.$$

$$\left. + \frac{(3k - 5)^{1/2}}{(\zeta^2 + 0.25k - 0.75)^{1/2}}\right) \qquad \text{for } 1/k < \zeta < 1 \qquad (5.135)$$

$$u_3(\zeta) = -\frac{F}{16\pi\mu r}\left(6 - \frac{(3k+5)^{1/2}}{(0.75 + 0.25k - \zeta^2)^{1/2}}\right) \qquad \text{for } 1 < \zeta < \gamma \quad (5.136)$$

$$u_3(\zeta) = -\frac{3F}{8\pi\mu r} \qquad \text{for } \gamma < \zeta. \qquad (5.137)$$

The normalized displacement $U(\zeta) = -u(\zeta)\pi\mu r/F$ is plotted in figure 5.46(b). Note that the Rayleigh arrival (R) again takes the form of a square root singularity (equation (5.134)), but this time of negative polarity, with a slowly rising 'wash' preceding the singularity. The compression arrival (P) is a broad positive pulse, and the shear (S) a change in slope. Mooney (1974) has extended the work of Pekeris (1955) to any Poisson's ratio. In this general case the solution is no longer analytical. A study of his results shows that the main difference caused by changing v is in the low-frequency 'wash' preceding the Rayleigh arrival, and that this is small if v is changed, for example, to 0.29 from 0.25.

The ablative source function usually takes the form of a force pulse rather than a step. Thus $G_{33}(r, t; 0, 0)$ is required rather than $G_{33}^{H}(r, t; 0, 0)$ as calculated above. All that is necessary is to differentiate equations (5.134)–(5.137) with respect to time. The result of this is shown in figure 5.47(c), where, as in the case of the thermoelastic source, the waveform has additionally been convolved with a Gaussian function to take into account effects such as finite temporal width of the source function and finite receiver bandwidth. As before, care must be taken at this stage to correctly represent the strength of the Rayleigh singularity. The effect of increasing the width of the Gaussian is to reduce the amplitude of the Rayleigh arrival compared with the rest of the waveform.

The agreement of theory with experiment (figure 5.47) is better for the line focus source than the point focus. This is believed to be because, in the former case, the ablation is relatively weak, so that the force function is indeed a short pulse. In the latter case, however, the plasma produced is so intense that the ablative force persists to produce a tail, in exactly the same way as discussed in §5.9.2. If the data were convolved with a force function of the type portrayed in figure 5.35 rather than a simple Gaussian pulse, then a much closer fit with the data of figure 5.47(a) would be obtained.

In the above treatment it was assumed that the source was either in the thermoelastic or ablation regime. In reality (as discussed in §5.9.2) there is often likely to be a combination of thermoelastic and ablative stresses, especially in intermediate power regimes. In figure 5.48(i), the energy of the incident pulse is steadily increased, while maintaining a constant source diameter (0.6 mm). At 16 mJ the source is purely thermoelastic, and the Rayleigh wave is first negative and then positive. As the power is increased, the pulse both increases in magnitude and changes in shape until, at 80 mJ, there is a relatively intense plasma, and a Rayleigh wave that is at first

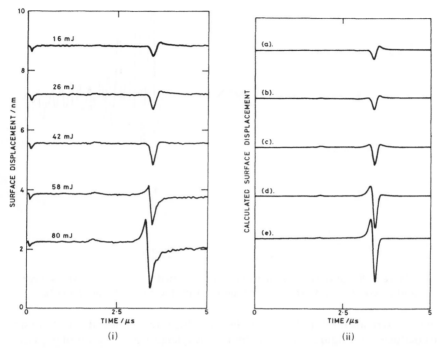

Figure 5.48 (i) Surface displacement measured as a function of incident optical energy for $r = 10$ mm. The source progressively changes from thermoelastic at 16 mJ to fully ablative at 80 mJ (Scruby and Moss 1985). (ii) Displacement calculated from combination of theoretical waveforms for the thermoelastic and ablative sources shown in figures 5.45(c) and 5.47(c).

positive and then negative. Note that there is an interesting intermediate regime (i.e. 42 mJ) where a virtually monopolar R pulse is generated. These waveforms can be modelled by a combination of both types of source (figure 5.48(ii)). At the extremes, good agreement is obtained for purely thermoelastic and ablative stresses (waveforms (a) and (e) respectively). For waveforms (b), (c) and (d) however a combination of both was required to fit the experimental data.

When the broadband receiver, such as the laser interferometer, is replaced by a narrower band contact transducer, then the discrete surface-wave arrivals are replaced by something approximating to a decaying sinusoid. Thus figure 5.49 shows the surface acoustic waveform on aluminium as received by a 3 MHz edge-mounted piezoelectric transducer. The source was a 20 mJ pulse from a Nd:YAG laser operating in the thermoelastic regime (Srcuby *et al* 1982). The Rayleigh arrival exhibits a long reverberation due to resonance in the transducer. The surface-skimming compression wave generates a much weaker arrival, consistent with the earlier work. Note that the true nature

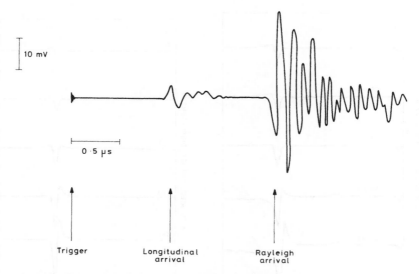

10 mV

0·5 µs

Trigger

Longitudinal
arrival

Rayleigh
arrival

Figure 5.49 Surface acoustic waveform generated by a thermoelastic source
and received by a 3 MHz edge-mounted transducer (Scruby *et al* 1982).

of both arrivals tends to be lost in the transducer resonance: it is not possible
to deduce from figure 5.49 that the true Rayleigh wave is a bipolar pulse.

Aindow *et al* (1982) have shown that similar scaling laws apply to surface
waves as to bulk waves. Thus in the thermoelastic regime the peak-to-peak
amplitude of the Rayleigh-wave arrival is directly proportional to the
absorbed laser energy per pulse. Aindow *et al* (1980) have investigated the
use of a cylindrical lens to produce a line thermoelastic source on the surface
of the specimen and hence generate directional Rayleigh waves. Figure 5.50
shows the experimental data obtained using a 3 MHz piezoelectric surface-
wave transducer. It can be seen that the cylindrical lens produces a highly
directional beam compared with the spherical lens. The principle by which
this focusing works is exactly the same as for bulk waves, discussed in, for
example, §5.8.1. The force dipole induced in the surface is the same
perpendicular and parallel to the line source. The spatial extent is, however,
different. Because the thermoelastic stresses must be located with the
$b = 0.1$ mm width of the line, the bipolar Rayleigh wave propagating
perpendicular to the line need only be convolved with a very narrow
broadening function (width of the order of $b/c_R \simeq 0.03$ µs, where c_R is the
Rayleigh velocity). In a direction parallel to the line (length a = 4.5 mm),
the broadening function should have a width of 1.5 µs, i.e. ~45 times greater.
Assuming a constant arrival strength in all directions means that the
amplitude of the Rayleigh arrival should be approximately 45 times greater
perpendicular to the line, which is reasonably consistent with experiment.

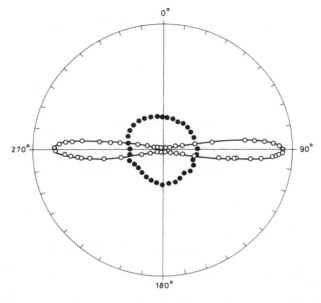

Key

o   5mJ pulse focused by cylindrical lens, thermoelastic
•   5mJ pulse focused by spherical lens, thermoelastic

Figure 5.50   Rayleigh directivity for a thermoelastic line (4.5 mm × 0.1 mm) source compared with a circular source, received by a 3 MHz surface-wave transducer (Aindow *et al* 1980).

Focusing the beam into a circular line to produce convergent surface waves for special applications is discussed in §6.1.4.

5.10.3   Lamb waves

As briefly described in §5.7, if two boundaries in a solid such as a plate are sufficiently close together, the elastic wave motions on each surface will interact to produce Lamb (or plate) waves, whose propagation characteristics are partly a function of the separation between the two boundaries. The two basic types of Lamb-wave mode, antisymmetrical and symmetrical, according to whether the displacements on the two surfaces are in phase ('flexural' modes) or antiphase ('breathing' modes), are shown schematically in figure 5.51. If the source of ultrasound is transient, then the wave motion in the plate close to the source consists of resolvable discrete wave arrivals, made up of reverberations of compression and shear waves between the two surfaces. Following each arrival is low-frequency 'wash', as shown for instance in figure 5.34. However, at larger distances r from the source (i.e. $r \gg h$), these individual arrivals are no longer resolved. Nevertheless, the lower frequency components continue to build up in amplitude to constitute the

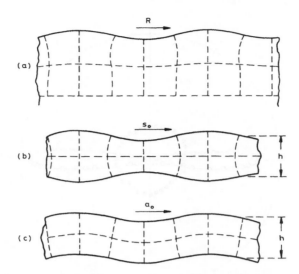

Figure 5.51 Schematic of particle motion in an elastic solid due to (a) Rayleigh surface wave, (b) lowest order (s_0) symmetric Lamb wave, (c) lowest order (a_0) antisymmetric Lamb wave (displacements greatly exaggerated).

Lamb-wave modes, which can alternatively be thought of as the normal modes of vibration of the plate. Thus the long-term response of the plate can be expanded as a series of normal modes.

In practice, certain modes will dominate, usually the lowest order modes. The lowest order symmetrical mode (s_0) is the fastest of the modes, being slightly slower ($\sim 90\%$ in steel) than the compression wave. On experimental waveforms it resembles the surface-skimming compression wave. The lowest order antisymmetric mode (a_0) is a few per cent faster than the shear wave in steel, and therefore tends to precede any Rayleigh wave that might also be present. Because of the proximity of the two boundaries, the Lamb-wave velocity is related to the thickness, and the waves are dispersive. For a more detailed discussion of Lamb waves the reader is directed to Viktorov (1967).

Dewhurst *et al* (1987) demonstrate that the pulsed laser can be used to generate symmetric and antisymmetric Lamb-wave modes. They focus the 7 mJ pulse to a line, so as to give directional ultrasonic radiation, and remain within the non-destructive thermoelastic regime. They employ a laser interferometer as a broadband receiver. They concentrate on thin plates, of thickness, h, in the range 27–425 μm. They also show that for very thin plates such as these, where typical ultrasonic wavelengths $\Lambda \gg 2h$, the symmetric modes tend towards a single well-defined 'sheet' velocity given by

$$c_{\text{sheet}} = \left[4\left(\frac{\lambda + \mu}{\lambda + 2\mu} \right) \frac{\mu}{\rho} \right]^{1/2} \tag{5.138}$$

For a material such as aluminium, $\lambda \simeq 2\mu$, so that substituting for the shear velocity, $c_2 = (\mu/\rho)^{1/2}$, we obtain

$$c_{\text{sheet}} = \sqrt{3}c_2 = \tfrac{1}{2}\sqrt{3}c_1. \tag{5.139}$$

Thus the sheet wave, which is shown so clearly in figure 5.52(*b*), travels at $\sim 87\%$ the compression speed. Figure 5.52 is, however, dominated by the lowest order antisymmetric mode, a_0. Figure 5.52(*a*) especially shows very clearly that this mode is dispersive, the higher frequency component arriving before the lower. Dewhurst *et al* (1987) show that that the group velocity of this antisymmetric mode is given by

$$c_g^2 = \frac{4\pi f h}{\sqrt{3}} c_{\text{sheet}}. \tag{5.140}$$

If the sheet velocity is known, the thickness, h, can be calculated from measurements of group velocity as a function of frequency for the a_0 mode. Dewhurst *et al* (1987) first measured the sheet velocity in their aluminium samples to be 5500 m s^{-1}. This is in good agreement with equation (5.139), since a typical shear velocity in aluminium is 3200 m s^{-1}, yielding a calculated sheet velocity of 5542 m s^{-1}. From a graph of the square of the group velocity measured from their data as a function of frequency, they were able to use equation (5.140) to deduce a value for the thickness h. The agreement was

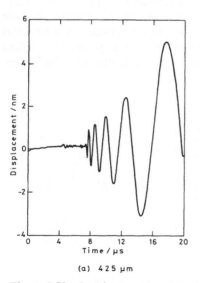
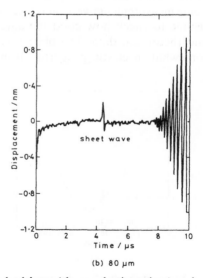

(a) 425 μm   (b) 80 μm

Figure 5.52 Lamb waves generated by pulsed laser (thermoelastic regime) and detected by an interferometer 24 mm from the acoustic source in aluminium samples of different thicknesses (Dewhurst *et al* 1987).

good for very thin samples, where the approximation $\Lambda \gg 2h$ held well, but it deteriorated for larger h.

In a more recent paper, Hutchins *et al* (1989) make a comprehensive study of laser-generated and laser-detected Lamb-wave propagation, using very thin specimens ($< 25\ \mu$m). They use their technique to deduce the thickness and longitudinal- and shear-wave velocities, using again the two lowest order Lamb-wave modes (s_0 and a_0).

Weaver and Pao (1982) have used a normal mode expansion technique to calculate the displacement generated by a point normal loading (with Heaviside-function time dependence) on a parallel-sided plate. They calculate each normal (Lamb-wave) mode for the plate, and then sum the modes with appropriate coefficients to satisfy the initial conditions and boundary conditions. Figure 5.53 shows the results of their calculation for a plate of thickness 9.6 mm, density 2300 kg m^{-3}, P-wave speed 5760 m s^{-1}, S-wave speed 3490 m s^{-1} and Poisson ratio 0.21. At a range of 100 mm, which corresponds to approximately 10 plate thicknesses, the calculated normal displacement (figure 5.53(a)) is dominated by the lowest order antisymmetric mode, which takes the form of a large-amplitude, low-frequency oscillation. The Rayleigh wave is also apparent, but neither the P nor S waves nor symmetric modes are resolved. At the larger range of 400 mm (figure 5.53(b)) the first antisymmetric mode still dominates, demonstrating clearly its dispersive nature, and the Rayleigh wave is still resolved.

These calculations are directly relevant to the ablation source, at higher power, when the time dependence of the source function resembles a step. Some measurements were therefore made using the laser source in the ablation regime to check how good the agreement would be between experimental and theoretical data. The plate was a 10 mm thick steel plate. This differs somewhat in elastic properties from those of the calculation, which were

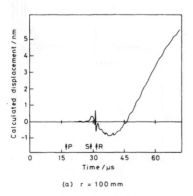
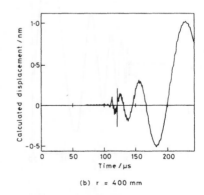

(a) $r = 100$ mm

(b) $r = 400$ mm

Figure 5.53 Surface displacement calculated by the normal-mode method (Weaver and Pao 1982) for a 9.6 mm plate at two different ranges in response to normal point force excitation on the same surface.

chosen to simulate glass, which could not be used here because it would be transparent to the laser pulse. A laser interferometer was not available for use as receiver, and a broadband, point-contact piezoelectric transducer was used instead.

The experimental data (figure 5.54) show some similarities and some differences compared with calculation (figure 5.53). Thus at ten plate thicknesses (figure 5.54(a)) the experimental data exhibit more high-frequency arrivals around the shear wave, which is itself larger in amplitude. The lack of higher frequencies in the theoretical data may be due to the top cut-off chosen in the summation over the normal modes. Conversely, the lower frequencies in the antisymmetric mode are somewhat attenuated in the experimental data, due to a roll-off in the low-frequency response of the transducer–amplifier combination. Similar observations can be made at 40 plate thicknesses (figure 5.54(b)). The high-frequency modes (absent from the calculated data) are particularly apparent here. The arrival times and mode frequencies do not agree between figures 5.53 and 5.54 because of the differences in the elastic constants between glass and steel. Nevertheless, the overall forms of the displacement waveforms are similar, and demonstrate the usefulness of the laser source for investigating Lamb-wave propagation.

5.10.4   Other guided waves

Very little reported work exists in the literature to demonstrate the generation by pulsed laser of the complex wave modes corresponding to propagation

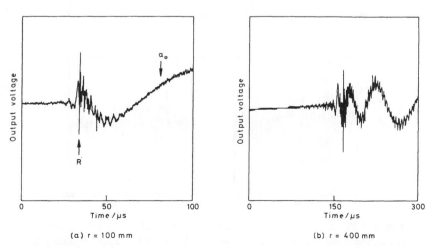

(a) r = 100 mm

(b) r = 400 mm

Figure 5.54 Experimental data for an ablation source on the same side as the receiver for a 10 mm steel plate at similar ranges to calculated data (figure 5.53).

through rods, bars and more complicated structures. However, Britton *et al* (1975) used a Q-switched ruby laser (50 ns pulse) to initiate a broadband stress pulse in a straight cylindrical steel rod waveguide of 10 mm diameter. By recording the ultrasonic pulse by means of a piezoelectric disc sandwiched between the main waveguide and a backing rod of the same silver steel at a distance of 1.8 m from the source, they were able to study the dispersion of the longitudinal (extensional) and flexural modes. The apparatus could not be used to study torsional modes for reasons of symmetry. The authors reported somewhat better agreement than in the data of Percival (1967) who had also used a laser to excite rod waves in a shorter specimen.

Figure 5.55 shows the rod waves that can be generated by the ablation source. The steel rod was 60 mm long and 47.6 mm in diameter. The detection was made by means of a piezoelectric transducer with a damped resonance at ~ 2.5 MHz, whose diameter was considerably larger than the rod. The laser was reasonably tightly focused to produce strong ablation, firstly at the centre and secondly towards the edge of the free end. In the former case (figure 5.55(a)) the laser predominantly generates extensional modes because of

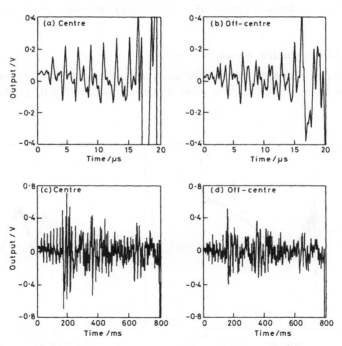

Figure 5.55 Showing the generation of rod waves by an ablation source (a) at the centre of the end of a steel rod, and (b) at approximately $\frac{1}{4}$ diameter from the centre; (c) and (d) show the same data respectively on a much longer time base.

symmetry. Note that when the source is moved to a lower symmetry position, the mode pattern becomes more complex (figure 5.55(c)). When the data are recorded at a longer timebase, yet more modes contribute, including flexural modes. There are also likely to be waves that propagate from the source the whole way along the surface.

5.11 LASERS FOR ULTRASONIC GENERATION

5.11.1 Choice of laser properties

Most of the discussion of the factors controlling the choice of laser to generate ultrasonic waves in solids has already been made in the preceding sections. The purpose of this section is therefore to summarize the main technical criteria for the choice of laser. The main effect of continuous-wave lasers is to produce steady heating, rather than the rapid temperature changes that generate stress and hence elastic-wave radiation. We note, however, that narrow-band ultrasonic waves can be generated by a modulated cw laser, provided a very sensitive lock-in amplifier system is used to enhance signal/noise. We shall restrict our attention to pulsed lasers. We shall first consider the principal laser characteristics of wavelength, pulse energy, pulse duration, beam profile and pulse repetition rate in turn, with respect to typical sample materials and ultrasonic requirements.

(a) Optical wavelength

Since for most purposes, the laser acts solely as a transient source of localized heat, the choice of optical wavelength is not very critical. The main effect of changing the wavelength of the incident electromagnetic radiation is to modify the reflectivity of the solid surface, and hence the efficiency with which the incident radiant energy is converted into elastic-wave energy. Outside the regime in which quantum effects are important, the efficiency of absorption (and hence ultrasonic generation) by a metal is inversely proportional to the square root of the optical wavelength since, from equation (5.7) and (5.5), substituting $\lambda = c/v$:

$$1 - R = 4/(\mu_0 \sigma c \delta) = 4\left(\frac{\pi \mu_r}{\mu_0 \sigma c \lambda}\right)^{1/2} \tag{5.141}$$

Aindow et al (1982) examine a wide range of laser wavelengths, and show that compression-wave generation in steel by a krypton fluoride laser in the near ultraviolet ($\lambda = 246$ nm) is 2.3 times more efficient per unit energy than by a Nd:YAG laser in the near infrared ($\lambda = 1.06\ \mu m$). This latter is in turn 3.0 times more efficient than a carbon dioxide laser ($\lambda = 10.6\ \mu m$). The factors for aluminium are only slightly different: 2.7 and 3.5 respectively. The factors calculated from equation (5.141) should be 2.1 and 3.2 respectively. The agreement with experiment is quite good in the latter case, but in error due to quantum effects in the former. As noted at the beginning of this chapter, absorption also depends upon surface condition, so that a rough and/or blackened surface increases efficiency at all wavelengths. The situation may be more complicated in certain non-metals, where the absorption of electromagnetic radiation is strongly frequency dependent. In such cases the depth of penetration will depend upon optical wavelength.

There may be other factors that influence choice of wavelength. Let us mention the important one of safety. High-power lasers are particularly hazardous to the eyes in the visible and near-visible parts of the spectrum. At 10.6 μm, the carbon dioxide laser generates radiation that is less dangerous, and which is attenuated by glass. Thus, although it is a less efficient generator, it may be a good candidate for industrial use on safety grounds. This is undoubtedly one of the reasons for the exploratory study by Edwards *et al* (1989).

(b) Pulse energy

Pulse energy is a very important factor in determining the characteristics of laser-generated ultrasound. There are two, often conflicting considerations. On the one hand, as equation (5.69) for instance shows, the ultrasonic amplitude in the truly non-destructive thermoelastic regime is proportional to optical power, i.e. proportional to pulse energy for a fixed duration pulse. Thus, to enhance ultrasonic amplitudes and thereby improve the sensitivity of, for example, laser-ultrasonic defect detection, as high a pulse energy as possible is desirable. On the other hand, however, increasing the pulse energy brings the source closer to the threshold for surface damage, and a certain amount of unwanted plastic deformation, melting or vaporization may occur. This threshold depends on power density rather than energy, so that it can sometimes be avoided by enlarging the irradiated area but, as discussed below, such an action may significantly change the characteristics of the ultrasound.

For common metals such as steel and aluminium, the damage threshold for a Q-switched pulse in the near infrared is 1–10 MW cm^{-2} at room temperature. At elevated temperatures, the threshold is inevitably less, since it requires a smaller temperature rise to raise the metal to its melting point. In non-metals, the wide variations in absorption coefficient and thermal conductivity make it impossible to quote a comparable threshold.

Some applications permit a small level of surface damage, so that higher incident pulse energies can be used and higher ultrasonic amplitudes generated. The user of the technique must be aware that small amounts of ablation can drastically change the form of the ultrasonic wave field, as discussed in earlier sections of this chapter.

(c) Irradiated area and beam profile

The area of the beam as it leaves the laser is less important because it can readily be changed by standard optical means. The area of the incident radiation on the sample surface is, however, much more important, because it affects both the power density (and hence the threshold for ablation and surface damage) and the dimensions of the ultrasonic source. Many commercial lasers produce beams with diameter of the order of a centimetre, which coincidentally is comparable with the diameter of many ultrasonic transducers. This tends however not to be particularly relevant since optical components such as lenses are often used to tailor the dimensions of the beam. Thus, as discussed in earlier sections, particular ultrasonic beam characteristics can readily be obtained by changing the irradiated area from a circular region, to a point, to a line, to a circular line. Phased arrays of small irradiated areas offer the maximum flexibility in ultrasonic directivity.

The variation of power across the beam may be important for a few applications, although for most it is satisfactory if the beam either has an approximately Gaussian or flat profile. Most commercial lasers meet this criterion. However, especially if the alignment of the optical components is not perfect, the beam may sometimes have more serious intensity variations which cause local 'hot spots', which may both damage the specimen and perturb the ultrasonic field.

(d) Pulse duration

This is a relatively important parameter since it influences both the generation mechanism and the character of the ultrasonic waveform. It was shown in §5.1 that the optical energy must be compressed into a very short pulse in order to obtain a sufficiently high instantaneous power ($> 10^5$ W) to generate useful ultrasonic amplitudes. The preferred pulse length depends upon the desired ultrasonic waveform. Thus, as also shown in the same section, a pulse duration of 20 ns or less is required to generate adequate ultrasonic energy up to ~ 20 MHz. By a happy coincidence, this is a typical pulse length for many commercial lasers.

Higher frequency ultrasonic generation requires sub-nanosecond (i.e. picosecond) pulses, which can be generated by means of mode-locked lasers. If lower ultrasonic frequencies (< 1 MHz) are required, it is arguably still better to employ a pulse duration in the range 10–50 ns than switch to a normal pulse laser (intermediate pulse lengths are not generally available). This is because the generation efficiency in metals falls with longer pulse duration and probably more than compensates for any advantage gained by compressing the energy into a narrower frequency band. In addition, the possibly complicating effects of thermal conductivity also increase with pulse duration.

(e) Pulse repetition rate

The desired repetition rate is likely to depend upon the intended application. The advantage of a high repetition rate for flaw detection, for instance, is that extensive, rapid signal averaging can be carried out to improve signal to noise ratios. Currently available recording instrumentation can average 1000 digital waveforms, each of 1024 points, in 1 s, using microprocessor hardware. Thus repetition rates ~ 1 kHz would be desirable for averaging purposes alone.

High repetition rates also mean that large areas of the sample can be rapidly scanned, making the most of the non-contact nature of the source. There are indeed lasers available with very high repetition rates. The copper vapour laser typically operates in the range 2–5 kHz. The restriction on average power imposed by various components in the laser system means that pulse energies tend to be smaller (e.g. 1–10 mJ). Another new and interesting possibility is the 'slab' laser (Kane *et al* 1986).

For many uses, especially in the laboratory, the more modest repetition rates of 1–100 Hz that are available from the most widely used and least expensive laser systems, are perfectly adequate. Indeed, in specimens where there is much ultrasonic reverberation, higher repetition rates are useless, because the ultrasonic waveforms from successive pulses, which last many milliseconds, may overlap. Power dissipation considerations tend to limit such lasers to energies of 1 J in the case of Nd:YAG or excimer lasers.

Table 5.4 gives examples of suitable lasers for ultrasonic generation.

5.11.2 Summary of characteristics of laser-generated ultrasound

Before concluding this chapter on ultrasonic generation by laser, it may be helpful to summarize its main characteristics.

(*a*) The incident energy is optical. Thus the laser can be remote from the specimen, the beam can readily be steered, and hostile environments and awkward geometries are accessible.

(*b*) Energy is transferred to the sample without couplant. Thus the sample surface remains unloaded. Although couplant variability is eliminated, variations in surface condition can significantly affect generation efficiency.

(*c*) The laser is used principally as a transient source of high-power localized heat. The coherent nature of the laser radiation is relatively unimportant.

(*d*) Optical wavelength is only important insofar as it affects reflectivity. Thus most short (< 100 ns) pulse lasers can be used.

(*e*) The laser-induced source is at the surface (i.e. within a surface layer a few micrometres thick) of a metal sample. It may extend into the bulk of a non-metallic sample.

(*f*) Source dimensions (except depth) can readily be varied, by for instance the use of lenses.

Table 5.4 Examples of lasers for ultrasonic generation.

Laser type	λ (μm)	E (J)	τ (ns)	rep (Hz)	Comments
Carbon dioxide (Q-switched)	10.6	0.1–1	200	400	Low generation efficiency, bulky, less hazardous
Nd:YAG (Q-switched)	1.06	0.01–10	10–50	<1000	Wide application, compact systems, available for low repetition rate.
Nd:glass (Q-switched)	1.06	1–10	10–50	<10	For highest power application, > 10 J with amplification.
Ruby (Q-switched)	0.694	1–10	10–50	<50	High-power applications.
Metal vapour (e.g. copper)	0.511 0.578	<0.03	10–60	2k–5k	High repetition rate, lower power applications.
Excimer (e.g. KrF)	0.249	0.1–1	10–20	10–200	High generation efficiency.

(g) Different source types exist: thermoelastic at lower power, ablation at higher power, more complex sources when the surface is constrained, and the 'air-breakdown' source.

(h) All types of elastic wave can be generated, including bulk waves (compression and shear), surface waves (including Rayleigh) and guided waves (e.g. Lamb waves in plates, extensional and flexural waves in rods).

(i) A wide range of ultrasonic amplitudes can be generated by varying the incident laser power, by for instance inserting neutral density filters in the beam. If generation is strictly to remain within the non-destructive thermoelastic regime, the maximum bulk compression-wave amplitude that can be generated in most metals is in the range 1–10 nm at a distance of ∼25 mm from the source, assuming no surface constraint. In the ablation regime, amplitudes up to ∼40 dB larger are realizable, depending upon the maximum laser power available. Similar amplitudes can be obtained from constrained surface sources.

(j) The directivity (beam profile) of the ultrasonic beam can be varied. In the ablation and constrained surface regimes, the directivity is similar to a small, normal incidence compression-wave probe used in short-pulse mode. The beam can be focused to a certain extent by increasing the area of the irradiated area, although a high pulse energy is required to maintain ablation over very large areas. In the thermoelastic regime the compression-wave beam is mainly directed in a conical lobe at ∼60° to the surface normal. Large-amplitude shear waves are also generated at various angles. More directional beams can in general be produced by means of the line source or phased arrays. However, directed shear waves at, for example, the most common testing angles of 45° and 60° are difficult to generate.

(k) In all three regimes, the ultrasonic pulses are short and of high bandwidth, extending from very low frequencies (from DC in the case of bulk waves) to typically 20 MHz, depending upon the optical pulse length and sample material properties. The bulk waves mostly take the form of single, δ-function-like pulses, although step-function arrivals can also be observed at some orientations or at very high ablative powers. The Rayleigh waves are mostly bipolar pulses.

5.11.3   Comparison with piezoelectric transducers

Finally, we briefly compare the laser-ultrasonic source with a conventional piezoelectric transmitter of ultrasound. As table 5.5 shows, there are significant differences between lasers and piezoelectric probes in three main areas, i.e. in their proximity and coupling to the sample, in the nature of the wave field (bandwidth and directivity) and in terms of safety and cost. As we shall show in Chapter 6, the non-contact nature of the laser source is nearly always a positive advantage, but this has to be weighted against cost penalties. The difference in wave-field characteristics tends to be less important for most applications than the other two factors.

Table 5.5 Comparison of laser and piezoelectric sources of ultrasound.

Physical parameter	Comparison of laser and piezoelectric probe
Proximity to sample	Laser non-contact and totally remote: probe must remain in contact via solid or liquid medium.
Couplant requirements	Laser requires no couplant: problems of couplant variability with piezoelectric probe.
Ultrasonic amplitudes	Laser generates amplitudes comparable with short-pulse piezoelectric probe.
Beam directivities	Laser source variable by optical means; ablation good as omnidirectional source. Directivity fixed for probe, better for highly directional beams.
Signal bandwidths	Laser generates wider band signals than probe.
Size and weight	Laser much larger and heavier than probe, but can be placed remote from sample. Actual laser ultrasonic source can be much smaller than probe.
Effect of environment	Laser can be situated remote from harmful environment whereas contact probe has limited use in very hot, corrosive or radioactive environments.
Economics	Laser technique more expensive than piezoelectric.
Safety considerations	Extra precautions needed for eye safety with high-power lasers.

Although review articles and books have been written summarizing the properties and uses of piezoelectric transducers (e.g. Sachse and Hsu 1979, Silk 1984) it has proved somewhat difficult to find absolute amplitude data for the displacement fields generated in solids by piezoelectric transmitter probes that can be compared with the calculations and measurements made for the thermoelastic and ablative laser sources which are the subject of this chapter. However, one recent study (Moss and Scruby 1988) has used a laser interferometer to measure the displacement wave fields generated by some typical ultrasonic transducers that are regularly used for the NDT of airframe components. Thus, by way of example, the peak displacement generated by a 5 MHz compression-wave probe at a range of 10 mm in aluminium was measured to be ~10 nm. A 5 MHz shear-wave probe (i.e. a compression probe attached to an angled perspex shoe to generate a 45° mode-converted shear-wave beam) generated a peak displacement of ~13 nm at a range of 7 mm in aluminium (plus 12 mm travel through the shoe).

Turning first to the amplitudes generated by the thermoelastic source, §5.8.4 deduces the peak compression-wave amplitude on the surface of a 100 mm aluminium hemisphere to be 0.88 nm for 6 mJ absorbed energy. Thus, for a range of 10 mm, the peak displacement would be ~9 nm, assuming that the amplitude is inversely proportional to the distance travelled. Note, however, that this is at an orientation of 64°, rather than on the axis of the source. Furthermore, an infinitesimally narrow source and an infinitely broadband detection system are assumed, so that there is no pulse broadening. In practice, peak thermoelastic amplitudes would be probably reduced by at least a factor of four. The maximum compressional amplitude for ablation is also calculated in §5.8.4. This time the maximum is on the axis as it is in the case of the piezoelectric probe. For a peak ablative force of 8 N (Dewhurst et al 1982) the peak displacement at a range of 100 mm in aluminium can be shown to be ~2.5 nm.

Thus it can be seen that the compression-wave amplitudes generated by the pulsed laser are only slightly smaller (say 12 dB) than those generated by a typical resonant piezoelectric probe. They are probably of the same order of magnitude as a short-pulse (i.e. damped) piezoelectric transmitter. The laser data assumed a 100 mJ incident laser pulse, which is an average size for a small–medium system. Much larger lasers are available, delivering energies in the range 1–10 J. It can be seen that these can easily overtake typical piezoelectric transducers in wave-field strength. These comparisons predict that the replacement of a piezoelectric transmitter by a pulsed-laser source should give negligible loss of sensitivity in most practical NDT situations, and may even give small improvements for broadband applications. This contrasts with laser reception of ultrasound, where it is very difficult to obtain sensitivities that even approach those of piezoelectric receivers (§4.8).

REFERENCES

Achenbach J D 1973 *Wave Propagation in Elastic Solids* (Amsterdam: North-Holland)
Aindow A M, Dewhurst R J, Hutchins D A and Palmer S B 1980 *Proc. Acoustics 80 Conf.* (Edinburgh: Institute of Acoustics) p 277
—— 1982 *Opt. Commun.* **42** 116
Aki K and Richards P G 1980 *Quantitative Seismology* vol 1 (San Francisco: Freeman)
Aussel J D, Le Brun A and Baboux J C 1988 *Ultrasonics* **26** 245
Berthelot Y H 1989 *J. Acoust. Soc. Am.* **85** 1173
Bleaney B I and Bleaney B 1965 *Electricity and Magnetism* 2nd edn (Oxford: Oxford University Press)
Bresse L F and Hutchins D A 1989 *J. Appl. Phys.* **65** 1441
Britton, Parkes, Burgum and Evans 1975 *Acustica* **33** 237
Burridge R and Knopoff L 1964 *Bull. Seismol. Soc. Am.* **54** 1875
Bushnell J C and McCloskey D J 1968 *J. Appl. Phys.* **39** 5541
Calder C A, Draney E C and Wilcox W W 1980 *Lawrence Livermore Laboratory Report* UCRL-84139
Carslaw H S and Jaeger J C 1959 *Conduction of Heat in Solids* (Oxford: Clarendon)
Chao C 1960 *J. Appl. Mech.* **27** 559
Cooper J A 1985 *PhD thesis* Hull University
Dewhurst R J and Al'Rubai W S A R 1989 *Ultrasonics* **27** 262

Dewhurst R J, Edwards C, McKie A D W and Palmer S B 1987 *Appl. Phys. Lett.* **51** 1066

—— 1988 *J. Appl. Phys.* **63** 1225

Dewhurst R J, Hutchins D A, Palmer S B and Scruby C B 1982 *J. Appl. Phys.* **53** 4064

Dewhurst R J, Nurse A G and Palmer S B 1988 *Ultrasonics* **26** 307

Doyle P A 1986 *J. Phys. D: Appl. Phys.* **19** 1613

Edwards C, Taylor G S and Palmer S B 1989 *J. Phys. D: Appl. Phys.* **22** 1266

Emmony D C 1985 *Infrared Phys.* **25** 133

Fox J A 1974 *Appl. Phys. Lett.* **24** 461

Gournay L S 1966 *J. Acoust. Soc. Am.* **40** 1322

Graff K F 1975 *Wave Motion in Elastic Solids* (Oxford: Clarendon)

Harada Y, Kanemitsu Y, Tanaka Y, Nakano N, Kuroda H and Yamanaka K 1989 *J. Phys. D: Appl. Phys.* **22** 569

Hsu N N 1985 *NBS Report* NBSIR 85-3234

Hutchins D A, Dewhurst R J and Palmer S B 1981a *Ultrasonics* **19** 103

—— 1981b *J. Acoust. Soc. Am.* **70** 1362

Hutchins D A, Lundgren K and Palmer S B 1989 *J. Acoust. Soc. Am.* **85** 1441

Hutchins D A and Nadeau F 1983 *IEEE Ultrasonics Symp.* vol 2, p 1175

Kane T J, Kozlovsky W J and Byer R L 1986 *Optics Lett.* **11** 216

Kaye G W C and Laby T H 1973 *Tables of Physical and Chemical Constants* 14th edn (London: Longman)

—— 1986 *Tables of Physical and Chemical Constants* 15th edn (London: Longman)

Kino G S and Stearns R G 1985 *Appl. Phys. Lett.* **47** 926

Knopoff L 1958 *J. Appl. Phys.* **29** 661

Knopoff L and Gangi A F 1959 *Geophysics* **24** 681

Krehl P, Schwirke F and Cooper A W 1975 *J. Appl. Phys.* **46** 4600

Ledbetter H M and Moulder J C 1979 *J. Acoust. Soc. Am.* **65** 840

Lee R E and White R M 1968 *Appl. Phys.* **12** 12

Lord A E Jr 1966 *J. Acoust. Soc. Am.* **39** 650

Miller G F and Pursey H 1954 *Proc. R. Soc.* A **223** 521

Mooney H M 1974 *Bull. Seismol. Soc. Am.* **64** 473

Moss B C and Scruby C B 1988 *Ultrasonics* **26** 179

Murphy J C and Wetsel G C 1986 *Mater. Eval.* **44** 1224

Opsal J and Rosencwaig A 1982 *J. Appl. Phys.* **53** 4240

Pao Y H, Gajewski R R and Ceranoglu A N 1979 *J. Acoust. Soc. Am.* **65** 96

Pekeris C L 1955 *Geophysics* **41** 469

Percival C M 1987 *J. Appl. Phys.* **38** 1967

Ready J F 1971 *Effects of High Power Laser Radiation* (New York: Academic)

Rose L R F 1984 *J. Acoust. Soc. Am.* **75** 723

Rosencwaig A and Gersho A 1976 *J. Appl. Phys.* **47** 64

Sachse W and Hsu N N 1979 *Physical Acoustics* vol XIV, eds W P Mason and R N Thurston (New York: Academic) p 277

Scruby C B 1985 *Research Techniques in Nondestructive Testing* vol 8, ed R S Sharpe (London: Academic) p 141

—— 1986 *Appl. Phys. Lett.* **48** 100

Scruby C B, Dewhurst R J, Hutchins D A and Palmer S B 1980 *J. Appl. Phys.* **51** 6210

—— 1982 *Research Techniques in Nondestructive Testing* vol 5, ed R S Sharpe (New York: Academic) p 281

Scruby C B and Moss B C 1985 *Rayleigh Wave Theory and Applications* eds E A Ash and E G S Paige (Berlin: Springer) p 102

Scruby C B, Wadley H N G, Dewhurst R J, Hutchins D A and Palmer S B 1981 *Mater. Eval.* **39** 1250

Sessler G M, Gerhard-Mulhaupt R, West J E and von Seggern H 1985 *J. Appl. Phys.* **58** 119

Sigrist M W and Kneubuhl F K 1978 *J. Acoust. Soc. Am.* **64** 1652

Silk M G 1984 *Ultrasonic Transducers for Nondestructive Testing* (Bristol: Adam Hilger)

Sinclair J E 1979 *J. Phys. D: Appl. Phys.* **12** 1309

—— 1982 Private communication

Spooner T 1927 *Properties and Testing of Magnetic Materials* (New York: McGraw-Hill)

Tam A C 1986 *Rev. Mod. Phys.* **58** 381

Tam A C and Leung W P 1984 *Appl. Phys. Lett.* **45** 1040

Viktorov I A 1967 *Rayleigh and Lamb Waves* (New York: Plenum)

Vogel J A and Bruinsma A J A 1988 *Non-destructive Testing* eds J M Farley and R W Nichols (Oxford: Pergamon) vol 4, p 2267

von Gutfeld R J 1980 *Ultrasonics* **18** 175

von Gutfeld R J and Melcher R L 1977 *Appl. Phys. Lett.* **30** 357

Wadley H N G, Simmons J A and Turner C 1984 *Review of Progress in NDE, Santa Cruz, 1983* vol 1, eds Thompson and Chimenti (New York: Plenum) p 683

Weaver R L and Pao Y H 1982 *J. Appl. Mech.* **49** 821

White R M 1963 *J. Appl. Phys.* **34** 3559

6 Applications Using Laser Generation of Ultrasound

In Chapter 4 we discussed applications of laser interferometry to the reception of ultrasound. The combination of laser generation with reception offers a completely remote ultrasonic inspection system, opening up a wide range of exciting applications to flaw detection and materials characterization. In this chapter we therefore propose to concentrate on applications where the ultrasound is both generated and received by means of lasers. We shall, however, also discuss some applications of laser-generated ultrasound alone at appropriate points in the chapter. Because of its great importance, we shall start with applications which relate to the main purpose of non-destructive testing, i.e. the detection and characterization of defects (flaws).

6.1 APPLICATIONS TO FLAW DETECTION

6.1.1 Introduction

Ultrasound propagates very efficiently through materials such as metals which are relatively opaque to electromagnetic and particle radiation. It is rapidly becoming therefore the main candidate for inspecting a wide range of materials, components and engineering structures to determine whether any significant defects are present, and if so to quantify their size, etc. Conventional ultrasonic techniques employing piezoelectric transducers are well established for flaw detection, and have made a major contribution to quality assurance and structural integrity which is now very well documented (see, for instance, Krautkramer and Krautkramer (1977), Szilard (1982), Farley and Nichols (1988)).

The piezoelectric transducer has proved itself to be a highly versatile device, transmitting and receiving ultrasound efficiently, having undergone many stages of development and refinement for an increasingly wide range of applications. Although other materials such as quartz, lead metaniobate and polyvinylidene fluoride are used for transduction, the most popular material for operation close to room temperature is still lead zirconate titanate (referred to universally as PZT from its chemical composition). A disc of PZT of suitable dimensions can be engineered into a practical probe with a resonance frequency in the range 1–10 MHz relatively simply and cheaply. Particular skills and experience are needed when special characteristics are required, e.g. a focused beam or very short pulse, but otherwise it is not difficult to produce probes with adequate efficiency as transmitters and good sensitivity as receivers of ultrasound, provided care is taken with the various interfaces and bonds during construction.

In view of this, the reader would be justified in asking why we are writing about replacing such a simple yet efficient device as the piezoelectric transducer with a far more complicated and expensive system using lasers. By way of reassurance, we do not see any need for an alternative to the piezoelectric transducer for the vast majority of ultrasonic applications, at least in the foreseeable future. Nevertheless, there are a small number of applications where the piezoelectric transducer is either completely ineffective or else extremely difficult to use. Furthermore, it seems likely that this number will increase with time, as industry seeks higher standards of quality and reliability, in a wider range of materials and components of increasing complexity and difficulty of inspection, and under greater extremes of temperature and other environmental conditions.

As already discussed in Chapter 1, the major drawback of the conventional transducer lies, not in the piezoelectric element itself, but rather in the need for a couplant to transmit the ultrasound to the specimen. The most commonly encountered problems during defect detection are changes in signal strength and frequency response due to couplant variations. However, the consequent changes in loading of the transducer front face also give rise to changes in performance. The couplant layer may itself introduce unwanted resonances, in addition to introducing errors in time-of-flight measurements. Finally, there are extra difficulties associated with the transmission of shear waves through the viscous fluid couplant. Because all these problems are exacerbated during scanning, as much ultrasonic inspection as possible is carried out in a water bath.

It was also pointed out in Chapter 1 that ultrasonic inspection becomes particularly difficult at elevated temperatures, mainly because the most common couplants tend to lose their viscosity, evaporate, and undergo chemical degradation. At temperatures much in excess of 100 °C it is very hard to find couplants suitable for scanning so that dry, semi-permanent bonding methods have to be considered. At elevated temperatures the materials used during probe construction (e.g. epoxy resins, plastics, solders) also begin to deteriorate, so that alternative construction procedures become necessary. At even higher temperatures (e.g. in excess of 300 °C) PZT must be replaced by other piezoelectric materials such as lithium niobate, but even these cannot be used much above 700 °C. Various different methods of probe construction and coupling, including the use of waveguides, wheels and water-jets, are all considered in Andrews (1982), but they all present problems of data interpretation, and may also be somewhat inflexible. Thus non-contact (and preferably remote) ultrasonic sensors would be most beneficial.

Although high temperature is probably the hostile environment of most interest, it should also be noted that there are major problems in using piezoelectric contact transducers in either corrosive or radioactive environments. Both types of environment cause the materials used in transducer construction to degrade slowly, so that some form of remote sensing is clearly to be preferred.

In recent years there has been a growing interest in the rapid, automatic scanning of transducers over specimens in order to detect various defects. If immersion in a water bath is not permissible, it is difficult to combine adequate scanning speeds with good reproducible sensitivity using contact transducers. The mechanical scanning of contact probes over complicated geometrical shapes or within confined spaces is especially difficult.

Thus, while it is recognized that the vast majority of defect detection will continue to be carried out by contact or immersion piezoelectric probes, there are clearly some application areas where a remote, non-contact technique such as laser generation and reception could be used to good advantage, i.e. as follows.

(a) For inspecting materials at very high temperatures, especially above 600–700 °C. A remote sensor such as the laser method would be far less temperature sensitive than a contact or near-contact probe (such as an EMAT).

(b) For inspecting materials in hazardous environments (e.g. highly corrosive or radioactive). A remote sensor behind a window would not be exposed to materials degradation in the way a contact or near-contact probe would. Only an optical method can be used in this manner.

(c) For inspecting materials or components with sensitive surfaces that must be kept clean, free of couplant and unscratched. A non-contact probe is essential, although it does not need to be remote. Thus EMAT and laser techniques would be options.

(d) For inspecting specimens that are very small or thin, and materials such as high-strength ceramics, where very small defects need to be detected in the region immediately below the surface. If use of a liquid is acceptable, focused immersion probes may be suitable, e.g. in an acoustic microscope. Otherwise contact probes may not have sufficient resolution, and in any case tend to have a dead zone just below the surface. This is not a problem for laser techniques, where the extremely small probe area leads to a dead zone currently estimated to be less than 100 μm, and which could certainly be reduced with further development.

(e) For inspecting awkward geometries and confined spaces. Conventional probes may be too large to follow surface geometry or to be introduced into restricted spaces, e.g. at the root of a weld, or inside a small-diameter pipe. Laser beams can be introduced into such regions by means of mirrors or optical fibres.

(f) For making area inspections at speeds that cannot be attained by mechanical probe scanning. Laser beams can be rapidly scanned by a number of methods.

(g) When quantitative, broadband, high-resolution ultrasonic data are required, for instance during calibration tests.

This is probably a suitable point to comment on the relative merits of laser techniques and EMATS, since for purely non-contact applications there will be a choice. EMATS may well have the advantage because of their simplicity and cheapness compared with lasers, although they offer lower bandwidths. However, if a remote technique or a very small probe area is also indicated, the EMAT no longer remains an option. We also note that there may be applications where the best inspection system involves a combination of laser and EMAT transducers.

Various configurations of laser source and receiver can be used when inspecting samples, which closely resemble those used for conventional probes. Thus the first configuration has the source and receiver on opposite surfaces of the sample (figure 6.1(a)) so that the probing ultrasonic wave field samples the material in transmission. This configuration is advantageous for laser ultrasonics since the sample isolates one optical system from the other. The transmission configuration is particularly useful for detecting the presence of planar defects (such as lamellar defects and inclusions in hot rolled steel) which cause a loss of signal when interposed between source and receiver. It is also a good geometry for inspection in the ablation regime, since this source radiates high-intensity compression waves normal to the source surface.

In the thermoelastic regime little energy is radiated normal to the surface (figure 5.20), so that the geometry of figure 6.1(a) would give disappointing results. The geometry of figure 6.1(b) would therefore be preferable for a transmission measurement in the thermoelastic regime. Both of these

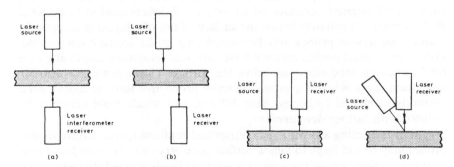

Figure 6.1   Four basic configurations for laser ultrasonic testing of solids. (a) Direct transmission, especially suitable for bulk measurements with compression waves in the ablation regime. (b) Oblique transmission, more suitable than (a) for the thermoelastic regime, and also for shear-wave measurements. (c) Single-sided measurement, especially suitable for surface-wave measurements and the time-of-flight technique. (d) Single-sided pulse-echo configuration.

configurations, however, require access to both sides of the specimen. This is regrettably not always possible in practice, so that the ability to inspect from one surface is essential. Two single-sided probe configurations are shown in figures 6.1(c) and (d). The former is obviously ideal for detecting surface-breaking defects, since it permits Rayleigh waves to be used in the inspection. However, it is also useful for time-of-flight measurements of the ultrasound diffracted by volumetric defects. In figure 6.1(d) the source and receiver are coincident, to emulate a pulse-echo arrangement. Whereas this is the simplest way to use a conventional probe, it is more difficult for laser ultrasound, since separate laser systems must be employed for generation and reception, and care must be taken to minimize breakthrough from the high-power source laser to the detection system. This configuration is useful for detecting echoes from subsurface or deeply buried defects.

In figures 6.1(a) and (d) in particular, the laser technique can be visualized as a non-contact replacement for a piezoelectric inspection system. This exemplifies one approach to the deployment of laser generation and detection for flaw detection, which is to tailor the performance of the laser system to emulate that of conventional pulse-echo, pitch-and-catch, transmission, etc, probe arrangements. Such an approach works well in the ablation regime, where the directivity of the source and receiver are both similar to that of a standard compression-wave probe. There is, however, an alternative approach in which the inspection arrangement is designed so as to make best use of the ultrasonic fields produced in laser ultrasonics. This is an especially important consideration in the thermoelastic regime, where the majority of the compression-wave energy is radiated obliquely to the surface. In figures 6.1(b) and (c), the material under inspection is close to the maximum ultrasonic intensity for bulk and surface waves from the thermoelastic source.

Much conventional ultrasonic inspection employs angled shear waves. At present, shear-wave inspection appears more difficult to carry out with laser ultrasonics for two main reasons. Firstly, the source always generates compression as well as shear, while secondly interferometers have mainly been developed to measure perpendicular surface displacement. Thus any shear-wave data are likely to be confused by compression-wave echoes which may precede or overlap them. The conventional option of using a mode-converting angled shoe to generate shear waves defeats the purpose of making an entirely non-contact technique. Some advances have been made in the use of phased array laser sources in the laboratory (Vogel and Bruinsma 1988), which may go part of the way to solving this problem.

The technical benefits of laser techniques which have been outlined above must always be weighed against disadvantages and costs. Although these have been discussed at various points in earlier chapters, they deserve brief reiteration here. Firstly, the capital cost of installing laser equipment is still much higher than that of piezoelectric probes: at least an order of magnitude higher. Furthermore, the running costs are also likely to be very much higher.

Secondly, although the actual probe areas on the specimen are much smaller than for piezoelectric transducers, the lasers themselves are far more bulky, and often need extra space in the form of a controlled environment, which is vibration free, dust free, and is temperature controlled, etc. We cannot, thirdly, ignore the safety hazards associated with high-power lasers, and the intrusion of the various safety requirements into the working environment, mainly to protect the eyes. Finally, it must be borne in mind that laser systems require highly trained staff for their installation, operation and maintenance. Industry is conservative, and the perceived costs of installation have been sufficient to prevent non-contact flaw detection from making a significant industrial impact, in spite of a number of laboratory demonstrations since 1973.

The main technical disadvantage of laser ultrasonic flaw detection lies in its sensitivity. Most laser interferometers are far less sensitive than piezoelectric receivers, as discussed in §4.8. Thus a completely remote laser-based system is likely to be less sensitive than a conventional system under most practical conditions. It is possible to raise its sensitivity by using higher power laser systems, but costs and operating problems escalate in consequence. Furthermore, although the laser source can be made to deliver ultrasonic pulses comparable in amplitude with those from broadband piezoelectric transmitters, this is generally only true at higher optical power densities, when ablation occurs, or when a liquid layer is applied to the specimen surface (negating the non-contact aspect of the laser source). In the truly non-contact and non-destructive thermoelastic regime, ultrasonic amplitudes are somewhat lower, although the line source, for instance, can be used to raise the ultrasonic amplitude in chosen directions. For applications where the specimen is to undergo further treatment or finishing, however, a limited amount of shallow surface damage is acceptable because it will later be removed, so that power densities can be raised into the ablation regime.

In the context of sensitivity, it is important to remember that the alternative non-contact receiver, the EMAT, is only slightly less sensitive than when an interferometer is used under ideal conditions (i.e. polished surface finish), as discussed in §4.8 (see table 4.1). It could, however, be a better choice in terms of sensitivity on very rough or oxidized surfaces. The EMAT is, generally speaking, less efficient than the pulsed laser as a broadband transmitter of compression waves. In the case of shear waves the situation is more complicated. One option, therefore, which may alleviate the problem of low sensitivity with an all-laser system, is to combine the laser source with an EMAT receiver.

6.1.2 Volumetric defect detection

The use of laser ultrasonic techniques for the detection of flaws in metals first appeared in the literature in the early 1970s. Thus Calder and Wilcox (1973) used a 1.2 J Q-switched ruby laser to demonstrate the detection and location of $\frac{1}{16}$ inch side-drilled holes (to simulate flaws) in a 1 inch thick aluminium plate. Using a 30 MHz piezoelectric transducer as receiver, they were able to detect ultrasonic echoes from the holes. About the same time, Giglio (1973) also demonstrated laser generation of ultrasound as well as laser interferometry as a means of detection, although he investigated these separately, and only discussed their combined use in a non-contact flaw detection system. Giglio used a relatively low-power (5 kW) laser for the source with a long pulse length (300 ns). It is likely that the ultrasonic signal amplitude would have been too low for interferometric detection.

One of the earliest published reports of a completely non-contact laser ultrasonic flaw detection system was made by Bondarenko *et al* (1976). The ultrasonic source was generated by a Q-switched ruby laser, emitting pulses of length 30–50 ns, and peak power up to 50 MW, so that the energy per pulse must have been in excess of 1 J. The beam diameter was approximately 1 mm and the incident power density on the steel sample was therefore as high as 5×10^9 W cm^{-2}. Thus the source would have been well into the plasma regime and far from non-destructive. The signals were detected with a Michelson interferometer, in which the minimum detectable surface displacement was 1 nm over a very wide frequency band extending from 5 kHz–150 MHz.

Bondarenko *et al* (1976) used this system to make transmission measurements of a simulated planar defect, which was produced by clamping together two thin polished steel blocks. Part of the interface between the blocks was well coupled ultrasonically by oil, while the other part was dry. Their apparatus detected compression waves of amplitude greater than 10 nm, which were transmitted through the oiled portion of the surface. They could locate the unlubricated area by the loss of ultrasonic amplitude. There was no report of the detection of ultrasound either reflected or scattered from the defective region.

A laser ultrasonic defect detection system was also described by Wilcox and Calder (1978) (cf Calder and Wilcox 1980). Their experimental arrangement (figure 6.2) was similar to Bondarenko *et al* (1976), except that they

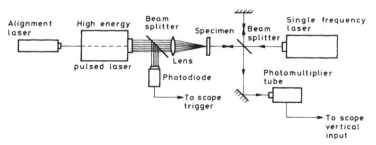

Figure 6.2 Schematic diagram of an experimental arrangement for non-contact flaw detection (Calder and Wilcox 1980).

used an even more powerful source laser, i.e. 15 J Q-switched neodymium–glass of pulse length 20–30 ns. The interferometer was also a Michelson type, employing a single-frequency argon laser to give an intrinsically more sensitive system than one incorporating a lower power helium–neon laser. The sensitivity was also maximized by using a lens to focus the interferometer beam onto the specimen and collect a larger fraction of the returned light from the surface. In addition, the photomultiplier was pulsed to switch on just prior to the arrival of required signal. However, the interferometer was unstabilized, so that the output amplitude was uncalibrated.

Calder and Wilcox (1980) demonstrated the potential of their system in an experiment in which they side-drilled a 1.5 mm hole into the side of a 25 mm thick aluminium plate. A single laser shot was fired and the displacement measured at the opposite surface (figure 6.3). The first arrival is a sharp pulse, consistent with the compression wave generated by ablation. There is also a compression-wave echo some 8 μs later. They also claim a flaw signal about 4 μs after the first arrival, which they interpret as an echo between the back wall and the flaw. However, the shear wave would also arrive at about the same time, although it is not clear how strong this would be in their experiment. It is also possible that the signal immediately following the compression arrival could be due to scattered ultrasound from the defect. It is unfortunately not possible to quantify these data because the sensitivity of an unstabilized interferometer varies with time. The signal/noise for the data as presented is exceptionally good, reflecting the very high power of the laser but, as the authors themselves point out, the specimen surface undergoes considerable damage due to melting and vaporization, so that the measurements were not non-destructive.

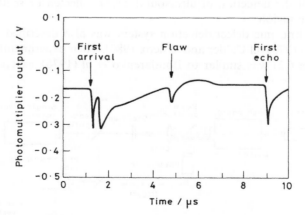

Figure 6.3 Flaw detection using the laser-ultrasonic technique of figure 6.2. The flaw echo is obtained from a 1.5 mm diameter side-drilled hole at the midpoint of a 25 mm aluminium plate (Calder and Wilcox 1980).

Budenkov and Kaunov (1979) discuss the potential use of laser-generated ultrasound for the on-line inspection of pipes and boiler tubes at elevated temperatures (700–900 °C). They planned to use an EMAT as ultrasonic receiver rather than an interferometer, but did not present any non-destructive testing data in their paper.

At the same conference, Krautkramer (1979) discusses non-destructive testing in which the ultrasonic source is generated by a pulsed laser, and the receiver is either a Michelson interferometer (as used in the other studies reported in this section) or a 'transit time' (i.e. long-path) interferometer. The light scattered from the specimen surface is divided into two parts, the first travelling directly to the photodetector, and the second travelling via a delay line before interfering with the first at the photodetector. This type of interferometer, which has been described in more detail in Chapter 3, is particularly well suited for displacement measurement on rough surfaces or at high temperatures, provided a sufficiently powerful laser is used. Wells Krautkramer (1984) have applied this laser inspection system to the detection of defects in continuously cast steel at 1000 °C. The application to carbon-fibre composite inspection is also briefly discussed.

Wellman (1980) used a Q-switched ruby laser (pulse energy 0.9–1.4 J, length 50 ns) and Michelson interferometer as receiver. Some of this work was carried out in transmission, using a similar experimental arrangement to Wilcox and Calder (1978). For the results presented on flaw detection, the source laser and interferometer were on the same face of the sample (figure 6.4(a)). Wellman varied the incident power density from 140 to 500 MW cm^{-2}, remaining within the plasma regime. Under these conditions, care must be taken first to prevent light from the source laser and the plasma itself from entering the interferometer and, secondly, to avoid airborne signals from the detonation wave created by the plasma.

The data from his first flaw-detection experiment indicate that his source and receiver were sufficiently well separated to receive the back-wall echo before any disturbance through the air. Wellman examined a 25 mm thick, 50 mm diameter polished steel disc, which contained two parallel side-drilled holes to act as simulated flaws (figure 6.4(b)). These holes were 2 mm in diameter, separated by 1 mm, and were 16.5 mm below the surface. The oscilloscope trace (figure 6.5(a)) shows a signal from the defects between what are presumed to be the surface wave and the back-wall echo. This relatively simple experiment also demonstrates the good resolution obtainable using laser ultrasonics, since the two defects were apparently individually resolved at higher laser power (figure 6.5(b)). Such data would be difficult to obtain using piezoelectric probes without extensive signal processing to remove the 'ringing' that would follow the wave arrivals.

The same technique was applied to the detection of defects in artillery shells. Because the surface was no longer specially polished, and was also curved, he found it necessary to modify his interferometer to obtain adequate

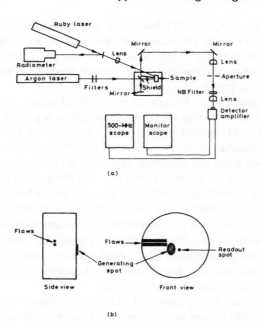

Figure 6.4 (*a*) Schematic diagram of an experimental arrangement for single-sided non-contact ultrasonic testing. (*b*) Flawed specimen showing positions of laser source and interferometric receiver. Flaws were 2 mm diameter holes separated by 1 mm (Wellman 1980).

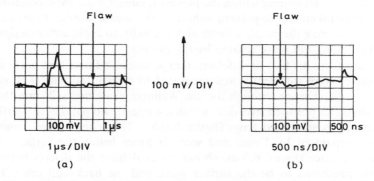

Figure 6.5 Oscilloscope traces of non-contact ultrasonic data from the specimen of figure 6.4 (Wellman 1980).

sensitivity. He used a lens to focus the beam onto the surface to collect more returned light, and a second lens to focus the fringe system onto the detector, which was itself changed to an avalanche photodiode. The system suffered from vibration problems, which he was unable to compensate for during the timescale of his programme. Figures 6.6(*a*) and (*b*) show some of the data

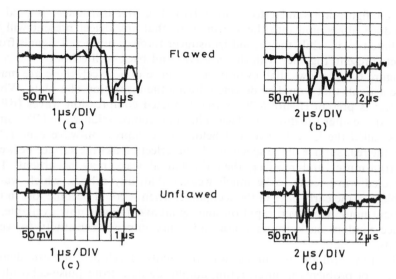

Figure 6.6 Oscilloscope traces comparing the ultrasonic signal from a 2 mm deep, 10 mm long by 0.5 mm wide flaw in an artillery shell casing with the signal from the unflawed area (Wellman 1980).

from a 2 mm deep, 10 mm long slit, machined in the shell to simulate a defect. When compared with the signals from the unflawed region, figures 6.6(c) and (d), some differences due to the flaw are apparent, especially at the beginning of the data. The defect was surface breaking, and would therefore be expected to interact with surface waves. The data are otherwise not easy to interpret. It was noted at the time that this defect was much smaller than the critical defect size, and also that the small amount of surface damage caused by ablation was not considered to be serious.

In more recent work, Hutchins *et al* (1989) have demonstrated that artificial defects can be imaged inside metallic cylinders using ultrasonic tomography. They compared a laser source – EMAT detector system with an EMAT – EMAT system, and using reconstruction based on compression and shear waves.

6.1.3 Time-of-flight technique

The time-of-flight ultrasonic technique was developed at Harwell over a number of years (Silk 1977). It differs from conventional ultrasonic testing in that it does not depend upon measuring an echo (reflection) amplitude from a defect. Rather, it makes use of the fact that the tips of defects act as diffraction sources of ultrasound, and it then measures the arrival times of these diffracted pulses relative to fixed arrivals, such as a back-wall echo. In its simplest form, the technique employs two short-pulse, broadband probes in a configuration similar to figure 6.1(c). The first transducer is used to introduce ultrasound into the specimen so that any defect present will be irradiated by it, while the second transducer receives diffracted signals from defects present, together with direct waves and back-wall echoes.

Figure 6.7(a) shows that, even in the absence of a defect, there are many wave arrivals that would be detected during the inspection of a plate. When a defect is present (figure 6.7(b)), the diffracted compression waves (dPP) must arrive after the direct, surface-skimming compression wave (P), sometimes called the lateral wave, but before the compression-wave echo (PP) from the rear surface of the specimen. If the defect breaks the surface between the transmitter and receiver, then the lateral wave may be absent. The diffracted wave tends to be much weaker than any echoes, which was a problem in the early days of the technique, when it could become lost in the ambient noise. However, rapid on-line signal averaging is now available, to reduce random background noise and hence improve sensitivity to weak signals.

The time-of-flight technique was originally developed to size defects because of problems in interpreting amplitude data from pulse-echo ultrasonics. This it does by accurately locating the tips of any defect present,

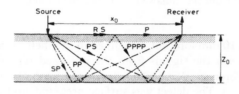

(a) Wave arrivals in parallel plate

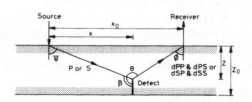

(b) Wave arrivals diffracted by tip of defect

Figure 6.7 Experimental arrangement for the time-of-flight technique showing wave arrivals referred to in text and in figure 6.8. For simplicity arrivals in absence of defect are shown separate from diffracted waves. P, S, R are direct compression, shear and Rayleigh waves, PP is P wave reflected as P wave, dPP is P wave diffracted as P wave, etc (Scruby *et al* 1987).

using the arrival times of the waves diffracted by each defect. If the two transducers are scanned together across the specimen so as to vary x while maintaining the geometry shown in figure 6.7(b), then the arrival time of the crack-tip diffracted compression wave is given by

$$T_{dPP} = \{(x^2 + z^2)^{1/2} + [(x_0 - x)^2 + z^2]^{1/2}\}/c_1 \qquad (6.1)$$

where c_1 is the compression-wave speed. As figure 6.8(a) shows, this dependence takes the form of a characteristic parabola on a B scan. The lateral wave and back-wall echo arrival times are independent of x, being given by

$$T_P = x_0/c_1 \qquad (6.2)$$

$$T_{PP} = [(x_0^2 + 4z_0^2)^{1/2}]/c_1. \qquad (6.3)$$

Figure 6.8(a) confirms that the diffracted wave lies between these two arrivals, in the centre of the range, so that it is easy to resolve. When the time-of-flight technique is used in practice, data similar to figure 6.8(a) are obtained, from which the depth and location of a crack tip can be calculated. If the defect is entirely buried within the specimen, and therefore does not break the surface, there will be diffraction from both its ends. The size of the defect can then be calculated from the difference in the coordinates deduced for each end.

Although the time-of-flight technique was originally developed for defect sizing, it is now also used for defect detection because it is sensitive to every form of scatterer within a specimen, and is unlikely to miss any significant defects. Considerable care is, however, needed in interpretation to avoid false alarms. Thus it has been applied in the defect detection trials which were used to assess the reliability of inspection methods for the primary pressure circuit of a pressurized water reactor (Curtis and Hawker 1983). The time-of-flight technique has also been developed for the inspection of offshore structures (Newton 1987).

For optimum performance, the time-of-flight technique requires transmitter and receiver transducers which are broadband and which generate a reasonably broad beam within the specimen. From theoretical studies (Ogilvy and Temple 1983, Scruby et al 1987), the maximum diffracted-wave amplitude for the type of geometry shown in figure 6.7 occurs for beam angles in the range 50–70° to the surface normal. The pulsed laser source and interferometer receiver fulfil the criterion for broad bandwidth better than most piezoelectric transducers. The interferometer receiver, acting virtually at a point, has a very wide reception beam as required, although it is not maximized around 60°. In the ablation regime the laser source generates a very broad compression-wave beam which, although not a maximum at 60°, has reasonable amplitude (0.53 × maximum) in this direction. However, in the thermoelastic regime, the compression-wave beam is a broad lobe centred

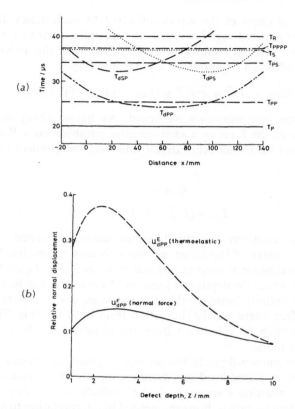

Figure 6.8 (*a*) Early wave arrival times calculated for $z_0 = 47$ mm, $z = 41$ mm, $x_0 = 120$ mm, as a function of x for compression and shear velocities of 5960 and 3240 m s^{-1} respectively (symbols defined in figure 6.7). (*b*) Comparison of diffracted compression-wave amplitudes from defects at a depth z, for normal force and thermoelastic sources. $x_0 = 20$ mm, $x = 10$ mm, $z_0 = 10$ mm. Displacement amplitudes are scaled to give equal back-wall echoes for each source (Scruby *et al* 1987).

on 60°, which is ideal for the time-of-flight technique. Thus, for equal back-wall echo amplitudes, the thermoelastic source generates a larger amplitude diffracted wave than the ablation source for depths of the crack tip, z, less than half the source–receiver separation, x_0, as shown in figure 6.8(*b*).

This theoretical model can be used to calculate simulated time-of-flight data for both the ablation (Gaussian impulse) source and thermoelastic (dilatation) source. The geometry is similar to figure 6.7, the receiving transducer being assumed to operate at a single point, and with a reasonably broadband frequency response. In the simulated B-scan data of figure 6.9, the diffracted compression wave which precedes the back-wall echo between

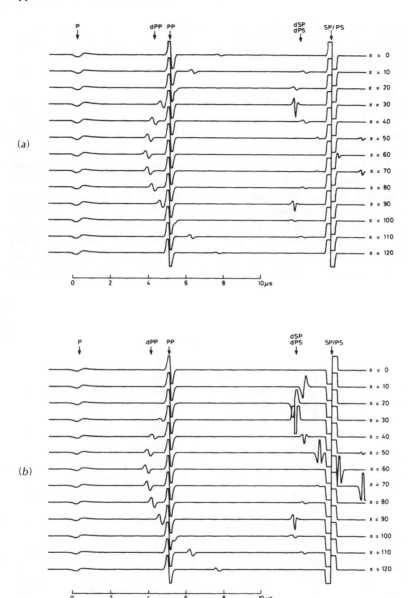

Figure 6.9 Computed B-scan data for (*a*) ablation (normal force), and (*b*) thermoelastic sources and receiver (e.g. interferometer) with Gaussian impulse response, showing ultrasound diffracted from 6 mm defect. Geometry and symbols defined in figure 6.7 with $x_0 = 120$ mm, $z_0 = 47$ mm, $z = 41$ mm (Scruby *et al* 1987).

$x = 30$ and 90 mm is of slightly greater intensity for the thermoelastic source. The direct P wave in both cases travels immediately below to the surface. Figure 6.9(b) shows that the thermoelastic source generates higher amplitude diffracted SP and PS waves than the ablation source. This is because the thermoelastic regime acts as very strong shear-wave source.

An experimental study of diffraction from defects in plates was carried out (Aindow *et al* 1985). The testing geometry was as close as possible to the one used in the above calculations, i.e. the 47 mm thick steel plate had a 6 mm slit to represent a defect on the opposite surface from the source and receiver. The laser line source generated a weak plasma and was separated from the broadband piezoelectric transducer by a fixed distance of 120 mm. The source and receiver were scanned together across the plate, so that the horizontal source–defect distance, x, varied in 10 mm steps. The agreement between the theoretical B scan (figure 6.9(a)) and these experimental data (figure 6.10) is remarkably good. The lateral wave (P) is similar in shape, although of larger amplitude in the experimental data. The diffracted compression wave (dPP) from the end of the defect can be seen ahead of

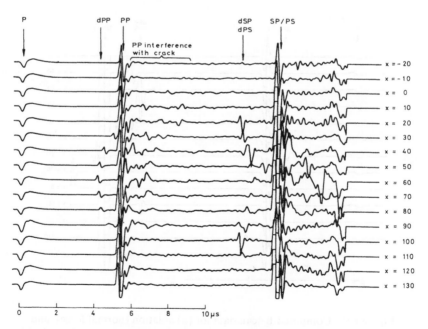

Figure 6.10 Experimental B-scan data for the same geometry as the computed data of figure 6.9. The laser line source generates weak ablation, and the receiver was a broadband piezoelectric transducer on a 60° angled compression-wave shoe.

the back-wall echo (PP) between $x = 30$ and $90 \, \text{mm}$, and is of similar amplitude to that calculated. There is, however, some structure following the back-wall echo that is absent from the calculated data. This is believed to be due mainly to ultrasonic energy that is both reflected from the back wall and diffracted at the tip of the defect, with mode conversion as an additional possibility. Some of the weaker structure may be due to scattering of the ultrasound from the microstructure of the steel.

Figure 6.11 shows the experimental arrangement for inspecting a welded joint that was suspected of having a defect present in the weld or associated heat-affected zone. The ultrasonic source was a Q-switched Nd:YAG laser delivering approximately $80 \, \text{mJ}$ of energy per pulse at the surface. The beam was focused by means of a cylindrical lens to form a sharp line parallel to the weld line on the surface. The incident power density was high enough to cause some ablation of the steel. The receiver, placed the other side of the weld, was a short-pulse piezoelectric probe (Krautkramer type G5KB) of $5 \, \text{MHz}$ nominal frequency, which was mounted on a perspex shoe so as to receive compression-wave beams at angles centred on $60°$ in steel. The optical system and the piezoelectric receiver remained fixed, while the welded specimen was slowly scanned in a direction parallel to the weld (i.e. perpendicular to the plane of the page). Although signal averaging was not used, the signal to noise ratio was very good even for the relatively weak diffracted wave, as the individual central A scan in figure 6.12 shows. This wave is the first arrival because the direct compression wave is heavily attenuated by the weld geometry.

When these individual A scans are combined to form a B scan, as in figure 6.12 (converting amplitudes into grey-scale levels), the diffracted signal from the crack tip is seen to vary in arrival time. This indicates that the depth of

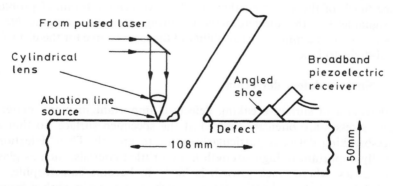

Figure 6.11 Apparatus used to scan an angled butt weld. The laser pulse was focused to a line in the ablation regime to generate ultrasound, and an angled piezoelectric transducer was used to receive ultrasound (Scruby 1989).

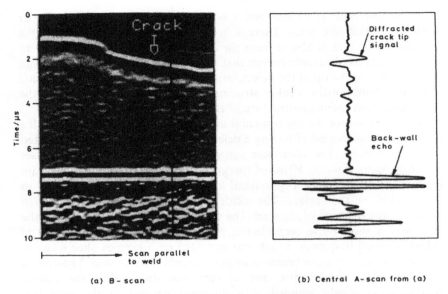

(a) B-scan   (b) Central A-scan from (a)

Figure 6.12 (a) Change in arrival time of wave diffracted from crack tip in a time-of-flight B scan shows the variation of depth of crack in the heat-affected zone of the weld. (b) Single A scan from centre of (a) showing more clearly crack-tip diffraction and back-wall echo (Scruby 1989).

the crack varies across the width of the plate. Similar data could have been obtained using an entirely contact transducer system, although it might have been difficult to get a standard piezoelectric transducer in quite as close to the weld as the laser pulse, because of the weld geometry. It is a pity that an entirely non-contact optical system was not used, but with the very poor surface finish of the specimen there could have been substantial problems with signal level in the interferometer, requiring extensive averaging. Nevertheless, these data confirm the suitability of the laser source for the ultrasonic time-of-flight technique.

6.1.4 Surface defect detection

The detection of surface-breaking defects is important in many materials because stresses are often maximized at the specimen surface, so that any surface-breaking defect is particularly prone to growth. Their detection is especially important in high-strength more brittle materials, such as glasses and ceramics, where crack growth is likely to become catastrophic. The critical defect size for a brittle material, such as a ceramic, with a fracture stress of 400 MPa is of the order of 100 μm. Advanced high-strength engineering ceramics such as silicon nitride are being developed with failure stresses even higher than this. Thus a ceramic with a fracture toughness of 10 MN m$^{3/2}$

and a service stress in excess of 1000 MPa cannot contain defects in excess of 30 μm. This imposes a considerable burden on the detection of all defects, but especially of those at the surface. In the case of ceramics, the strong preference is for a technique which is non-contact, since some of these materials are slightly degraded by fluids such as water, and mechanical contact may scratch and damage the surface.

Laser ultrasonic techniques clearly have potential for the detection of surface defects in such sensitive materials. They are capable, as shown in Chapter 5, of generating high-amplitude Rayleigh waves that probe the surface region. As will be discussed shortly, published data show that defects as small as 100 μm can be detected on metals using entirely non-contact laser techniques. While this resolution is inadequate for the high-strength ceramics, there would appear to be no fundamental reason why defects as small as 10 μm cannot be detected by using very short (picosecond) lasers and exceptionally broadband detection. Figure 6.13(a) illustrates the manner in which a completely non-contact system based on laser ultrasonics can be used to detect a surface-breaking defect. The type of data obtained from this measurement is shown in figure 6.13(b). First there is a direct Rayleigh wave, whose time delay corresponds to the source–receiver separation. Then almost 2 μs later there are two signals linked to the presence of the defect (in this case simulated by a narrow slot).

A very detailed study of the detection of such surface defects has been carried out by workers at Hull University. Thus Cooper et al (1986) studied the interaction of laser-generated ultrasound with a series of slots machined in the surface of an aluminium block. A 30 ns pulsed YAG laser was used to generate the ultrasound, while detection was carried out by means of a small capacitance probe (a non-contact device). The slots varied in depth between 0.3 and 5 mm, being 0.1 mm wide. They show that such slots both transmit and reflect Rayleigh waves (figures 6.14(a) and (b)). Furthermore, because

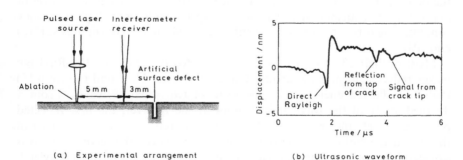

(a) Experimental arrangement   (b) Ultrasonic waveform

Figure 6.13 (a) Experimental arrangement to demonstrate scattering of a Rayleigh wave by a surface-breaking defect (Scruby 1989). (b) A direct Rayleigh wave is followed by a Rayleigh wave reflected from the defect. The second scattered signal is associated with diffraction from the crack tip.

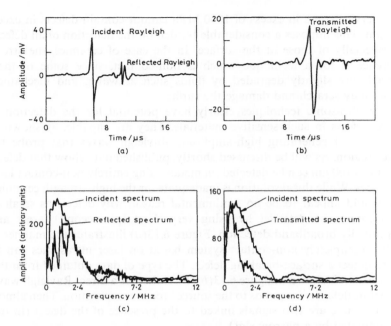

Figure 6.14 (*a*) Interaction of laser-generated Rayleigh pulse with a 0.75 mm deep slot in an aluminium block; source and receiver are on the same side of the slot; source–slot and receiver–slot distances are 25 mm and 6 mm respectively. (*b*) As (*a*), but with receiver on opposite side of slot; receiver–slot distance 10 mm, to show transmission of Rayleigh wave. (*c*) Comparison of frequency spectra of incident and reflected Rayleigh waves in (*a*). (*d*) Comparison of frequency spectra of incident and transmitted Rayleigh waves in (*b*) (Cooper *et al* 1986).

the depth penetration of the Rayleigh wave into the surface is a function of frequency, the transmitted Rayleigh-wave spectrum loses its high-frequency content, which is a major component of the reflected wave (figures 6.14(*c*) and (*d*)).

Most of their effort is, however, concentrated on analysing the Rayleigh waves reflected from these artificial surface defects. They find (figure 6.15) that the reflected energy takes the form of two separate waves. From a detailed study of the wave-arrival times at the receiver, the first reflected pulse has the characteristics of a Rayleigh wave reflected from the corner of the slot at the surface. As figure 6.15 shows so clearly, the second arrival is slot-depth dependent. Their data are consistent with scattering of the Rayleigh wave at the bottom of the slot into shear waves, which propagate back to the surface to produce a Rayleigh wave again (Cooper *et al* 1986). They also show that their ultrasonic technique could be used to measure the depth of the slot with comparable accuracy to the mechanical gauge used to check

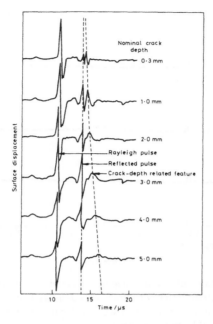

Figure 6.15 Laser-generated ultrasonic waveforms showing variation of second scattered wave from crack tip with depth (Cooper *et al* 1986).

their data, i.e. ± 0.1 mm. The shallowest slot that could be detected with their technique was 0.3 mm.

The same group of workers (Cooper *et al* 1986, Dewhurst *et al* 1986b) use a laser interferometer as a receiver instead of the capacitor, which gives better spatial and temporal resolution. This entirely remote, non-contact surface defect-detection system provides even better sensitivity. Using the laser source in the plasma regime, they were able to detect even shallower slots of 100 μm depth using single-shot data (i.e. without signal averaging). Hutchins *et al* (1986) have also used laser generation and reception in a parallel study of the interaction of ultrasound with slots in the surface of a plate. However, they focus more attention on transmitted waves. They observe that the Rayleigh wave is strongly mode converted into bulk waves by the corners of the slot, in a manner consistent with the results of Cooper *et al* (1986).

In a separate publication, Dewhurst *et al* (1986a) apply the laser-generated Rayleigh-wave technique to the detection of a real crack in a welded steel specimen. The surface was rusty, and an EMAT was used as an alternative non-contact receiver. Figure 6.16 shows that the crack produced a clear reflection of the Rayleigh wave in a similar manner to the artificial slots. The Rayleigh-wave reflection (*b*) is followed by some additional features which

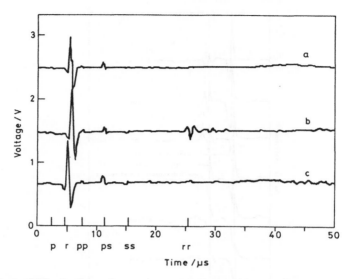

Figure 6.16 Surface ultrasonic waves detected on welded specimen (Dewhurst *et al* 1986a). (*a*) On plane surface away from weld; (*b*) with EMAT receiver 30 mm from cracked weld; (*c*) with EMAT receiver 30 mm from uncracked weld (with undercut).

the authors consider may be due to scattering from the crack tip. Data taken from an uncracked region of the weld ((*a*) and (*c*)) give a similar Rayleigh-wave signal to the parent plate, without any strong scattering or reflection from the rough weld surface, which even included some undercut at the edge. Hutchins and Hauser (1987) have also used laser generation (227 mJ ruby laser, unfocused to give a 6 mm × 4 mm beam, either in the thermoelastic regime or with an oil coating) and EMAT detection, in their case to study the ultrasonic echoes from a 7 mm surface-breaking slot in aluminium.

Many of the above data have been obtained using a small circular source of ultrasound, generated by focusing the incident light with a spherical convex lens. The source is thus of the ablation type, with or without the addition of oil or grease to enhance ultrasonic generation and reduce damage to the surface. One of the beauties of an optical source is that it can be easily changed in size and shape by means of standard optical components. As Aindow *et al* (1982) show, the laser light can be focused by means of a cylindrical lens into a line on the surface, which is ideal for generating directional Rayleigh surface waves. Such Rayleigh waves should be ideal for detecting linear surface defects, provided that they are parallel to the source line. Cielo *et al* (1985), however, focus the incident laser light into a circle on the surface of the specimen. In their non-contact surface inspection technique (figure 6.17) the pulse from the YAG laser passes through an axicon lens that focuses the light into a narrow circular ring. Thermoelastic stresses

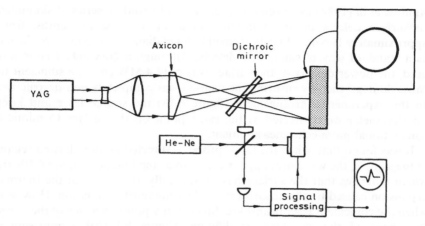

Figure 6.17 Experimental arrangement to focus incident laser pulse into annulus on specimen surface, and detect surface waves at centre of annulus by means of a laser interferometer (Cielo *et al* 1985).

in this ring launch circular Rayleigh waves that converge onto the centre of the circle, where they produce a very high surface displacement, which can be detected by means of a Michelson laser interferometer directed at the centre. The amplification factor for the Rayleigh waves is of the order of 20, and this is obtained without crossing into the damage-inducing ablation regime. Thus small defects can be detected by the relatively insensitive interferometer with much better signal to noise than otherwise. The authors describe the application of their technique to crack detection, the measurement of coating thickness, and the quality of adhesive bonding. In the first case they were easily able to detect cracks 1 mm deep and 0.1 mm wide. A much shallower crack of depth 0.1 mm was only just detectable. It is interesting to note that the detection threshold is similar to that of Dewhurst *et al* (1986b) using a similar interferometric detection system, but a focused point source in the ablation regime. One further advantage of the convergent-wave technique is that it is not preferentially sensitive to cracks lying in any given orientation on the surface. The indications are, however, that the alignment of the axicon lens with respect to the surface is very critical, which may restrict its broader application.

6.1.5 Subsurface defects

Defects that are immediately below the surface of the sample can sometimes prove a problem with contact transducers, because they are located within the 'dead zone' of a compression- or shear-wave probe, and yet do not break the surface to enable ready detection by surface waves. Defects of this nature lend themselves therefore to investigation by a laser ultrasonic technique.

Aindow *et al* (1984) use a remote laser system to study a series of simulated subsurface defects in an aluminium plate at a range of depths, from approximately 1–5 mm. Flat-bottomed holes of diameter 6.35 mm are chosen as a standard way of simulating defects. Although a Q-switched laser was used, the power density at the surface was only 4 MW cm^{-2}, insufficient to cause damage, thereby ensuring the method was completely non-destructive. In the experimental arrangement (figure 6.18) the source laser and laser interferometer were directed at the same point on the surface to mimic a conventional pulse-echo measurement.

It was found that, when the laser 'probe' is directed at an unflawed region of the sample, the waveform approximates to a step function (figure 6.19(a)), which indicates that the surface has risen locally as a result of the thermal expansion caused by the absorption of electromagnetic radiation. However, when the coincident laser beams are directed at a point over one of the flaws, the shape of the waveform is different (figure 6.19(b)), comprising a small-amplitude high-frequency (2.65 MHz) oscillation superimposed on a larger amplitude lower frequency (152 kHz) oscillation. In their analysis of the data, the depth of the flaw is calculated directly from the higher frequency oscillation, while the diameter is deduced from the lower frequency, provided the depth has first been calculated. The expression for the lower frequency is based on an expression from Morse (1936), which predicts that the frequency is proportional to the depth but inversely proportional to the square of the radius.

The experimental data of figure 6.19 are in good general agreement with the results of a finite-element calculation reported by Cielo (1985). He models the elastic response of a 0.3 mm thick aluminium layer of radius (a) 13 mm

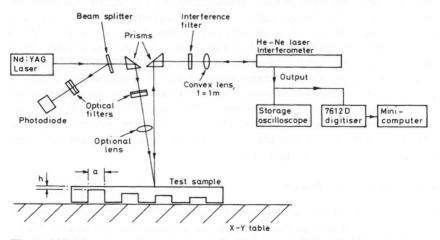

Figure 6.18 Apparatus to study detection of subsurface flaws. Note that the laser source (thermoelastic regime) and interferometer are coincident (Aindow *et al* 1984).

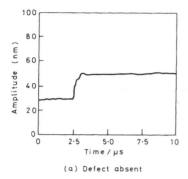 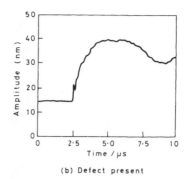

(a) Defect absent   (b) Defect present

Figure 6.19 (a) Initial elastic response when probing an unflawed region.
(b) Response above 6.35 mm defect, 1.2 mm deep. Note low-frequency
oscillatory signal, and low-amplitude high-frequency modulation (Aindow
et al 1984).

and (b) 4 mm, subjected to a short (5 μs) duration heating pulse over a
3.5 mm radius area. The calculated displacement at the centre of the heated
layer is predicted to oscillate in a similar manner to figure 6.19. If the
expression from Morse (1936) is used to extrapolate the period of this
oscillation from the start to the first minimum (6.5 μs as measured from figure
6.19) to the situation modelled by Cielo (1985), then a period of the order
of 30 μs is obtained, which is consistent with his calculation.

Aindow et al (1984) also demonstrate an alternative and arguably more
reliable method for sizing the defect by scanning the laser probe across one
of the defects. As figure 6.20 shows, the size can be estimated reasonably
accurately from the points where the oscillations first begin to appear. This
method would seem to have particularly good potential for lamellar flaws
close to the surface, since the oscillation is a very obvious feature. One
problem is that the sensitivity of the technique falls rapidly with increasing
depth for a given defect diameter: the oscillations are very weak for depths
greater than 3–4 mm for a diameter of 6 mm. Although not investigated in
this study, it would be expected that appreciable defect roughness or
variations in defect orientation relative to the surface could markedly
attenuate the signals.

Subsurface defects can also be detected by the perturbation they cause to
surface waves. They are expected to have a number of effects. Firstly, they
might cause attenuation and/or distortion of the Rayleigh-wave pulse
itself, depending upon the ratio of the defect depth to the Rayleigh wavelength.
Secondly, some echoes or scattering by the defect might be observed, while
thirdly (if the defect is large enough) a short-range guided wave (cf Lamb
wave or flexural mode) might be set up between the defect and the surface.
The study of subsurface defect detection by surface waves reported in Aindow
et al (1984) confirms that all these effects are indeed present. In their

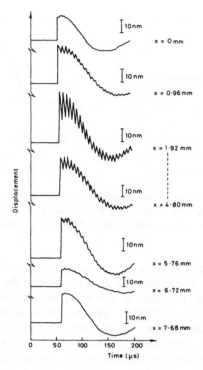

Figure 6.20 Waveforms obtained from linear scan of coincident source and receiver across region of specimen containing 6.35 mm diameter subsurface defect (Aindow *et al* 1984).

experiment the source and receiver are separated by 10 mm, so that a weak surface-skimming compression wave and strong Rayleigh wave are both observed, as shown in the first trace of figure 6.21. When the source–receiver combination is incrementally scanned across one of the defects, the Rayleigh is considerably distorted over the defect (see the central traces of figure 6.21), and a low-frequency oscillation is observed, whose period is of the same order as the flexural mode observed in the pulse-echo configuration discussed earlier. There are, furthermore, indications of echoes between the direct compression and Rayleigh waves. These waveforms are, however, more difficult to interpret than the pulse-echo waveforms, so that defect depth cannot as readily be deduced. Furthermore, the changes in the waveforms are less sudden making it more difficult to determine defect size.

6.1.6   Bonding defects

Without doubt one of the most important types of subsurface defect is a disbond or delamination of a surface layer. This may be some form of coating

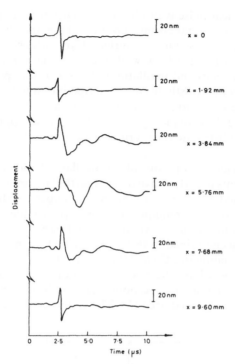

Figure 6.21 Detection of 6.35 mm subsurface flaw by perturbation in surface-wave waveforms (Aindow *et al* 1984).

(such as a ceramic plasma-sprayed onto a metal), or one ply of a composite. There is a growing interest in protective coatings and in many different types of composite, and techniques for detecting unbonded areas are currently being sought. As for some monolithic ceramics, non-contact inspection is often preferred to avoid contamination. Thus one candidate technique is thermography. Laser ultrasonics is a possible alternative. The principle upon which laser ultrasonic inspection would operate is very similar to that described in §6.1.5 for subsurface defects.

There are two approaches to estimating the size of the surface displacement. The first is more appropriate for thermal pulses whose length is much greater than the time taken for an ultrasonic pulse to propagate from the top surface to the back wall and return again to the top surface (of the order of $1-10 \mu s$ for surface layers of the order of 1 mm). In this low-frequency limit (which is the regime of thermography and photoacoustics), the thermoelastic strains induced at and just beneath the surface when it is subjected to a heat pulse cause it to expand. Because it is constrained at the ends, the surface 'buckles' outwards, and can be detected by an enhanced displacement of the surface over the defect.

If, however, the heat pulse is produced by a Q-switched laser, then shorter timescale effects must be taken into account. Thus the thermoelastic effect launches ultrasonic waves into the surface layer, which are reflected back strongly from the disbonded back wall to cause large surface displacements and a build-up of echoes and oscillations. After a certain period of time (i.e. $\gg 10 \,\mu s$) the energy has mostly been transferred into low-frequency components, which produce the long-term displacement of the surface, which is determined by the low-frequency limit of the first approach. Thus the ultrasonic waves redistribute the thermoelastic strain throughout the surface layer.

The work discussed in §6.1.5 on subsurface flaws demonstrated some large surface displacements over defective regions, so that it is interesting to examine the results obtained by Cielo (1985) in a number of different potential applications. One of his techniques (an example of the second approach noted above) is to focus the Q-switched pulsed laser into a circle by means of an axicon lens, as described in §6.1.4. Before making measurements on an aluminium–epoxy laminate, Cielo (1985) reports preliminary studies of a 0.125 mm copper–beryllium sheet bonded to a Plexiglass substrate by epoxy resin. Figure 6.22 enables the response of a well-bonded area (a) to be compared with an unbonded area (b). The radius of the annular ultrasonic

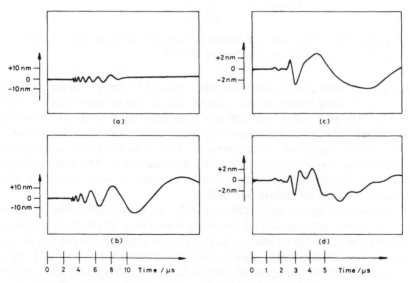

Figure 6.22 Response of 0.125 mm copper–beryllium sheet epoxy-bonded to Plexiglass substrate to ultrasonic impulse-loading technique: (a) well-bonded area, (b) unbonded area. Diameter of annular heating beam is 14 mm. Response of $0.020''-0.020''-0.016''$ aluminium–epoxy laminate to ultrasonic impulse-loading technique: (c) unbonded area (d) well-bonded area. Diameter of annular heating beam is 16 mm (Cielo 1985).

source is 7 mm, while the unbonded area has a considerably larger radius of 50 mm, so that to all intents and purposes the sheet behaves as a free membrane. As might be expected, therefore, the waveform received by the interferometer shows that the source sets the thin sheet into large-amplitude oscillations. In the well-bonded region ultrasonic energy is coupled into the substrate to give lower amplitude oscillations that are more highly damped.

The aluminium–epoxy laminate consists of a multiple sandwich of three layers of aluminium separated by epoxy. A bonding defect is simulated by a thin 10 mm diameter Teflon disc inserted immediately below the surface layer. As figure 6.22 also shows, differences between good and defective regions can be observed. Over the defective, unbonded region (c) the signal is dominated by a normal-mode type of oscillation in the surface layer, whereas the well-bonded region (d) exhibits higher frequency structure that the author believes is due to multiple ultrasonic echoes in the various layers. We note that there are some similarities between these signals and the surface-wave data from the subsurface defects shown in figure 6.21.

Cielo (1985) reports an alternative method for detecting and characterizing areas of disbonds which is an example of the first general approach noted above. This technique involves depositing energy by means of a pulsed laser over a small surface area, and detecting surface displacements at the centre of this area by means of a laser interferometer. The apparatus (figure 6.23(a)) is therefore very similar to figure 6.18. However, the Nd:YAG laser is used in the free-running rather than Q-switched mode, so that the pulse duration is 1 ms. Thus, although the details of the bandwidth of the displacements measured by the interferometer are not given, they seem likely to fall mainly within the sonic rather than ultrasonic spectrum. He reports that when the sample (copper–beryllium sheet epoxy-bonded to Plexiglass) is scanned across the laser beams, they detect a large increase in surface displacement over defects varying in size from 10 mm × 30 mm down to 3 mm diameter. The results from an aluminium–epoxy honeycomb panel are equally encouraging. An artificial defect (20 mm diameter Teflon disc) was inserted between the 0.3 mm thick outside aluminium sheets and 3 mm cell-size honeycomb core prior to curing of the epoxy bond. As figure 6.23(b) shows, the unbonded region is clearly visible for all diameters of laser source beam. However, the honeycomb cell structure also becomes visible for the 1 mm diameter beam.

Bresse et al (1989) have applied laser-generated ultrasound to the characterization of layered composites, consisting of a sandwich of an aluminium sheet and an epoxy adhesive layer, either bonded to a second aluminium sheet or to a foam substrate. They detected the ultrasonic waves by means of an EMAT transducer situated on the same surface as the source. They investigated both transient Lamb waves and resonances caused by reflections from the boundaries between the layers of the composite. They were able to conclude that both approaches show promise for the detection of disbonds between the aluminium and epoxy layers, and possible advantages over

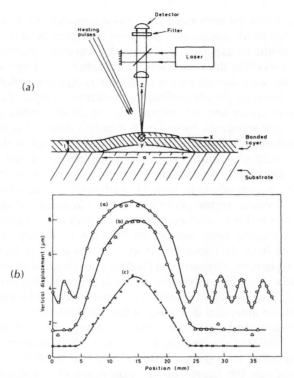

Figure 6.23 (*a*) Principle of thermoelastic interferometric method for detecting disbonds. Surface-heating pulse causes unbonded layer to bend, while vertical displacement is detected by interferometer. (*b*) Interferometric scan of 20 mm diameter delamination on aluminium–epoxy honeycomb laminate. Heating beam diameter: (*a*) 1 mm, (*b*) 3 mm, (*c*) 6 mm (Cielo 1985).

conventional techniques because of their high spatial resolution and wide bandwidth.

Cielo (1985) and Cooper *et al* (1985) also report preliminary work on bonding defect detection in plasma-sprayed coatings, but it is more convenient to discuss this in §6.2, together with other surface treatments and coatings.

6.1.7 Weld defects

The most important method for joining steel components is by welding. In order to ensure that the weld has sufficient strength for the application, it must be inspected to check that it is free of any defects that might otherwise reduce its strength. The most important potential defects include lack of side-wall fusion, porosity, slag, hot cracking and delayed cracking. Conventionally, the welded joint is inspected by ultrasonics or radiography after fabrication. The problem is that, if any defects are found, they have to be ground out and the weld repaired by filling in with more weld metal. Not only is this expensive, but the repair may weaken the weld or introduce more defects. The greater the number of weld passes that are laid after the pass in which the defect occurs, the more serious the repair. In the case of multi-pass welds especially, there is a clear need for an on-line technique to detect the defects as they form, or as soon as possible after the weld bead has been laid down. There are two techniques relevant to the present subject matter, acoustic emission (which was discussed in §4.6.2) and ultrasonics.

On-line ultrasonic inspection should in principle be able to detect most of the serious defects. By definition it cannot readily detect delayed cracking, since this occurs some time after welding has been completed. Some of the most serious defects, including incomplete penetration and lack of fusion, are directly caused by poor control of the size and position of the weld pool. There is thus an additional interest in sensing the position of the solid–liquid interface around the perimeter of the pool. Several research teams worldwide are investigating ultrasonic methods for monitoring the solid–liquid interface, including the detection of defects on line.

One practical problem is that the metal in the vicinity of the weld becomes very hot, so that conventional piezoelectric transducers have to be placed some distance away. This is therefore a potential application for laser

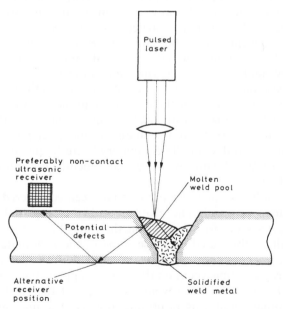

Figure 6.24 Schematic of apparatus to detect defect formation during weld fabrication. Potentially serious defects which may form at weld-pool/parent plate interface during solidification can be detected by change in transmission of P or S waves through interface. Advantage of laser is ability to generate sound in weld-pool.

ultrasonics. Thus a laser (e.g. pulsed YAG) can be directed at the weld pool itself in order to generate the ultrasound as close as possible to the interface and potential defects, as shown in figure 6.24 (Scruby and Wedgwood 1986, Carlson and Johnson 1987). A non-contact receiver of ultrasound is ideal. However, the poor quality of the surface, etc, during welding is rather unsuitable for a laser interferometer, and the group at EG and G (Carlson and Johnson 1987) have therefore used a piezoelectric transducer. In a more recent study they have replaced the piezoelectric transducer with an EMAT (Carlson and Johnson 1988). In a simplified experiment, in which they melted a pool of metal on a plate, they detect changes in the wave arrivals that appear to be related to the extent of the molten zone.

6.2 APPLICATIONS TO MATERIALS PROPERTY MEASUREMENT

6.2.1 Introduction

The second main application area for laser generation and reception is in materials characterization. The measurement of mechanical and materials parameters, such as internal stress, grain size, elastic constants and temperature, is itself a growing field for any technique such as ultrasonics that can probe the interior of solids non-destructively. One reason is the increasing industrial requirement for high-quality materials. Even in the case of well-established engineering materials, such as aluminium and steel, customers are continually tightening their specifications in terms of composition and performance. Manufacturers are consequently faced with the need to monitor their products more closely. In the case of many new materials that have been developed for specialized applications in, for instance, the aerospace industry, the required specifications can only be met after extensive tests and monitoring during production. These materials are frequently designed for high strength combined with low weight, and include ceramics, composites and special alloys.

In the past, off-line testing of a product has been considered an adequate method for quality control. The problem with this approach is that, if a given process goes wrong for some reason, there may be considerable wastage before the fault is discovered. Because some of the new materials such as metal matrix composites are expensive in terms both of raw material costs and of the energy and time spent on the process, appreciable wastage cannot be tolerated. There is thus a growing awareness of the need to introduce continuous on-line monitoring of product quality. The final step is then to close the loop by feeding back the information from the sensors into the process control instrumentation.

Ultrasonics is an obvious candidate for making measurements on most engineering materials. In the case of metals, which are opaque to light and highly attenuating to most other radiations, ultrasonics may be the only available technique for interior measurement. Suitable ultrasonic techniques have been under laboratory development for some years. They include the measurement of elastic constants via ultrasonic velocity, the characterization of microstructure via attenuation and the measurement of residual (internal) stress via a combination of velocities. Already some of these measurements are being proposed for production control, with some preliminary trials in progress.

For on-line measurements the limitations of conventional ultrasonic transducers, especially in regard to couplants, are likely to be more serious than for off-line flaw detection. This is firstly because many engineering materials are fabricated at higher temperatures than can be accommodated by conventional probes. However, a second reason is that the product is likely to be travelling through the plant at a relatively high speed, making it difficult to maintain consistent coupling between probe and product. It seems likely therefore that much on-line materials property monitoring would benefit from non-contact ultrasonic sensors such as those based on lasers. Again EMATs are also candidates which may prove simpler and cheaper to install (they have indeed already been installed for certain quality control measurements (Wilson *et al* 1977)).

We next summarize some of measurements that are likely to be of interest to manufacturers and customers, and which are candidates for measurement by laser ultrasonic techniques.

(a)   Dimensional properties, such as thickness and density.
(b)   Composition, including impurity levels, homogeneity, segregation.
(c)   Microstructural properties, including grain size, anisotropy, texture, number and distribution of phases present, fibre distribution in composites, areas of phase transformation.
(d)   Mechanical properties, such as strength, ductility, fracture toughness, magnitude of residual stresses.
(e)   Surface properties, roughness, surface treatment quality, coating quality and thickness.
(f)   Presence and size of all defects and discontinuities, such as cracks, inclusions, porosity.
(g)   Quality and strength of interfaces, bonds, joints, including welds.
(h)   Progress of thermo-mechanical and other treatments, solidification, extrusion, forging, including measurement of internal temperature.

Laser ultrasonics can, in principle, be applied to a good number of these measurement problems. One of the simplest (and most versatile) measurements is the time of flight of an ultrasonic pulse. If the ultrasonic velocity in the material is known, then its dimensions (e.g. thickness) can readily be deduced. Ultrasonic thickness gauges are readily available. Conversely, time-of-flight data can be used to deduce the velocity if the thickness is known. The velocity is an extremely useful quantity since elastic constants and density can be calculated from it, so that it is, for instance, possible to characterize different phases during a transformation. By measuring the velocity of two or more propagation modes as a function of orientation, the magnitude of internal stresses and texture (grain anisotropy) can also be estimated. Velocity is also a function of temperature, so that measurement of ultrasonic time of flight can in principle be used to deduce internal temperature distribution. According to whether bulk- or surface-wave velocities are measured, then bulk or surface properties can be deduced. The use of surface waves should in principle be especially useful for characterizing surface layers and coatings.

Laser ultrasonics has the added advantage of being able to give extremely accurate time-of-flight data. This is firstly because the elastic waves propagate as very fast rise-time monopolar pulses and, secondly, because no couplant or shoe correction is necessary. For some parameters (e.g. residual stress) the time of flight needs to be measured with an accuracy of 1 part in 10 000. This accuracy is attainable using laser generation and reception, especially if a picosecond laser is used for generation and state-of-the-art digitizers or clocks are used for timing. Q-switched lasers can enable ultrasonic propagation times to be measured to an accuracy of typically a few nanoseconds.

The other common parameter measured from an ultrasonic signal is its attenuation as a function of distance travelled. Work in various laboratories has shown that frequency-dependent attenuation is linked to scattering from microstructural features such as dislocations and grain boundaries. Attenuation is, however, considerably more difficult to measure than time of flight, if it is to provide data that can be interpreted with confidence. Again laser ultrasonics should be ideal for attenuation measurement, since the broadband pulses cover the whole frequency range of interest. Microstructural features attenuate the wavefronts by scattering the energy into other modes and other directions, and we shall show later in this chapter that laser ultrasonics is well suited for the measurement of the scattered radiation also.

We shall now discuss these applications in turn, and summarize published work reporting the application of laser generation and reception of ultrasound to materials characterization.

6.2.2 Thickness gauging

Thickness measurement by ultrasonic means has been used for some years. One approach has been to measure the time of flight of a compression-wave pulse using a pulse-echo transducer in contact with the sample. In addition to being limited to modest temperatures, these techniques are restricted to thicknesses above about 0.5 mm because of transducer dead-times and problems in resolving overlapping echoes. Laser ultrasonics has the potential for surmounting these problems. The first reported demonstration was by Bondarenko et al (1976) who generated and detected the ultrasound on the same side of a steel plate and steel rod. The data from the plate (approximately 10 mm thick) show a train of echoes from which the round trip time can be measured, and hence thickness can be deduced if the velocity is known. Using an ultrasonic pulse length of 50 ns, they show the minimum measurable thickness to be 0.14 mm in steel. The error in measured thickness is likely to be much less than this figure.

Turning to much thinner specimens, to exploit the extremely high bandwidths attainable with lasers, Krautkramer (1979) reports work in which a mode-locked laser generator is used, with pulse length ~ 2 ns, and rise time ~ 1 ns. The published data show a train of echoes in a thin metallic shim (a razor blade). He proposes the laser ultrasonic method for very-high-resolution thickness, etc, measurement. More recently, Tam (1984) has reported using laser-generated ultrasound coupled with a high-frequency zinc oxide piezo-electric receiver (rise time < 1 ns) to measure the thickness of stainless steel films. Using a nitrogen laser, with a pulse duration 0.5 ns, he measures round-trip echo times to an accuracy of 1% from steel films of varying thickness (figure 6.25). Thus, for the thinnest film, the ultrasonic data gave a thickness of 12.6 μm, which compares favourably with a much less accurate

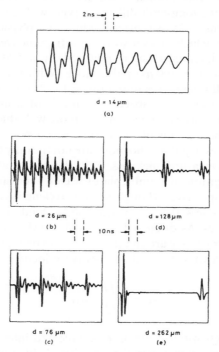

Figure 6.25 Laser-generated ultrasonic pulses multiply reflected in type 302 stainless steel films of thickness d (as measured by micrometer) (Tam 1984).

micrometer measurement of (14 ± 2) μm. Dewhurst and Al'Rubai (1989) have used a picosecond laser and polyvinylidene fluoride transducer to study the ultrasonic echoes in duralumin plates of thickness between 250 μm and 3 mm. The plate thickness could be deduced from the ultrasonic transit time measured, provided the velocity of sound was known. They also repeat the experiment reported by Krautkramer (1979) and study the ultrasonic echoes in a razor blade (thickness 108 μm). The first nine compression-wave echoes are well resolved. They also observed the lower frequency 'wash' that is predicted theoretically between the echoes, that is due, for instance, to the shear-wave arrival. This was unaccountably absent from the Krautkramer (1979) data.

Interestingly, one of the applications of laser ultrasonics which is closest to industrial exploitation, in that it has undergone trials under realistic plant conditions, is the dimensioning of hot steel tubes (Keck et al 1987). They employ a krypton fluoride excimer laser (energy 700 mJ, pulse duration 15 ns, repetition rate 8 Hz) to generate compression waves in the ablation regime on the surface of the hot (~ 1200 °C) tube blank. In one configuration, the interferometric receiver is directed at the same point on the tube surface so as to simulate the conventional pulse-echo probe configuration. Their interferometer is of the long-path-difference type, with an optical time delay between the two arms of 25 ns (corresponding to a physical path difference of the order of 5 m), which gives a working ultrasonic frequency maximum of ~ 20 MHz. The interferometer is powered by a 500 mJ Nd:YAG laser operating in the long (i.e. 100 μs) pulse mode. With this system they were able to measure wall thicknesses in the range 15–25 mm while the tube was travelling past the system at a speed of 2 m s^{-1}. Measurements were thus made every 250 mm. This was a genuinely remote wall-thickness gauge since the distance from the instrument to the tube was 5 m.

Compression waves in pulse-echo mode are not the only form of ultrasound that can be used to measure thickness. Thin sheets of material in particular support the propagation of Lamb (plate) waves over large distances. The Lamb waves are dispersive, although if the wavelength is very much larger than the plate thickness the symmetric modes tend to single frequency-independent velocities. As discussed in §5.10.3, Dewhurst et al (1987) show that the antisymmetric mode group velocity (c_g) can be expressed as a function of frequency (f), thickness (h) and sheet-wave velocity (c_{sheet}) thus:

$$c_g^2 = \frac{4\pi f h}{\sqrt{3}} c_{\text{sheet}}. \tag{6.4}$$

Then they use a Q-switched YAG laser source and interferometer receiver, separated by 24 mm, to measure the velocity of different frequency components of the dispersive antisymmetric mode. They obtain a measurement accuracy of about 2% on sheets as thin as 27 μm. The errors in this method for measuring thickness increase for thicknesses of the order of 100 μm or more because it became progressively more difficult to satisfy the condition that the wavelength is much greater than the thickness. We note with interest that this Lamb-wave technique enables a longer pulse (20 ns) to be used than in Krautkramer's (picosecond) or Tam's (0.5 ns) work, so that a cheaper laser can be used in any future commercial exploitation.

6.2.3 Elastic constants measurement

Bondarenko *et al* (1976) also mention the use of laser ultrasonics to measure longitudinal velocity and hence deduce the elastic modulus of a steel cylinder, but do not give details. Prior to this, Brammer and Percival (1970) showed how a laser-pulse technique can be used to measure elastic moduli at elevated temperatures. More recently, Wilcox and Calder (1978) have extended this work to a wider range of materials, including both metals and plastics. They use a Q-switched laser source and a 10 MHz contact piezoelectric receiver. By making use of mode conversion at boundaries, they are able to measure shear (c_2) as well as compression (c_1) wave speeds. From the ratio of the speeds, $k = c_1/c_2$, and the density ρ, they calculated the following elastic constants:

Shear modulus $\qquad\qquad \mu = \rho c_2^2$ $\qquad\qquad\qquad\qquad\qquad$ (6.5)

Lamé constant $\qquad\qquad \lambda = \rho c_2^2 (k^2 - 2)$ $\qquad\qquad\qquad\qquad$ (6.6)

Young's modulus $\qquad\quad Y = \rho c_2^2 (4 - 3k^2)/(1 - k^2)$ $\qquad\qquad$ (6.7)

Bulk modulus $\qquad\qquad B = \rho c_2^2 (k^2 - \tfrac{4}{3})$ $\qquad\qquad\qquad\qquad$ (6.8)

Poisson's ratio $\qquad\qquad \nu = (2 - k^2)/(2 - 2k^2).$ $\qquad\qquad\qquad$ (6.9)

There was agreement with published data to within 1–2% for most of their results. They commented that the laser source was very good at generating sufficient energy to propagate through the more attenuating materials.

In a more recent publication, Calder and Wilcox (1980) combine the laser source with an interferometer receiver to make a non-contact materials characterization system, which is capable of operation at elevated temperatures. By way of example, they deduced Young's modulus for tantalum at 200 °C, their value agreeing with published data to within 2%. It is difficult to measure ultrasonic velocities in liquid metals at high temperatures and pressures. However, this was also demonstrated using laser ultrasonics, with samples 1 mm in diameter and 25 mm long. The metal is subjected to transient heating, and the measurement of velocity is made only a few microseconds after melting. For example, they find that the compressional velocity for lead decreases on melting from 2.21 to 1.80 mm μs^{-1}, which is in good agreement with other data.

Calder *et al* (1980) report the application of laser techniques to the measurement of elastic constants as a function of temperature. They study plutonium and a plutonium–gallium alloy over a temperature range extending from 40 to 500 °C. They are regrettably unable to make their measurements above 120 °C in the pure metal, because the phase change from alpha to beta plutonium destroys the mirror-like surface on the specimen that is essential for the operation of their interferometer. A totally non-contact laser method is ideal for this study, not only because of its suitability for elevated temperatures, but also because it can be carried out through windows in the type of sealed container that is necessary for toxic plutonium. A schematic of apparatus similar to theirs is shown in figure 6.26. Using samples of length 101.0 mm and diameter 9.53 mm for the alloy, they obtain an accuracy of about 2% in their elastic constant values, which are shown in figure 6.27. It is noted that a 10–15 J laser pulse is used, and focused to an area of 5 mm^2, which must have caused some surface damage. In this work shear velocities are calculated from wave arrivals that undergo mode conversion at the boundaries. Ledbetter and Moulder (1979), however, combine a pulsed laser source with compressional and shear receiving transducers, and hence measure both velocities more directly. The data they obtained for 2014 aluminium agree with conventional pulse-echo data to better than 1%.

In the most recent work at the time of writing, Aussel and Monchalin (1989a) have carried out a comprehensive study of the precision with which ultrasonic velocities and hence elastic constants can be measured using laser techniques. In their experimental arrangement, the specimen is enclosed in an evacuated oven in a similar manner to that illustrated in figure 6.26. The ultrasound

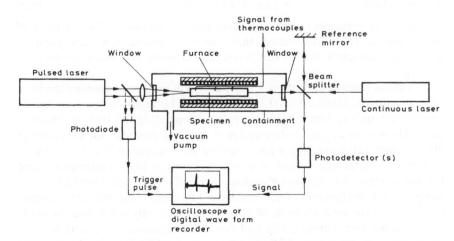

Figure 6.26 Experimental arrangement for measuring elastic constants of a solid as a function of temperature by non-contact ultrasonic means.

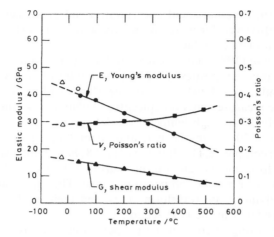

Figure 6.27 Variation of elastic moduli of Pu–1 wt% Ga with temperature. Solid data points from laser-ultrasonic measurements; open points from conventional measurements (Calder *et al* 1980).

is generated by means of a 750 mJ Nd:YAG laser, delivering 10 ns pulses, focused typically to 0.5 mm diameter, to give a source with both thermoelastic and ablative mechanisms in operation. The ultrasound was detected at the epicentre by means of a displacement interferometer employing a 1 W argon laser. They describe applications of their technique to PZT ceramic, metal–ceramic composites and single-crystal germanium. In the case of the composites, they measured longitudinal and shear velocities over a wide temperature range from ambient to almost 1000 °C.

As a result of their analysis of errors, they were able to deduce values for ultrasonic velocity in single-crystal germanium to an absolute accuracy of 0.1%. They note that this is as accurate as classical pulse-echo and resonance techniques. In order to ensure as accurate a result as possible they included an analysis of the diffraction correction for the finite area of the laser-generated source. They showed that if the source has a radius a, and the velocity is measured between two points a distance z_1 and z_2 from the source, then the longitudinal velocity c_1 must be reduced by $\Delta c_1/c_1 = a^2/(4z_1z_2)$ in the ablation regime and by $\Delta c_1/c_1 = a^2/(2z_1z_2)$ in the thermoelastic regime. For the shear wave the reductions are $\Delta c_2/c_2 = a^2/(3z_1z_2)$ and $\Delta c_2/c_2 = a^2/(2z_1z_2)$ in the ablation and thermoelastic regimes respectively. Thus, in the case of their measurements on the 6.066 mm thick PZT sample, using a spot diameter of 0.5 mm, the diffraction correction for a longitudinal velocity measurement from the P and PPP arrival times is calculated to be 0.03%. They point out that this is of the same order as the errors resulting from the thickness and time measurement uncertainties and must therefore be taken into account. However, for a smaller source and/or a thicker specimen the correction would be negligible.

Monchalin *et al* (1987) have used a similar experimental system to study two metal–ceramic composites based on titanium boride and aluminium, and also PZT ceramic, as a function of temperature. Unlike the above work, they also made measurements off-epicentre to characterize the shear arrival more accurately. Their high-temperature data show transitions due to the Curie point at $\sim 300\,°C$ in the PZT, and to melting of the aluminium at $\sim 660\,°C$ in the composite. Dewhurst *et al* (1988) have built a similar, entirely non-contact ultrasonic system for measuring velocity, and hence elastic constants also, over a wide temperature range. They have also used their system for studying phase transformations, and their work will, for convenience, be considered in the next section.

There is a growing interest in the properties of composite materials and in their ultrasonic characterization. Some preliminary work that has been carried out on the measurement of elastic constants using laser-generated ultrasound in plastic matrix (Buttle and Scruby 1988) and metal matrix (Buttle and Scruby 1989) composites will be discussed separately in §6.2.10.

All of these studies have relied on the measurement of time of flight from which ultrasonic velocities and hence elastic constants can be deduced. Bresse *et al* (1988) have adopted a different approach which should, in principle, be capable of greater accuracy, because it relies upon the whole recorded ultrasonic waveform. Because they use a pulsed laser to generate the ultrasound and a laser interferometer as receiver, the measurement system response can be approximated to a simple impulse response, so that their measured waveforms are an accurate representation of the ultrasonic field in the specimen. They record the whole waveform and use a first-principles theoretical model to calculate a waveform for comparison based on an initial guess for the elastic constants. The analysis then follows an iterative procedure in which the theoretical waveform is made to fit the experimental waveform as closely as possible by varying the elastic constants. In some preliminary results obtained on 12.6 mm aluminium plate they were able to calculate the compression and shear velocities with estimated uncertainties of 1.5 and 1% respectively.

In concluding this section on the measurement of elastic constants, we note that the stated errors in most of the above results are between 1 and 2%, which is inferior to conventional ultrasonic methods. However, the very careful work of Aussel and Monchalin (1989a) permits ultrasonic velocities to be measured to an absolute accuracy of 0.1%, which is comparable with classical methods. There is still some room for further developments to improve accuracy, especially if it is required to measure non-linear effects in order to deduce residual stress (see §6.2.6). It should be possible to attain a relative, if not an absolute, accuracy approaching 1 part in 10 000.

6.2.4   Monitoring changes of phase

Velocity and elastic constants measurements can readily be applied to the monitoring of various phase changes and structural transformations. Laser techniques bring the benefit of a wide temperature range, and of being suitable for very small samples, or regions of a sample. However, as Calder *et al* (1980) observed, some structural transformations distort the surface, making interferometric detection more difficult.

Rosen *et al* (1981) have applied laser-generated ultrasound to the measurement of Young's modulus in metallic glass ribbons, in order to monitor the percentage of amorphous metal that has transformed to the crystalline phase. The palladium–copper–silicon alloy is melt-spun into ribbons 1.25 mm wide and 40 μm thick. It is then cut into 250 mm long strips. A Q-switched Nd:YAG laser (pulse energy 20 mJ and duration 15 ns) is employed to generate extensional waves in the ribbon. In order to radiate ultrasonic energy preferentially along the ribbon, the laser light is focused into a 4 mm line, 0.25 mm wide, perpendicular to the ribbon length. The receiver is a 5 MHz quartz transducer, 1 mm wide, situated at a distance of 200 mm from the laser source. With this arrangement the transit time and distance were both measured to an accuracy of 1 part in 1000, to give a predicted error in Young's modulus of 0.2%. Because the Young's modulus is known to be about 7.4% higher in the crystalline than amorphous phase, the authors are therefore able to estimate the percentage of crystallization following annealing. A technique such as this could ultimately be used on line for the quality control of these materials, especially if the ultrasonic receiver were an interferometer. Greenhough *et al* (1987) have also shown that ultrasonic velocities in a metallic glass ribbon (Metglas 2605 SC) vary from 4940 to 5830 m s^{-1} as the material becomes progressively less amorphous. They used a similar arrangement for generation, i.e. a Nd:YAG laser focused by means of a cylindrical lens to give a line the width of the ribbon. The weak plasma source generated 'sheet' waves which were detected by means of a non-contact magnetomechanical sensor.

Dewhurst *et al* (1988) have built a non-contact ultrasonic materials characterization system, based on a Q-switched YAG laser to generate the ultrasound, and with a Michelson interferometer as receiver (compare figure 6.26). The iron, steel and aluminium alloy samples under investigation (19 mm diameter, 8 mm thick) are situated inside a vacuum furnace, the laser light entering and leaving through suitable windows. In the aluminium alloy measurements are made almost up to the melting point, while for the iron and steel measurements are made up to 1000 °C. The accuracy of the velocity data thus obtained is limited by the digitizer used to 1%. This is, however, good enough to resolve both the ferromagnetic to paramagnetic phase transformation at the Curie point (768 °C), and the martensitic phase transition (910 °C) in the iron (figure 6.28). The authors note that their

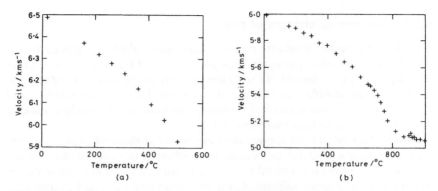

Figure 6.28 Temperature dependence of longitudinal ultrasonic wave velocity in (*a*) Dural aluminium alloy and (*b*) iron (Dewhurst *et al* 1988).

ultrasonic waveforms can also be analysed to give frequency-dependent attenuation data as a function of temperature.

6.2.5 Measurement of texture

Materials that have been manufactured by rolling or extrusion, for instance, invariably have anisotropic microstructures (e.g. grain elongation), which in turn give rise to anisotropic elastic properties. For certain applications there is a need to monitor and thence control the texture, because of its effect upon the mechanical properties of the material. Ultrasonic velocity measurement offers one method of achieving this.

Tam and Leung (1984) use laser-generated ultrasound to examine the texture of an extruded 6061-T6 aluminium disc (figure 6.29(*a*)). The ultrasonic source is generated by a 1 mJ, 8 ns duration pulse from a nitrogen laser, which is focused to approximately 4 mm^2 at the sample surface. The receiver is a PVDF transducer with a rise time of < 10 ns. They measure times of flight with an accuracy of 1 ns and the diameter of the 46.970 mm disc with an accuracy of 0.005 mm. This enables them to calculate the ultrasonic velocity with an accuracy of 0.02%. To date this is the closest published figure to the desired accuracy of 1 part in 10 000 discussed earlier in this chapter. The compression-wave velocity (and hence elastic modulus) was a maximum at 45° to the extrusion direction (figure 6.29(*b*)), which can be explained on the basis of a single-phase aggregate model for the grains (Ward 1962). Their laser technique could have considerable potential for production line use if combined with laser reception of comparable time resolution.

Lesne *et al* (1987) also discuss the use of a laser ultrasonic system employing a *Q*-switched laser source and heterodyne interferometer receiver to determine the elastic anisotropy induced in rolled steel sheet. They show that the Rayleigh-wave velocity differs according to whether the measurement is made parallel or perpendicular to the rolling direction.

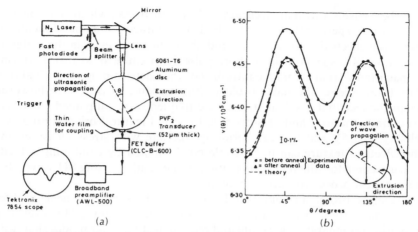

Figure 6.29(*a*) Schematic diagram of apparatus to measure texture in extruded aluminium alloy. θ is the angle between extrusion direction and ultrasonic propagation. (*b*) Longitudinal ultrasonic wave velocity v as function of θ in extruded 6061-T6 aluminium alloy. Solid lines join experimental data points, while the broken line is theoretical data for the single-phase aggregate model (Tam and Leung 1984).

6.2.6 Residual stress measurement

The authors do not know of any published work on the application of laser ultrasonics to the measurement of residual (internal) stress. However, because of the growing industrial interest in this measurement we shall discuss the possibilities briefly. Residual stresses are generally not desirable because they may take structural materials up to quite large fractions of the yield stress even in the absence of external stresses. Thus, when the structure is loaded, it may fail at a lower load than expected, and any defects present may grow at a higher rate than predicted. There are a number of common causes of residual stress, including gross deformation during rolling and pressing, thermoelastic effects during uneven cooling following heat treatment or welding, and martensitic phase transformations. It should be added that some internal stresses are designed for, e.g. in fasteners and steel tyres.

Ultrasonic velocities show a small dependence on stress (or to be more precise, on the difference between the principal components of the strain tensor), and a number of studies using piezoelectric and EMAT probes have shown how residual stress levels can be deduced from ultrasonic velocities (e.g. Sayers *et al* 1986). The effect of stress is only second order, being due to departures from linearity of the elastic constants, so that it only produces very small changes in ultrasonic velocity. Thus, to measure a change in stress of the order of 10 % of the yield stress in steel, the ultrasonic velocities must be measured with an accuracy of 1 part in 10 000, which is at the limit of present experimentation. Any higher accuracy is made difficult by dispersion effects due to scattering and absorption processes in the material. The effect of dispersion on a short pulse is to cause it to distort during propagation so that errors are introduced when determining the arrival time of the leading edge. The main problem with ultrasonic measurement, however, lies in the fact that residual stresses are frequently accompanied by micro-structural variations and anisotropies. Preferred textural orientation affects velocity (§6.2.5), and this can seriously interfere with stress measurement. The effects of texture can in principle be eliminated by combining velocity measurements for different propagation modes and by using SH waves, but the practical procedures are very difficult.

The short, broadband pulses of laser ultrasonics should be near ideal for non-contact stress measurement, especially on hot samples. However, it is more difficult reliably to generate and receive the two shear modes than with EMATS. In the authors' opinion stress measurement is one of the most difficult applications for laser ultrasound.

6.2.7 Internal temperature measurement

The steel industry especially would like to be able to measure temperatures inside hot products. One of the few possible practical techniques is ultrasonics, because ultrasonic velocities are known to be temperature dependent. Thus Wadley *et al* (1986) have used ultrasonic velocity measurements employing a laser source and EMAT receiver to compute the internal temperature distribution within a steel billet (figure 6.30). A tomographic reconstruction method is employed. In comparison with thermocouple data, errors of $\sim 20\,^\circ\mathrm{C}$ are found for a cylindrical billet, and $\sim 10\,^\circ\mathrm{C}$ for a test on a rectangular bar. Data are obtained at temperatures as high as $750\,^\circ\mathrm{C}$. This method could in principle be made entirely non-contact by using an interferometer as receiver.

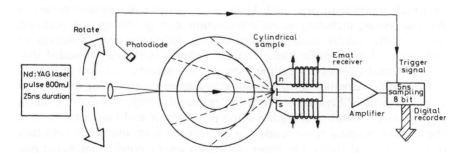

Figure 6.30 Schematic diagram of apparatus used to measure ultrasonic time of flight in order to deduce internal temperature distribution in cylindrical specimen (Wadley *et al* 1986).

However, there would be significant problems in obtaining sufficient sensitivity on the rough and dirty (e.g. oxide scale) surfaces typical of metal billets in production, so that EMAT transducers could ultimately prove the more practicable alternative as receivers. Although this is a difficult application for laser ultrasound, it is of great industrial importance, so that further results are awaited with interest.

6.2.8 Microstructural monitoring

It was mentioned in §6.2.1 that another common method of characterizing an ultrasonic signal is in terms of its attenuation after passing through the medium of interest, and that this is proving a valuable method for making non-destructive measurements of microstructure. The parameter to measure is the frequency dependence of the attenuation, since it is this which characterizes the source of the losses. Interfaces such as grain boundaries scatter ultrasonic energy in directions different from the main beam; they also convert compression waves into shear and vice versa. This leads to attenuation which increases as the fourth power of the ultrasonic frequency in the Rayleigh regime (scattering length parameter < ultrasonic wavelength). However, the dislocations present in a metal that has undergone plastic deformation remove energy from the ultrasonic beam by an absorption process, i.e. the ultrasonic energy is dissipated as heat, rather than scattered. The frequency dependence is, not unexpectedly, different for absorption: the attenuation increases as the square of the frequency.

Laser ultrasonics should be well suited to attenuation measurement for two reasons. First, the bandwidth of both laser source and inferferometer cover the entire frequency range of most current attenuation studies, without the need to change transducer frequencies. A typical pass band for single transducer measurements has been 10–50 MHz. A standard Q-switched laser and interferometer can readily achieve this frequency range: indeed the response also extends down to sub-megahertz frequencies, to enable dislocation absorption processes to be studied simultaneously with grain scattering. The bandwidth can also be made to extend above 100 MHz, by using shorter pulse source lasers.

Secondly, because this is a non-contact technique, there is no correction necessary to take into account couplant losses, the impedance of the couplant, and the loading of the transducer on the surface. Furthermore, provided the areas of the laser excitation and the interferometer spot are both very small, it is unnecessary to make the type of diffraction correction required for conventional transducers. This highlights one important difference when using the laser method. Conventional measurements use approximately plane waves from reasonably large diameter piezoelectric probes; a diffraction correction is necessary to take into account the finite probe area which causes edge effects around the circumference of the piezoelectric element. The laser ultrasonic technique works best with a very small (e.g. point) source and receiver, so that the wavefronts should be spherical. Thus, in a typical attenuation experiment, where a series of echoes is recorded from a parallel-sided sample, the amplitude of the compression-wave arrivals falls off linearly with propagation distance due to the inverse square law, in the absence of any material-dependent attenuation.

Scruby *et al* (1986) demonstrate the measurement of frequency-dependent attenuation in steels of different microstructure. The steel samples are 8 mm thick, with accurately parallel and polished faces. The source laser (*Q*-switched Nd:YAG, delivering pulses of maximum energy of 100 mJ and duration ~30 ns) is focused to a 1 mm spot at the centre of one face. A 100 MHz bandwidth laser interferometer receiver is accurately located to measure surface displacement directly opposite the source. The experimental arrangement is shown in figure 6.31, further details being given in the reference. They apply a small amount of volatile liquid to the source surface to enhance the amplitude of the ultrasonic pulse (as discussed in Chapter 5) without unduly damaging the sample surface.

The ultrasonic signal from the finer grained steel is dominated by a series of short pulses which are the compression-wave arrivals (figure 6.32(*a*)). Their amplitudes decay approximately inversely with distance travelled through the steel. However, when the time axis is expanded, the rise times of the echoes are observed to increase (figure 6.33) as higher frequencies are progressively attenuated. This is a consequence of scattering, principally by the grain boundaries in the steel. The second steel sample has a considerably coarser microstructure. The ultrasonic data still exhibit a train of echoes (figure 6.32(*b*)), but their rise times increase more rapidly than in the finer grained steel (figure 6.33). The pulses have also been analysed by Fourier

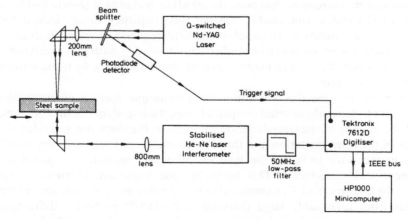

Figure 6.31 Apparatus to characterize microstructure by measuring ultrasonic attenuation and forward scattering (Scruby *et al* 1986).

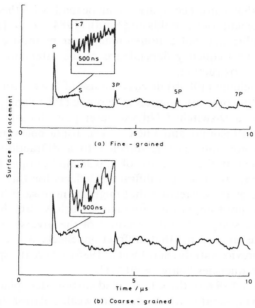

Figure 6.32 Data obtained from (*a*) finer-grained, (*b*) coarser-grained steel. Note successive reflections of P wave, from which frequency-dependent attenuation can be calculated, and ultrasonic forward scattering which is of large amplitude and longer periodicity in (*b*) (Scruby *et al* 1986).

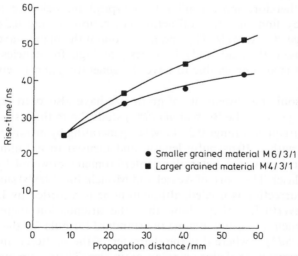

Figure 6.33 Rise times measured from P-wave echoes in figure 6.32. Greater increase in M4/3/1 due to enhanced frequency-dependent scattering by coarser microstructure (Scruby *et al* 1986).

transformation so that a direct comparison can be made with the conventional ultrasonic measurements of Reynolds and Smith (1984, 1985). The agreement is satisfactory, confirming that a non-contact laser method can indeed be used to measure the frequency dependence of the attenuation and hence characterize the microstructure.

Aussel and Monchalin (1989b) discuss the use of laser ultrasound for the measurement of ultrasonic attenuation in ceramic (PZT) and steel samples. In the former case, a Q-switched Nd:YAG laser pulse focused to a 0.5 mm spot generates the ultrasound, and a heterodyne displacement interferometer powered by an argon ion laser is directed to a diffraction-limited spot (≈ 0.1 mm diameter) at the epicentre of a 6.38 mm sample to act as the receiver. The authors then discuss diffraction effects for this configuration and the magnitude of any correction that might be necessary. They find that the spherical-wave limit applies for frequencies below ≈ 60 MHz, and the only correction necessary is to take account of the inverse square law for intensity. We note in passing that diffraction effects are comparable in magnitude to the previous study (Scruby et al 1986), where the spherical-wave limit applied for frequencies below ≈ 40 MHz.

In their study of a 4.19 mm thick hot-rolled carbon-steel plate, Aussel and Monchalin (1989b) generate ultrasound from weak ablation by means of a frequency-doubled Q-switched Nd:YAG laser over an irradiated spot of 8 mm diameter. The receiver is a Fabry–Perot velocity interferometer powered by a long-pulse Nd:YAG laser focused to a 10 mm spot on the surface. Both generating and receiving spots are larger than the specimen thickness so that the situation approximates to the plane-wave rather than spherical-wave limit, and is therefore more similar to a typical piezoelectric transducer geometry. They find that the diffraction correction is only 0.5 dB in the frequency range 20–40 MHz. They point out that if the plane-wave approximation is chosen, then not only is it best for high frequencies and thin samples, but it is also preferable to have the generating and receiving spots of similar size.

Laser ultrasonic measurements of grain size have also been reported by two other groups. Thus Le Brun et al (1988) study 2 mm thick stainless steel sheet over the frequency range 0.2–45 MHz, generating by means of a 240 mJ Nd:YAG laser with a 10 ns pulse length, and focused to a 1 mm diameter spot. Their receiver was a heterodyne interferometer powered by a 2 mW helium–neon laser. The work of Aussel and Monchalin (1989b) suggests that a diffraction correction is needed, although none is recorded by Le Brun et al (1988). Nevertheless they show that the attenuation increases with ultrasonic frequency (f), and assume the standard formula for the scattering coefficient, i.e. Sd^3f^4, where d is the diameter of the scatterer and S is the dispersion parameter, to deduce a mean grain size. Their estimate of 35 μm agrees reasonably well with the optical metallographic value of 25 μm.

Faßbender et al (1988) have carried out somewhat similar measurements of the grain size in steel plates varying in thickness between 0.3 and 20 mm. They use a 400 mJ ruby laser focused to a spot of diameter ≈ 1.5 mm to generate the ultrasound. Their receiver was either a heterodyne interferometer powered by a helium–neon laser or a time-of-flight (i.e. LPD) interferometer. They note problems on thinner sheets (<2 mm) using the heterodyne instrument due to large-amplitude lower frequency oscillations, probably due to thickness vibration modes of the specimen and near-field terms. These are eliminated in the data from the LPD interferometer because it is velocity sensitive and therefore attenuates lower frequencies. The grain sizes of the specimens are deduced from the standard approach used by Le Brun *et al* (1988). The results, which cover the range ≈ 20–70 μm, agree well with optical metallography. It is noted that the geometry approximated to the spherical-wave regime for the thicker specimens only, but how diffraction effects were taken into account was not discussed.

As an alternative to measuring the attenuation of the transmitted ultrasound, it is also possible to characterize the microstructure by analysing the ultrasound scattered out of the transmitted beam. Scruby *et al* (1986) consider this aspect by analysing the portions of their ultrasonic data that follow the main wave arrivals. They show that the ultrasonic 'noise' observed to follow both the compression and shear arrivals (figure 6.32) is a characteristic of the microstructure. Thus the finer microstructure is responsible for smaller amplitude, higher frequency data than the coarser microstructure. Although the authors discuss various ways in which this ultrasonic forward scattering might originate, the most probable seems to be mode conversion at grain boundaries lying reasonably close to the transmitted beam. Thus signals immediately preceding the shear arrival would be due to scattering close to the incident surface, because of the slower velocity of the shear wave, while signals immediately following the compression wave are likely to have undergone scattering close to the receiver surface. The authors conclude that, especially since this laser method is non-contact, it should have applications to the on-line monitoring of microstructure in metals at room or elevated temperatures. Such applications are, however, still some way off, partly because industrial interest in such types of on-line monitoring is only just beginning, and partly because of the cost of laser technology.

One type of sample where there is interest in microstructural gradients is the welded joint. Depending upon the success of the post-weld heat treatment, there may be considerable variations as one progresses from the weld metal itself, through the heat-affected zone into the parent plate. A welded mild steel plate has been investigated using laser-generated and laser-received ultrasound. The plate had undergone no post-weld heat treatment to relieve stresses in the vicinity of the weld. In addition to measurements in which the ultrasonic beam was perpendicular to the plate surface, the scanning arrangement was used to measure ultrasound that had propagated parallel to the boundary between the heat-affected zone and the parent plate. It was found that variations in microstructure as evidenced by changes in forward scattering were restricted to a very narrow region at the weld/heat-affected zone/parent plate interface.

6.2.9 Characterization of surfaces and coatings

Laser ultrasonics is also useful for measuring surface properties, because of the efficient generation and reception of surface waves. From surface-wave velocity measurements, information can be deduced about surface texture, residual stress, and the thickness and quality of surface treatments such as case hardening and protective coatings. From surface-wave attenuation measurements, surface microstructure, roughness, coating quality, etc, can in principle all be characterized. A non-contact technique would have potential for making these measurements under production line conditions. Although only a small amount of work has been published to date, it is hoped that the following discussion will give the reader an appreciation of the capability of laser ultrasonic measurements in this field.

One industrially important method of modifying surface properties is by means of a plasma-sprayed protective coating. Thus ceramic materials such as alumina are sprayed onto steel to improve high-temperature properties, such as corrosion resistance. Typical applications therefore include the coating of internal combustion engine components such as exhaust valves to give enhanced life at higher exhaust gas temperatures. Plasma-sprayed metallic coatings are also used to give protection: thus an aluminium coating may be sprayed onto mild steel to give added corrosion resistance in a sea-water environment. The most common problems encountered with plasma-sprayed coatings include variations in coating thickness and density and the presence of defects in the form of cracks.

The thickness of the coating is important since its protective properties will be much reduced where the layer is thin. Variations in thickness tend to occur in the vicinity of corners and other abrupt changes in shape because of inhomogeneities in the electrostatic field at these points. As dense a coating as possible is usually desired, since porosity permits diffusion of corrosive species through the coating to the substrate. Cracks perpendicular to the coating surface are a problem in that they may permit some localized corrosion of the substrate. Parallel cracks (laminar defects) may be produced by poor bonding of the first sprayed layer to the substrate (although the use of intermediate layers such as cermets is intended to avoid this) or by poor bonding between subsequent layers. These defects may be more serious than the perpendicular cracks since they may lead to the decohesion of a large area of coating from the substrate, leaving it unprotected.

There are two major benefits of laser ultrasonics that could be exploited for the inspection of plasma-sprayed coatings. Firstly, the method is non-contact. This is especially important in the case of ceramic coatings, some of which might be slightly degraded by contamination with ultrasonic couplants. Secondly, the method could be applied remotely on hot surfaces, giving the potential for *in situ* application during spraying. Not only could quality be monitored, but also any deficiencies in an early layer could be detected and put right before later layers are applied.

In a preliminary study of plasma-sprayed metal coatings, a 10 mm aluminium plate was plasma sprayed with aluminium coatings of variable thickness. One strip was bare substrate, but was roughened by abrasive grit to give a surface of the same roughness as the plasma-sprayed material. The other five strips were sprayed with respectively 1, 2, 3, 4 and 5 layers, each nominally 0.005″ (130 μm) thick. The ultrasound was generated by a Nd:YAG laser, Q-switched to produce pulses of approximate duration 25 ns and energy 80 mJ (before transmission through various optical components). The pulse was focused by a cylindrical lens to give a 1 mm wide line for the surface-wave measurements and by a spherical lens to give a 2 mm diameter spot for the bulk-wave measurements. The ultrasonic displacements were detected by a phase-locked reference-beam interferometer. While it was found satisfactory to use a relatively long focal length lens to focus the interferometer beam onto the back surface (reasonably flat and polished) for transmission data, a much shorter focal length lens (50 mm) was needed for detection on the rough coated surface. It was found that a short focal length lens of reasonably large numerical aperture could collect sufficient light as it was scattered from asperities on this surface to lock the phase of the interferometer signal.

For the single-sided, surface-wave measurements the source and receiver were positioned side by side on each coated surface in turn, as shown in figure 6.34. The data shown in figure 6.35(a) were obtained with a source–receiver separation of 33 mm. The most notable feature on the waveform

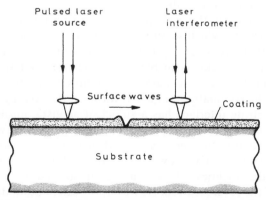

Figure 6.34 Apparatus for non-contact characterization of surface coatings (Scruby 1989).

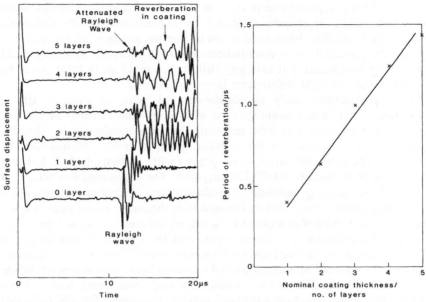

Figure 6.35 (*a*) Surface waveforms from uncoated material, and material covered by 1–5 layers of plasma-sprayed aluminium. Note attenuation and delay of Rayleigh wave and appearance of guided wave in coated samples. (*b*) Period of reverberation is proportional to coating thickness (Scruby 1989).

from the uncoated aluminium is the Rayleigh-wave arrival. On the thinnest coating (one layer) the Rayleigh wave is slightly attenuated, delayed in time and followed immediately by some reverberation. As the coating thickness is increased there is progressively more attenuation of the Rayleigh wave, and a progressively longer period of reverberation with the dominant frequencies shifting to lower values.

The interpretation of these data is that scattering in the coating is dominated by porosity, which selectively attenuates those wavelengths in the Rayleigh wave that are small by comparison with the coating thickness. Thus, as the coating thickness increases, a progressively larger part of the energy spectrum is lost. A more marked effect, however, is the reverberation, which is due to the stimulation of guided waves in the coating. It seems that the coating–substrate interface has a sufficient acoustic mismatch for this. Figure 6.35(*b*) shows that the period of these guided waves is directly proportional to coating thickness, as would be expected assuming a simple model. These results clearly show great promise for practical application, especially since the thickness of the coating can so readily be deduced from the frequency of the reverberation. However, the attenuation data could also be useful for characterizing the degree of porosity.

Single-sided measurements are likely to be of more practical importance since the rear surface of a sample is not always accessible, but some bulk measurements were also attempted. Two effects were observed. Firstly, there was a progressive delay in the bulk-wave arrivals, due to the increase in coating thickness. Such a measurement as this would only be useful if it could be ensured that the thickness of the substrate remained constant. The second effect was some ultrasonic 'noise' following the compression-wave arrival. Although it was tempting to interpret this again in terms of reverberation in the coating, the signals appeared rather random in amplitude and phase, and more reminiscent of the forward scattering studied in Scruby et al (1986). In this case the scattering centres would probably be a combination of various discontinuities, including layer–layer boundaries and porosity.

Cielo (1985) also reported some measurements made on 250 μm plasma-sprayed aluminium coatings on steel, to give improved corrosion protection. He studied the sensitivity of his thermoelastic testing technique (as described above in §6.1.6) to adhesion defects by masking a 4 mm wide strip of the steel with a Teflon strip (which probably melted in the process) during spraying. The signal from the laser interferometer showed a marked rise in surface displacement over the defective region. He also applied his convergent-wave technique to the same sample. The annular region heated by the Q-switched YAG laser was 8 mm in diameter (larger than the 4 mm wide defective region). Both well-bonded and unbonded areas exhibited more irregular ultrasonic signals than for other types of coating, which was put down to the inhomogeneity of the plasma coating. However, he did detect a shift to lower frequency in the oscillations following the surface-wave arrival for the unbonded region compared with the well bonded (figure 6.36). This result is consistent with the data of Scruby et al (1990b).

Cooper et al (1985) discuss a preliminary set of measurements carried out with a ceramic-coated plate. The coating was 0.6 mm alumina, and the plate was 3 mm thick steel. The ultrasonic source was generated by a 30 mJ, 30 ns pulse from a Nd:YAG laser, which was focused onto the plasma-sprayed alumina surface so as to cause ablation, although the authors report negligible surface damage. They employed two different types of receiver, a thick piezoelectric transducer, coupled by grease to the back face, and a non-contact EMAT transducer. They measured variations in wave amplitude and arrival time as the transmitter/receiver combination was scanned across well-bonded and disbonded regions. There was a big difference in waveform between the well-bonded and defective regions, doubtless because of the different modes of propagation necessary. This makes it difficult to interpret changes in amplitude across a disbonded region, and the authors conclude that measuring the increase in the time of flight is the best method for detecting disbonded regions.

Another common form of surface coating is by electroplating, and Cielo et al (1985) investigate the potential application of laser ultrasonics to electroplated coating thickness measurement. For this work they used the laser-generated convergent surface-wave technique, concentrating on ultra-

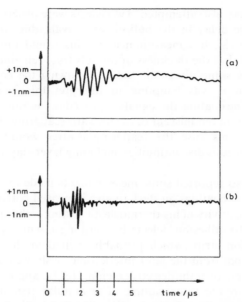

Figure 6.36 Response of 250 μm thick plasma-sprayed aluminium coating on steel to ultrasonic converging-pulse technique: (*a*) unbonded area, (*b*) well-bonded area (Cielo 1985).

sonic velocity measurement. The waves that are guided within the surface layer are frequency dispersive when their wavelengths are of the same order as the coating thickness. The authors carefully chose to electroplate copper with silver and chromium respectively, because the velocity in the copper substrate is lower than in the silver coating, but higher than in the chromium coating. They show that dispersion causes the higher frequencies to follow the lower frequencies in the silver-plated sample but to precede them in the chromium-plated sample. These effects are similar to those observed above in the plasma-sprayed aluminium coatings. The physical explanations of the data are equivalent, since the dispersive nature of the guided waves referred to in Scruby *et al* (1990b) gives rise to oscillations of varying period. Cielo *et al* (1985) also show that the time of flight of the main pulse increases with coating thickness of silver, and decreases with chromium, and propose this as a method for thickness determination.

Scruby *et al* (1990b) have applied the technique of figure 6.34 to the characterization of the surface of a steel component that had been hardened by a high-powered laser. Some of the observed effects were similar to those in the plasma-sprayed coating, e.g. attenuation and delay of the Rayleigh-wave arrival. It was, however, much more difficult to identify a guided wave from which the thickness could easily be measured for the coating, probably because the interface between hardened surface layer and parent metal is less sharply defined and also undergoes some variation from point to point.

6.2.10 Characterization of composites

Although the majority of work using laser ultrasonics has been carried out on metallic specimens, there is a growing interest in its use for non-metallic materials, especially composites, which reflects the growing importance of composites themselves as engineering materials. Some work on composites has already been discussed, in particular studies of lamellar and bonding defects in §6.1.6. In this section we shall therefore restrict ourselves to a discussion of the application of laser ultrasonics to the characterization of the composite material itself, by making measurements of velocity and attenuation. Because much of the information required by materials scientists about composites is different from metals, it is appropriate to discuss them separately.

As discussed in Chapter 5, the mechanisms whereby ultrasound is generated in a non-metal are different from a metal, so that the form of the source may be more difficult to model, especially in composites in which one phase is conducting and the other phase insulating. Because of such complications, the following work mostly only treats the laser generation as a broadband localized source. Nevertheless, except in more transparent materials, laser generation acts as an efficient and intense ultrasonic source. Because the absorption tends to be high, and the conductivity low, it is much harder to remain in a thermoelastic regime and avoid surface damage.

Optical detection of ultrasound is more difficult in most composites than metals, because many of these materials (e.g. carbon-fibre-reinforced plastic) have low reflectivity. The second phase also produces surfaces that are difficult to polish flat. Except for metal matrix composites of reasonably high conductivity, EMATs are not likely to be a very successful option. The work to be described therefore used a piezoelectric receiver for the ultrasound.

A final complication of fibre composites is that the materials are highly anisotropic with regard to ultrasonic propagation. They are furthermore susceptible to the propagation of guided waves along the fibres. In addition, the fibre composites to be studied are often in sheet form so that, even in the absence of fibres, most energy propagates in the form of plate (Lamb) wave modes. Because of all these complications only the first ultrasonic wave arrival was measured in the following work, and no attempt made to identify its mode. The anisotropy inevitably affects the radiation pattern of the source, whether in the thermoelastic or ablation regime. Care must be taken to take such variations into account.

In the first study, Buttle and Scruby (1988) used a pulsed laser source and a point-contact piezoelectric transducer (figure 6.37) to study elastic-wave propagation in glass fibre (GRP) and carbon fibre (CFRP) composites with an epoxy matrix. Although the main purpose of this study was to establish an acoustic emission location method suitable for these highly anisotropic materials, it also produced interesting data on the ultrasonic properties of these materials. By making a series of measurements, systematically scanning

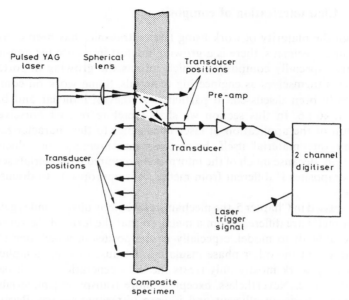

Figure 6.37 Diagram of apparatus to measure ultrasonic properties of composite materials using a pulsed laser source. Although not shown, the laser was also scanned systematically across the specimen (Buttle and Scruby 1988).

the laser source in directions parallel and perpendicular to the fibre axes, the velocities were measured with good accuracy. The velocities parallel to and perpendicular to the fibre directions were respectively 4770 and 2980 m s^{-1} in unidirectional GRP and 9310 and 2450 m s^{-1} in unidirectional CFRP. The measurement accuracy was estimated to be 0.5%.

Because of the ease with which it can be scanned, the laser was also used to plot ultrasonic velocity and attenuation as a function of angle in these sheet materials (e.g. figure 6.38). Effects due to source directivity variation were eliminated by measuring the attenuation between two receivers. Nevertheless, considerable care was still needed in making the attenuation measurements because of shot-to-shot variability in the intensity of the source, and the errors were therefore much larger than for the velocity data. One fascinating observation was the maximum in attenuation at approximately 60° in CFRP, rather than perpendicular to the fibres as might have been supposed. This was discussed in terms of a critical-angle effect at the fibre–matrix interface causing total reflection and therefore poor transmission.

The same technique was applied to a metal matrix composite, in which the matrix was 2014 aluminium alloy and the reinforcing phase was small particles (mean diameter 10 μm) of silicon carbide. Because this is a relatively new material, the purpose of the study (Buttle and Scruby 1989) was to

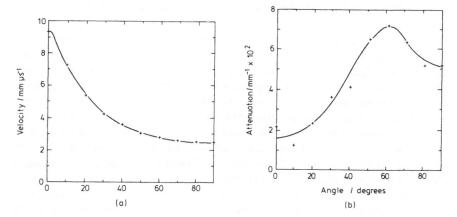

Figure 6.38 Ultrasonic velocity (a) and attenuation (b) as functions of angle between propagation direction and fibre axes for first wave arrival in unidirectional CFRP (Buttle and Scruby 1988).

deduce its elastic constants, and compare these with the pure alloy (table 6.1). The errors in the compression (P) wave velocity were less than 0.5%, but difficulties in making the shear velocity measurement led to an estimated error of 3–5%. The attenuation measurements were accurate to 3%. As expected, the elastic moduli (and hence velocities) are higher in the composite than the parent alloy (table 6.1). The attenuation is slightly higher, but this may be partly because these measurements were made using a surface-skimming compression wave. The increase in attenuation due to the silicon carbide particles should only be small at frequencies below 5 MHz. A result such as this should be useful to the manufacturers of the material because it indicated that this specimen had been well fabricated, with few defects such as voids or porosity to scatter and absorb the ultrasound.

Table 6.1 Ultrasonic properties and elastic constants deduced for metal matrix composite and parent aluminium alloy, assuming densities of 3200 and 3000 kg m^{-3} respectively.

Material Property	Composite Material	Parent Alloy
P velocity (m s^{-1})	7080	6350
S velocity (m s^{-1})	3490	3150
P Atten (nepers mm^{-1})	0.0176	0.0109
Bulk modulus (N m^{-2})	3.9×10^{10}	3.0×10^{10}
Shear modulus (N m^{-2})	1.1×10^{11}	8.1×10^{10}
Poisson's ratio	0.34	0.34

6.3 APPLICATIONS TO ACOUSTIC MICROSCOPY

Scanning acoustic microscopy (SAM) has been under serious development for a number of years. The most popular type of SAM is based on a design of Lemons and Quate (1979). The piezoelectric element is a thin film of zinc oxide coated on one end of a sapphire rod, the other end of which is ground into a small spherical concave (converging for compressional acoustic waves) lens. Water is used to couple the ultrasound into the sample. To be useful as a microscope, it is necessary to employ relatively short wavelengths, and so acoustic microscopes have been built with frequencies covering as wide a range as 50 MHz to 2 GHz. Both 'reflection' (working in 'pulse-echo mode') and 'transmission' instruments have been researched. The former dominates at the high-resolution (high-frequency) end of the range partly because it only incorporates one transducer/lens assembly and is therefore very much easier to operate. In the reflection SAM it is only necessary to align the lens and specimen, whereas in the transmission SAM it is also necessary to align the second lens accurately in all three dimensions.

Most of the interesting contrast at high frequencies arises because the large numerical aperture lens launches and receives leaky Rayleigh waves on the specimen surface which interfere with directly reflected waves. Thus the resolution is controlled by the Rayleigh-wave velocity. A velocity of 3000 m s^{-1} in steel leads to a wavelength-controlled resolution of the order of $1 \mu m$ at 2 GHz. This is comparable with optical microscopy. However, the importance of the SAM to materials science lies not so much in its resolution, but in the contrast mechanism. Thus it is sensitive to local variations in elastic constants, variations in microstructure, and extremely sensitive to all types of discontinuity, including cracks and interfaces (e.g. fibre/matrix interfaces in composite materials). Furthermore, unlike optical microscopy, SAM is able to probe beneath the surface in opaque materials such as metals to a depth of 1–2 Rayleigh wavelengths.

One of the problems of reducing sensitivity is the high insertion loss of all the components of the microscope at gigahertz frequencies. A large part of this loss is due to ultrasonic attenuation in the water couplant. However, attenuation is also high in the sample, making penetration through thicknesses of many wavelengths difficult. Any method for reducing attenuation and thereby improving sensitivity would therefore be most welcome. The laser source generates the ultrasonic wave field at the surface of the sample, eliminating the need for water couplant and the attendant losses. Its potential application would be in the transmission SAM, where reduction of insertion loss increases the capability to penetrate a thicker specimen.

Wickramasinghe *et al* (1978) have demonstrated the application of a pulsed laser source to SAM, combining it with a conventional zinc oxide/sapphire lens receiver (figure 6.39). In order to generate ultrasound at sufficiently high frequencies, they used a mode-locked Nd:YAG laser. The mode-locked pulse train consisted of 200 ps pulses repeating at 210 MHz. The mode-locked

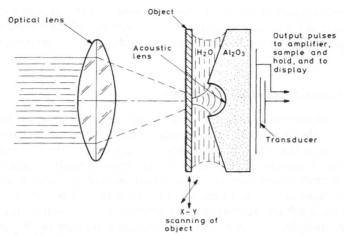

Figure 6.39 Diagram of scanning acoustic microscope where ultrasound is generated by a mode-locked pulsed laser and received by a piezoelectric lens assembly (Wickramasinghe *et al* 1978).

pulse train came within the envelope of a 200 ns Q-switched pulse, which was itself repeated at a frequency of 2.7 kHz. The incident pulse train generated ultrasonic signals in the sample not only at the fundamental of 210 MHz, but also at higher harmonics such as 420 and 840 MHz. Wickramasinghe *et al* (1978) selected the 840 MHz harmonic, and used a reasonably well matched detector centred around 800 MHz. Shorter picosecond pulses at higher repetition rates are available from Nd:YAG lasers, so that higher resolution should be attainable.

The full potential of laser ultrasound would only be achieved if the receiver used also a non-contact optical technique, such as a laser interferometer. Not only would there be further gains because of the elimination of the second water couplant, but also the optics could be scanned instead of the sample. It has always been something of a disadvantage to have to scan mechanically either the sample or lenses in the SAM, since it is difficult to move anything other than a very light sample with sufficient acceleration to produce a raster image in an acceptable time. Although lens scanning is standard in the reflection SAM, it is more difficult in the transmission instrument because of the need for synchronism between the transmitter and receiver. The authors do not know of any published application of a fully non-contact laser ultrasonic microscope. One problem is likely to be the low sensitivity of interferometry.

6.4 CALIBRATION

The laser ultrasonic source has a number of calibration uses, either separately (as in the case of acoustic emission), or else in combination with a laser interferometer receiver. The reason for proposing calibration as an application area is that the spatial and temporal characteristics of the ultrasonic wave field generated by the laser source closely reproduce theoretical delta-function sources.

6.4.1 Acoustic emission transducers

Absolute calibration is more difficult for the acoustic emission technique than most other non-destructive testing methods such as ultrasonics. The main reason is that the ultrasonic source is internal to the material, being for instance a growing crack, which is extremely difficult to reproduce or quantify for calibration purposes. Historically users of acoustic emission have tended to calibrate the electronic part of their detection systems (amplifiers, etc), but have left the acoustic part (propagation from source through the specimen, transducer response, etc) uncalibrated because of this difficulty. Lack of calibration causes problems when comparing results obtained in different laboratories or with different transducers, and the technique has remained largely qualitative. Measurements have shown that defect growth can be quantified, provided special broadband sensors and recording instrumentation are used (Wadley *et al* 1981, Scruby *et al* 1986). A reliable and practical calibration method would enable much more of the acoustic emission data obtained with conventional systems to be quantified and interpreted with confidence.

The ideal calibration method measures the response of the complete acoustic emission system to the type of signal it is to be used to measure. Since acoustic emission is a passive technique, relying on the material itself to generate the ultrasound, the calibration method must involve some form of acoustic source. It is insufficient to use an electrical calibration signal injected at the transducer output, since this does not measure acoustic factors such as couplant between specimen and transducer and the response of the transducer itself. It is also important that the calibration is as relevant as possible to the measurement being made. Thus the acoustic source needs to be applied to the specimen in a location as similar as possible to the location of the defect source. The reason is that the strength of the signal recorded by the system depends critically on how the ultrasonic waves propagate from source to transducer, which in turn depends upon the exact location of source and receiver. Although it is often difficult to predict in advance where the acoustic emission is going to operate, most advanced acoustic emission systems include a source location capability, so that calibration can be carried out afterwards.

Finally, the calibration source needs to reproduce as closely as possible the nature of the source events to be measured. Thus for most acoustic emission applications the source needs to simulate a growing crack in terms of the stress field generated at the source, in terms of its amplitude and in terms of its temporal characteristics and bandwidth. Defect sources in solids are best represented by combinations, or distributions, of force pairs (force dipoles) rather than single forces. They are also transient, lasting typically only fractions of a microsecond, so that their bandwidths extend to many megahertz. They also cover a very wide dynamic range in terms of amplitude.

A number of different methods for the calibration of acoustic emission systems have been used (which have been reviewed by Sachse and Hsu (1979)). They include the use of an ultrasonic pulser, gas jet, electrical spark, fracture of a pencil lead (or glass capillary), and pulsed laser. We shall discuss the merits and demerits of each briefly in turn. The ultrasonic pulser is simple and easy to use. The chief drawbacks are that it is subject to variability in the couplant used to attach it, making it difficult to obtain reproducible data. Furthermore, it is unlikely to have sufficient bandwidth to simulate correctly a real acoustic emission event. Finally, the stress field it generates is very different from a defect source. The helium gas jet generates continuous white noise at the surface, which is quite different in character from transient acoustic emission events. It is too variable for absolute amplitude calibration.

The fracture of a pencil lead (Hsu–Nielsen source) or glass capillary, on the other hand, constitutes a transient event, and offers absolute calibration if the fracture force is known. It is also extremely cheap and easy to apply. The drawbacks are that the stress field is strongly monopolar, the amplitude rather too large, it is difficult to vary systematically, and the bandwidth is rather less than the majority of real acoustic emission events.

The electrical spark causes rapid localized heating at the surface of the specimen, and generates an ultrasonic source that has approximately the correct stress field to simulate a crack. Although relatively cheap, it has the disadvantages of damaging the surface of the specimen, being somewhat irreproducible, and being difficult to apply in many field situations for safety reasons. It is also difficult to filter out the electrical transient induced by the large electrical discharge. Reasonably high bandwidths are attainable with carefully designed electrical circuitry.

Several authors (Egle and Brown 1976, Davidson *et al* 1976, Bentley *et al* 1978) have advocated the use of a pulsed laser as a simulated source of acoustic emission, mainly on the grounds of the extremely short duration ultrasonic pulse that is obtainable. There are a number of other technical advantages to this method, one of the most important of which is that the thermoelastic source generates a dipolar stress field in the surface of the specimen that is very similar to the stress field of a crack. Figure 6.40 illustrates this point. The principal stress generated by the thermoelastic source (*a*) can be represented by two pairs of force dipoles parallel to the surface. If the laser is focused by a cylindrical lens to give a line source, then at high frequencies one dipole dominates. The principal stresses in a surface-breaking crack (*b*) can also be represented by a force dipole. Thus, to a first approximation, the thermoelastic source is a good simulation of a surface-breaking crack, especially if the laser is focused to a line parallel to the crack direction.

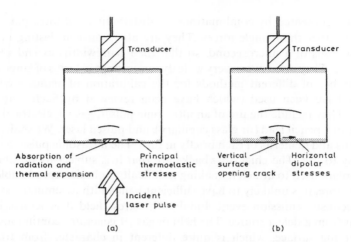

Figure 6.40   Schematic diagram to show how thermoelastic stresses from a laser line source approximate to stresses generated by a surface crack as it grows.

Users of acoustic emission are, however, also interested in monitoring defect sources that are buried within the bulk of a material. This is more difficult to simulate with the pulsed laser. One possibility is to bury a small absorbent 'inclusion' inside a block of transparent material (e.g. glass), as shown in figure 6.41(a). Absorption of electromagnetic radiation by the inclusion causes it to expand suddenly, generating a centre of expansion, which can be represented by three perpendicular force dipoles of equal strength. Such a specimen would, however, be difficult to manufacture, and in any case is strictly only relevant to acoustic emission measurements made on the same transparent material.

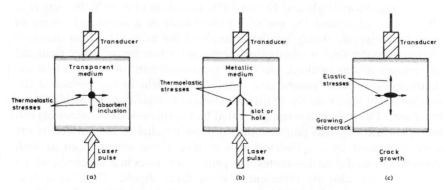

Figure 6.41   Schematic diagram showing two methods for using a pulsed laser to stimulate the stress field generated by growth of a buried microcrack.

A simpler alternative (figure 6.41(*b*)) is to drill or machine a hole or slit into the base of the specimen. This again acts as a centre of expansion, although the presence of the boundaries distorts the wave field close to the source. Theoretical studies (Achenbach and Harris 1979) have shown that this distortion is not too serious for the portion of the elastic wave field above the crack. In any case, the acoustic emission technique is frequently used to monitor the extension of a pre-existing macro-crack, in which case the emission event itself would be modified by the crack boundaries. A buried microcrack source is best represented by a combination of three perpendicular force dipoles. However, unlike the buried thermoelastic source, these dipoles are unequal in strength. Thus for a material such as aluminium with a Poisson's ratio of $\frac{1}{3}$, the vertical dipole is twice the strength of the two horizontal dipoles (figure 6.41(*c*)). This difference in relative dipole strengths is not serious, and good agreement between the waveforms from the thermoelastic source used as in figure 6.41(*b*) and from crack extension as in figure 6.41(*c*) is obtained, which is illustrated by figure 6.42 (Scruby *et al* 1981). Figure 6.42 also shows that the rise time (and hence the bandwidth) of the thermoelastic laser source is comparable with a real acoustic emission event.

Thus the thermoelastic source generated by a pulsed laser appears to be almost perfect as a simulation of the acoustic emission generated by a crack.

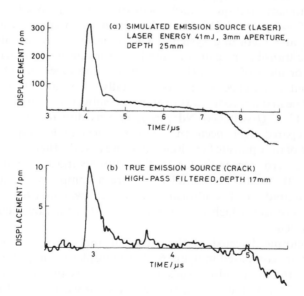

Figure 6.42 Comparison of ultrasonic waveforms generated by (*a*) absorption of pulsed laser radiation, and (*b*) cleavage microcrack propagation in mild steel (Scruby *et al* 1981).

The technique has the added benefits of being non-contact and readily adjustable in strength by means of attenuative filters inserted in the laser beam. The main disadvantages are those which apply generally to laser techniques, i.e. the need for safety precautions and much greater cost than alternatives such as the Hsu–Nielsen (pencil fracture) source. Although it is possible in principle to use the thermoelastic source as an *absolute* calibration device by accurately monitoring the quantity of heat deposited at the surface of the specimen, this is made difficult in practice by factors such as variable reflectivity of the specimen surface. However, as we shall discuss below, it can be used as part of an absolute transducer calibration technique.

We now turn to practical considerations for acoustic emission transducer calibration. First it is important to choose the type of laser. Most of the criteria are the same as for ultrasonic testing. Thus a short-pulse laser is required. One of the most convenient is the Q-switched Nd:YAG laser, since its pulse length and pulse energy are well matched to acoustic emission requirements. For laboratory use, any one of a wide range of bench lasers can be used. However, for field use, or for application where the apparatus must be portable, the bench laser is too large and unwieldy. In response to this need, miniature pulsed laser systems have been built (e.g. by Dewhurst (1983)). They are small enough to be held in the hand, fixed to an optical bench, or clamped onto a standard photographic tripod. Their power consumption is low enough for them to run off a battery.

In order to carry out quantitative acoustic emission work (e.g. Wadley *et al* (1981), Scruby *et al* (1986)) it has been essential to use calibrated transducers and recording systems. The approach was to calibrate each piezoelectric transducer against a standard transducer of known calibrated response. For most of the work a capacitive transducer was chosen as the standard, but cross checks were occasionally made against a laser interferometer. The first approach was to employ a substitution method. As shown in figure 6.43, a Q-switched Nd:YAG laser was directed at one face of a specially prepared steel specimen. The transducer under test was mounted at the centre of the opposite face. Remembering that the thermoelastic far-field amplitude is zero perpendicular to the source surface, the top face was oriented at 30° to the horizontal to ensure a compression-wave pulse of reasonable amplitude. The ultrasonic wave field was first measured by the standard transducer, which was then substituted in turn by each transducer to be calibrated.

The waveforms detected by the capacitor and one of the point-contact transducers are compared in figure 6.44. The displacement pulse that arrives 4.28 μs after the laser pulse struck the specimen is the compression wave. The step at 7.75 μs is due to the arrival of the shear wave. From the capacitive transducer the peak amplitude of the compression wave was calculated to be 41.5 pm. The piezoelectric transducer recorded an output signal of 18.5 mV, so that its sensitivity was computed to be 0.45 V nm^{-1}. Since the thermoelastic

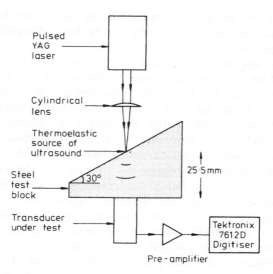

Figure 6.43 Apparatus to carry out acoustic emission transducer calibration by substitution, using a pulsed laser to simulate the crack growth event (Scruby 1985).

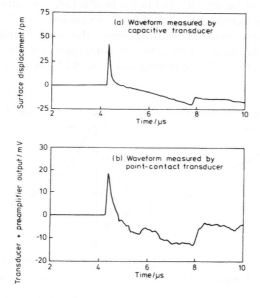

Figure 6.44 Surface waveforms measured using the apparatus of figure 6.43 when calibrating a point-contact piezoelectric transducer against a capacitor (Scruby 1985).

source is such a good simulation of an acoustic emission event, this calibration should be accurate enough for the compression waves detected from events in laboratory specimens.

It is sometimes required to determine the frequency response of an acoustic emission transducer. Because the capacitive transducer is known to have a flat frequency response to at least 10 MHz, this can be done by Fourier transformation of figure 6.44. The result (figure 6.45) indicates that the point-contact transducer under test also has a reasonably high fidelity response, although it falls off more rapidly with frequency than the capacitor. Spectral calibration such as this, extending from 0–10 MHz, is much more difficult without a broadband ultrasonic source such as the pulsed laser.

The substitution method has some practical disadvantages, and an alternative calibration block was designed (figure 6.46) to enable the capacitive transducer to be permanently mounted. The two symmetrical faces were machined at 45° which is close to the maximum of the thermoelastic directivity pattern. The distance from source to each transducer was 50 mm, making the calibration particularly suitable for specimens of approximately this thickness.

It is not always possible in acoustic emission testing to arrange for the emission source to be directly below the transducer. It is therefore necessary to calibrate the directivity of the transducer response. This relative rather than absolute calibration is again well suited to the pulsed laser. Using the apparatus shown in figure 6.47, the beam profile (angular directivity) of the response of a point-contact transducer was measured. The hemisphere was either steel or aluminium depending upon the final application of the transducer. In order to generate a strong compression wave normal to the irradiated surface, the ablation regime was used, with the laser pulse (12 mJ)

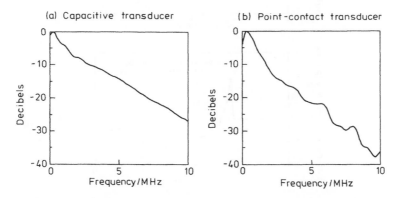

Figure 6.45 A thermoelastic source generates a sufficiently broadband ultrasonic pulse to enable spectral analysis of transducer response (Scruby 1985).

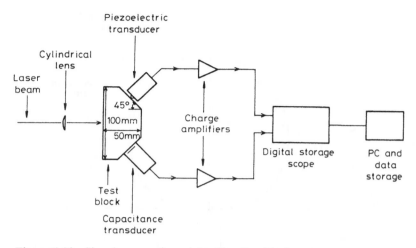

Figure 6.46 Showing use of special calibration block to compare response of piezoelectric transducer with absolute capacitive transducer.

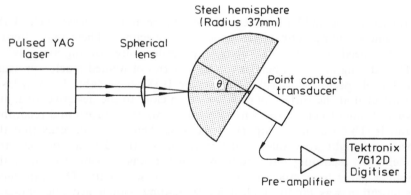

Figure 6.47 Apparatus to measure the directivity of a piezoelectric transducer by scanning a pulsed laser source around a hemispherical block (Scruby 1985).

focused to a small circular spot (0.5 mm diameter). After several pulses to 'clean' the surface at each orientation, it was found that ultrasonic waves of reproducible amplitude were generated. The output from the transducer was measured every 10° and the response plotted as in figure 6.48. It can be seen that the experimental directivity is in good agreement with theory.

6.4.2 *In situ* acoustic emission tests

The pulsed laser source can also be used to check the sensitivity and response of an acoustic emission system when it is installed on a specimen or structure. There may be a need, for instance, to check that the transducer sensitivity

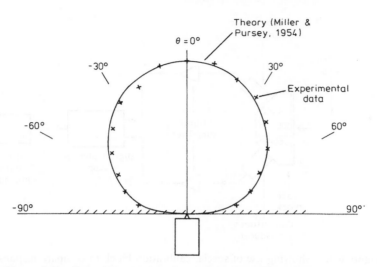

Figure 6.48 Experimental data for the directional response of a point-contact receiver (Scruby 1985).

is still acceptable and has not deteriorated. There may be a need to check the ultrasonic coupling between transducer and specimen. There may alternatively be a need to check that the analysis procedures are working correctly.

Thus during an experiment involving a series of welded reactor pressure vessel steel specimens (Scruby and Stacey 1988), a portable Nd:YAG laser was directed at the surface of each specimen close to the defective area, just prior to testing to check the sensitivity of the six transducers and recording channels. This was most important, because there were concerns that the sensitivity might vary due to deterioration of the delicate piezoelectric element. By recording data from the laser it was possible to adjust the amplifier gains to compensate for any loss of sensitivity. The transducer configuration was far from ideal for three-dimensional acoustic emission source location, and considerable errors could arise. By using the pulsed laser it was possible to check the accuracy of the location algorithm. Finally, some source radiation pattern analysis was carried out, which relied on the correct relative strengths of the wave field being measured at the various transducers. Because the radiation pattern of the thermoelastic source is known, small correction could be made for differences in gain between the recording channels.

The same portable pulsed laser has also been used to calibrate a set of acoustic emission transducers *in situ* on an experimental reactor pressure vessel at JRC, Ispra, Italy (figure 6.49) (Scruby *et al* 1990a). The purpose in using the laser was similar to the previous example. First it was necessary to record the sensitivity of each transducer, and that there was good coupling to the vessel. Each transducer had previously been calibrated on à steel test

Figure 6.49 Photograph showing a pulsed laser in use to check the response of an array of piezoelectric transducers mounted on a pressure vessel.

block of similar thickness to the vessel, and a sample waveform recorded at a source–receiver separation of one plate thickness (figure 6.50(*a*)). For each transducer it was checked that the waveform obtained at the same range agreed with the laboratory waveform (figure 6.50(*b*)). The laser was also used to determine the accuracy of source location procedures on the vessel.

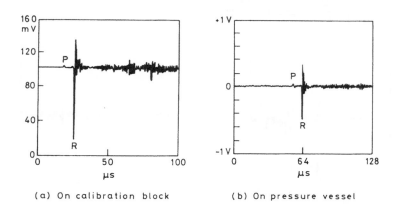

(a) On calibration block   (b) On pressure vessel

Figure 6.50 Surface waveforms from one transducer used on the Ispra vessel, obtained from a 1 mm diameter thermoelastic laser source (Scruby *et al* 1989a). (*a*) On a calibration block of similar thickness to vessel, at one plate thickness from source; (*b*) on vessel, at the same distance, in order to ensure transducer response is unchanged from laboratory.

A number of interesting acoustic emission events within the vessel were recorded, but it was not possible to locate them accurately, or to characterize them in terms of their radiation patterns, because it was only possible to site one, or occasionally two, transducers close to each defective area. In order to learn more about their possible origin, a laboratory simulation was set up, using a steel plate of similar thickness and composition and firing a pulsed laser at various points on the surface. In this manner it was found that some of the recorded events could be simulated surprisingly accurately. Thus the simulation of one particular event (figure 6.51) indicated that it was most probably due to a crack at or close to the rear surface of the vessel. The location of this event turned out to be where the vessel later failed during a prolonged period of cyclic fatigue. Such a simulation relies heavily on the fact that the thermoelastic source generates a local strain field which reproduces that of a surface-breaking crack.

6.4.3 Ultrasonic transducers

In Chapter 4 we discussed the application of laser interferometry to the calibration of ultrasonic transducers. Because the interferometer is a receiver of ultrasound, it is best suited to calibrating ultrasonic transmitters, although it can be used to calibrate the response of transducers as receivers by, for instance, a substitution method. An alternative method of calibrating the

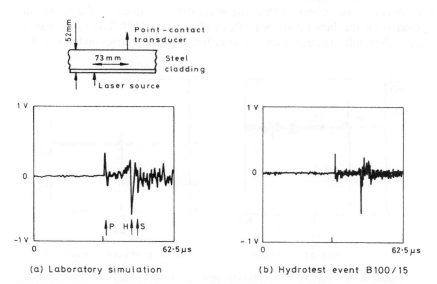

Figure 6.51 Characterization of an emission event recorded during test of vessel, by laboratory measurement using thermoelastic source to simulate defect (Scruby *et al* 1989). (*a*) Simulation, showing P, S and H (head waves); (*b*) real event recorded during vessel test.

response of a transducer as a receiver is by using the pulsed laser as a standard ultrasonic source, exactly as in the case of acoustic emission transducers discussed in the previous sections. Because it is difficult to quantify the pulsed laser source, this method is more applicable to relative than absolute calibrations.

In the study (Moss and Scruby 1988) of ultrasonic shear-wave probes referred to in §4.5.4, there was a particular interest in beam profiles, both when transmitting and receiving. The pulsed laser source is well suited to measuring the beam profile of an ultrasonic receiver. A photograph of the apparatus is shown in figure 6.52. On the right is a Nd:YAG laser operating in Q-switched pulse mode. Because damage to the specimen could not be tolerated, neutral density filters were used to reduce the energy per pulse to 5 mJ. When focused to a 1 mm diameter spot on the surface, the power density was low enough to prevent ablation and ensure a thermoelastic source. Specimens of aluminium of varying thickness were scanned in a raster through the light beam.

In figure 6.53 the receiver beam profile of a 5 MHz transducer fitted with 45° and 60° angled shear-wave shoes is plotted. Every effort was taken to ensure that as far as possible these measurements were the reciprocal of the earlier interferometric measurements of the same transducer. However, the

Figure 6.52 Photograph showing ultrasonic probe calibration by pulsed laser.

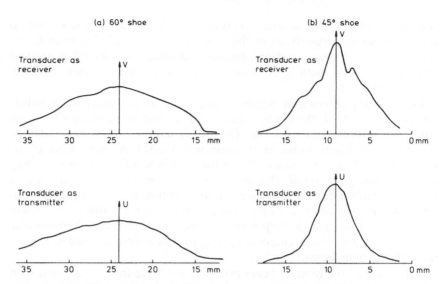

Figure 6.53   Directivity of a 5 MHz ultrasonic probe as a receiver of ultrasound, as measured on a 10 mm aluminium plate using a thermoelastic laser source. Lower plots enable directivity as transmitter to be compared (Moss and Scruby 1988).

thermoelastic source, generating stresses parallel to the surface, is not the reciprocal of the interferometer receiver, which measures perpendicular displacement. The correct reciprocal would have been the ablation source but this was not used because it would have damaged the specimen surface. Another problem with the thermoelastic source is that it has a zero in the shear directivity at 45°, and this probably explains the perturbation in the data for the 45° shoe. It is therefore somewhat remarkable that the directivity data for this transducer acting as a receiver agree as well as they do with the data (also presented in figure 6.53) for its performance as a transmitter.

Moss and Scruby (1988) also studied the ultrasonic waveforms produced by this 5 MHz shear-wave transducer. The waveform transmitted by the transducer as measured by a laser interferometer on the same 10 mm aluminium plate is presented in figure 6.54(a). For comparison purposes, the 'reciprocal' shear waveform is given in figure 6.54(b). This was obtained by recording the output voltage from the transducer when receiving the ultrasound generated by a pulsed laser located at the same point as the interferometer in the earlier measurement. These two waveforms are generally similar, but with small differences in low-frequency content, especially at the commencement of the waveform.

When this probe is used for pulse-echo work, the echo waveform should in theory be the convolution of these two waveforms. The result of this

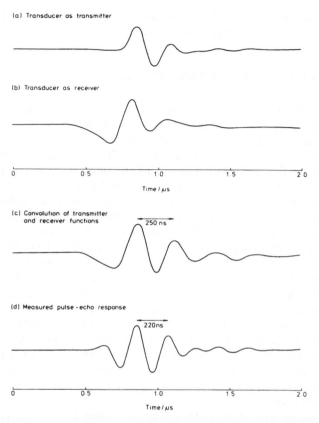

Figure 6.54 (*a*) Waveform at centre of 45° beam from 5 MHz shear transducer, as measured by laser interferometer; (*b*) output voltage from same transducer when excited by laser pulse at the same point on a 10 mm plate as used for interferometer measurement; (*c*) convolution of (*a*) and (*b*) to simulate pulse-echo response; (*d*) measured pulse-echo response of the same transducer using a 1.5 mm side-drilled hole as reflector (Moss and Scruby 1988).

convolution is shown in figure 6.54(*c*). This simulated pulse-echo waveform can be compared with figure 6.54(*d*) which is an experimental pulse-echo waveform, obtained with this transducer from a 1.5 mm side-drilled hole at a slightly larger range of 15 mm in a standard ultrasonic test block. Apart from a small shift in frequency, the agreement is good. These differences were thought to be at least partially due to the fact that the thermoelastic source is not the reciprocal of the interferometric displacement measurement. The thermoelastic source is known to generate a very broadband ultrasonic signal with significant energy at low frequencies. Another important factor is the frequency characteristic of the driver edge pulse, which is unlikely to be as broadband as the Q-switched laser pulse.

6.5   WAVE PROPAGATION STUDIES

Because of their unique properties, laser generation and reception of ultrasound can also be used in experimental studies of elastic (ultrasonic) wave propagation. Elastic wave theory has been progressively developed, especially over the past ten years, to enable ultrasonic and acoustic data to be interpreted. For a full account of such theory, the reader is referred to such books as Achenbach (1973), Graaf (1975) and Harker (1988). The theory of ultrasonic propagation in solids is more difficult under transient conditions (i.e. short pulses), and when the waves undergo significant amounts of reflection, refraction and diffraction. While full analytical or numerical solutions exist for wave propagation in an infinite body (Aki and Richards 1980), a half space (Pekeris and Lifson 1957) and an infinite plate (Pao et al 1979), only approximate solutions are available for more complex bodies. Only recently have solutions, for instance, been published for the diffraction of ultrasonic pulses by ideal cracks (Chapman 1981, Ogilvy and Temple 1983).

In the ablation regime, the pulsed laser source of ultrasound is very similar to the point force used when obtaining theoretical solutions to the wave equation. Similarly, the laser interferometer is very close to the ideal point receiver. Thus laser ultrasonics can be used, either to confirm wave propagation theory, or to extend it to problems that are not amenable to easy theoretical solution.

The source of ultrasound (which is usually some form of active defect) in an acoustic emission measurement is usually some distance away from the transducers, in a location that cannot be predicted beforehand. Thus the waves must propagate through the structure, undergoing various reflections, etc, before being converted into electrical waveforms by the transducers. It is not possible to interpret the acoustic emission data in terms of the characteristics of the source unless the effects of wave propagation are understood. Because one of the most important applications of acoustic emission is to the monitoring of plate-like structures, such as pressure vessels and pipes, considerable effort has gone into theoretical studies of propagation through plates (e.g. Pao et al 1979, Weaver and Pao 1982). Various workers have used the pulsed laser and broadband receivers to confirm experimentally the waveforms predicted by the theory.

Theoretical solutions are particularly difficult to derive for thin plate structures, where the waves have propagated so far that much of the energy is in the form of normal modes, e.g. at a range of 20 plate thicknesses. Although Weaver and Pao (1982) have derived the analytical solution for the vertical point force (which necessarily has cylindrical symmetry), solutions for generalized sources, e.g. horizontal and vertical force combinations which lack this symmetry, are not yet available. Because there is interest in applying acoustic emission to the monitoring of defects in offshore structures, which are fabricated from tubular sections that can be approximated for present purposes to thin plates, a study was carried out to measure experimentally the waveforms that are generated in such plates (Eriksen and Thaulow 1988). Because of the similarity of the stress field generated by the pulsed laser to real crack sources of acoustic emission, it was one of the ultrasonic sources chosen for this study.

The time-of-flight technique for defect detection and sizing described in §6.1.3 depends upon the diffraction of ultrasound from the edges of defects. Theoretical studies (Chapman 1981, Ogilvy and Temple 1983) have been carried out to predict diffraction coefficients. The pulsed laser source of ultrasound has been used in conjunction with a broadband receiver to confirm the angular dependence of the diffraction coefficient (Ravenscroft et al 1989). These measurements had a special poignancy, because theory predicted a zero in the diffraction coefficient, together with a small lobe of opposite sign, under certain conditions. An example is shown in figure 6.55(a). If this were true, then there was a danger of a crack-tip diffracted signal being missed and the defect going undetected. Experimental data using conventional short-pulse piezoelectric transducers had initially, however, failed to detect this zero, or even a significant minimum. This then raised the question of a possible error in the assumptions underlying the theory.

The laser was used in conjunction with various transducers using the apparatus shown in figure 6.55(b). It was found that, around the angles at which theory predicted zero amplitude and change of phase, there was a progressive change in waveform shape accompanied also by a phase change (figure 6.56), although not a zero in amplitude. As figure 6.56 shows, there appear to be two components in the waveform, separated slightly in time. If these were coincident then, since they are of opposite polarity, there would indeed be an angle at which they cancelled out. Thus the laser measurements appeared to confirm the theory, except for the separation of the two components. Further studies showed that this small difference in arrival times was due to the non-zero width of the tip of the slit in the test block (the theory had assumed zero width). Following these results, further very accurate measurements were made using piezoelectric probes. When the waveforms were carefully analysed, it was found that these data too showed the presence of two components of opposite phase which did not quite cancel out. Further work on fatigue cracks, rather than slots, showed that an ideal open crack does produce a zero in the diffracted energy consistent with the theory. From the point of view of practical ultrasonic testing, the experimental data confirmed that the zero in the diffraction coefficient takes the form of a narrow cusp which is likely to cause only a small perturbation in the output signal from most conventional transducers.

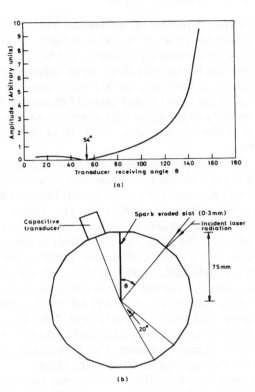

Figure 6.55 (*a*) Theoretical amplitude of compression wave diffracted by crack, for an incident angle of 20°; (*b*) apparatus to use a pulsed laser to measure diffracted waveforms (Ravenscroft *et al* 1989).

There are several other possible applications of laser generation and reception to the study of wave propagation. One is the study of ultrasonic propagation through anisotropic materials that are very difficult to model theoretically. Thus the pulsed laser can be used to generate broadband elastic waves in composite materials, either in combination with a laser inter-ferometer receiver or an alternative broadband receiver (e.g. Buttle and Scruby 1988). Another application might be in scaled-down measurements, e.g. of geological structures. Sub-sea exploration for oil and gas deposits depends on acoustic measurements using a broadband source (e.g. a small explosion) and hydrophone detectors. Wave propagation theory for anisotropic media is needed to interpret the results. An alternative might be to carry out model experiments in the laboratory using the pulsed laser source and a broadband ultrasonic receiver, with a suitable scale factor. There is no reason why the same approach could not be adopted to interpret the ultrasonic wave propagation in a complex structure, such as a welded node.

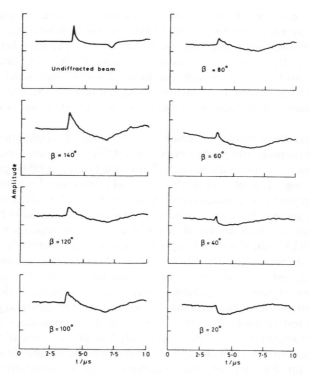

Figure 6.56 Diffracted ultrasound from a 0.3 mm slit, generated by a pulsed laser and received by a capacitive transducer (Ravenscroft *et al* 1989).

6.6 MEDICAL APPLICATIONS

It was pointed out in Chapter 1 that medical diagnostics is one of the most important application areas for ultrasonics. Laser ultrasonic techniques have made little direct impact in this field for the reasons touched upon in that introductory chapter. The contact transducers currently in use are very adequate for the purpose, so that there is no real need for alternatives. Couplants are in any case less of a problem in medical ultrasonics, since the acoustic impedance (which controls any mismatch at the interface) of much body tissue is similar to water and other commonly used fluid couplants. Furthermore, there are rarely any contraindications to the use of an external couplant, nor are any measurements made outside normal body temperatures, which are ideal for piezoelectric materials. Finally, because the velocity of sound is a factor of four smaller in water than the compression waves in most structural metals, the spatial resolution is correspondingly better.

Alternatively, comparable resolution is attainable at lower frequencies, so that it is rarely necessary to exceed 5 MHz for medical imaging. The high bandwidths of laser ultrasonic systems therefore appear to have little to offer either. Finally, the much higher cost of a laser probe compared with a piezoelectric probe, together with the safety hazards associated with any laser, cannot be neglected. It is hard to envisage replacing a safe and effective medical ultrasonic transducer with one which has a higher risk of damaging the patient.

Nevertheless there could be some very specialized applications where conventional piezoelectric probes prove unsuitable, and a few workers have carried out some preliminary studies. Thus von Gutfeld (1980) proposes a scheme for using a pulsed laser as the source of ultrasound for computerized axial tomographic (CAT) measurements, in order to take advantage of the high-speed scanning that can be achieved optically. This can now be applied to lasers, some of which are listed in Chapter 1, whose repetition rates are so high (> 10 kHz for the metal vapour lasers, and substantially higher for semiconductor lasers) that the limiting factor becomes the ultrasonic time of flight and reverbation. In his scheme (which, incidentally, is not recorded as having been engineered) a pulsed nitrogen laser 'fan beam' is directed via a rotating mirror onto the interface between a rubber boundary and the water in the testing tank. An array of piezoelectric transducers is used to receive the ultrasound. Although a relatively low-power laser is envisaged, the power must still be high enough to ensure that the ultrasonic amplitude is adequate for the required measurements, and so substantial optical shielding is necessary.

One of the other advantages of laser ultrasonic systems is that the probe 'head' can be made extremely small (< 100 μm diameter if required), especially if optical fibres are employed. It is interesting to speculate therefore whether in the long term there might be potential application to internal ultrasonic measurements. What is envisaged is a form of ultrasonic endoscopy. Perhaps a fibre-optic ultrasonic system could be used in association with laser micro-surgery, which already uses some of the same technology (e.g. fibre optics and pulsed metal vapour lasers).

REFERENCES

Achenbach J D 1973 *Wave Propagation in Elastic Solids* (Amsterdam: North-Holland)
Achenbach J D and Harris J 1979 *J. Appl. Mech.* **46** 107
Aindow A M, Dewhurst R J and Palmer S B 1982 *Opt. Commun.* **42** 116
—— 1984 *NDT Int.* **17** 329
Aindow A M, Dewhurst R J, Palmer S B and Scruby C B 1985 Unpublished
Aki K and Richards P G 1980 *Quantitative Seismology* vol 1 (San Francisco: Freeman)
Andrews K W 1982 *Ultrasonic Testing* ed J Szilard (New York: Wiley) ch 10

Aussel J D and Monchalin J-P 1989a *Ultrasonics* **27** 165
—— 1989b *J. Appl. Phys.* **65** 2918
Bentley M N, Denham A G and Dukes R 1978 *Br. Patent* no 1507486
Bondarenko A N, Drobot Y B and Kruglov S V 1976 *Sov. J. NDT* **12** 655
Brammer J A and Percival C M 1970 *Exp. Mech.* **10** 245
Bresse L F, Hutchins D A, Hauser F and Farabakhsh B 1989 *Review of Progress in Quantitative NDE* vol 8, eds Thompson and Chimenti (New York: Plenum) p 527
Bresse L F, Hutchins D A and Lundgren K 1988 *J. Acoust. Soc. Am.* **84** 1751
Budenkov G A and Kaunov A P 1979 *9th World Conf. NDT, Melbourne, Australia* paper A4A-14
Buttle D A and Scruby C B 1988 *J. Acoust. Emission* **7** 211
—— 1989 *Harwell Laboratory Report* AERE-R13273
Calder C A, Draney E C and Wilcox W W 1980 *Lawrence Livermore Laboratory Report* UCRL-84139
Calder C A and Wilcox W W 1973 *Lawrence Livermore Laboratory Report* UCID-16353
—— 1980 *Mater. Eval.* **38** 86
Chapman R K 1981 *CEGB Report* NW/SSD/RR/145/81
Carlson N M and Johnson J A 1987 *Review of Progress in Quantitative NDE* vol 7, eds Thompson and Chimenti (New York: Plenum)
—— 1988 *Proc. 3rd Int. Symp. on Nondestructive Characterisation of Materials, Saarbrucken* (Saarbrucken: IzfP) paper 33
Cielo P 1985 *International Advances in NDT* vol 11, ed W J McGonnagle p 175
Cielo P, Nadeau F and Lamontagne M 1985 *Ultrasonics* **23** 55
Cooper J A, Crosbie R A, Dewhurst R J, McKie A D W and Palmer S B 1986 *IEEE Trans. Ultrasonics, Ferroelect. Freq. Control* **UFFC-33** 462
Cooper J A, Crosbie R A, Dewhurst R J and Palmer S B 1985 *Proc. Ultrasonics International 85* (Guildford: Butterworths) vol E25 p 207
Cooper J A, Dewhurst R J and Palmer S B 1986 *Phil. Trans. R. Soc.* A **320** 319
Curtis G J and Hawker B M 1983 *Br. J. NDT* September 240
Davidson G, Emmony D C and Maloney T 1976 *2nd Acoustic Emission Conf. AML, Dorset*
Dewhurst R J 1983 *NDT Commun.* **1** 93
Dewhurst R J and Al'Rubai W S A R 1989 *Ultrasonics* **27** 262
Dewhurst R J, Edwards C, McKie A D W and Palmer S B 1987 *Appl. Phys. Lett.* **51** 1066
—— 1988 *J. Appl. Phys.* **63** 1225
Dewhurst R J, Edwards C and Palmer S B 1986a *Appl. Phys. Lett.* **49** 374
Dewhurst R J, McKie A D W and Palmer S B 1986b *Appl. Phys. Lett.* **49** 1694
Dobbs E R 1973 *Research Techniques in Nondestructive Testing* vol 2, ed R S Sharpe (New York: Academic) p 419
Egle D M and Brown A E 1976 *J. Test. Eval.* **4** 196
Eriksen M and Thaulow C 1988 *Non-destructive Testing* vol 4, eds J M Farley and R W Nichols (Oxford: Pergamon) p 3001
Farley M F and Nichols R W (eds) 1988 *Non-destructive Testing* vols 1–4 (Oxford: Pergamon)
Faßbender S, Kulakov M, Hofmann B, Paul M, Peukert H and Arnold W 1988 *Proc. 3rd Int. Symp. on Nondestructive Characterisation of Materials, Saarbrucken* (Saarbrucken: IzfP) paper 26

Giglio D A 1973 *Harry Diamond Laboratory Report* HDL-TM-73-32

Graff K F 1975 *Wave Motion in Elastic Solids* (Oxford: Clarendon)

Greenhough, Dewhurst R J and Edwards C 1987 *J. Appl. Phys.* **62** 4728

Harker A H 1988 *Elastic Waves in Solids with Applications to Nondestructive Testing of Pipelines* (Bristol: Adam Hilger)

Hutchins D A and Hauser F 1987 *Nondestructive Characterisation of Materials II* eds J F Bussière, J-P Monchalin, C O Rudd and R E Green Jr (New York: Plenum p 725

Hutchins D A, Hu J K, Young R P, Stoner R, Jansen D and Zhang Q L 1989 *J. Acoust. Soc. Am.* **85** 747

Hutchins D A, Nadeau F and Cielo P 1986 *Can. J. Phys.* **64** 1334

Keck R, Krüger B, Coen G and Häsing W 1987 *Stahl und Eisen* **107** 1057

Krautkramer J 1979 *9th World Conf. NDT, Melbourne, Australia* plenary lecture

Krautkramer J and Krautkramer H 1977 *Ultrasonic Testing of Materials* 2nd edn (Berlin: Springer)

Le Brun A, Lesne J-L, Cassier O, Goncalves F and Ferriere D 1988 *Proc. 3rd Int. Symp. on Nondestructive Characterisation of Materials, Saarbrucken* (Saarbrucken: IzfP)

Ledbetter H M and Moulder J C 1979 *J. Acoust. Soc. Am.* **65** 840

Lemons R A and Quate C F 1979 *Physical Acoustics* vol XIV, eds W P Mason and R N Thurston (New York: Academic) p 1

Lesne J-L, Le Brun A, Royer and Dieulesaint 1987 *Industrial Optoelectronic Measurement Systems using Coherent Light, SPIE* vol 863 (Bellingham, WA: SPIE) p 13

Miller G F and Pursey H 1954 *Proc. R. Soc.* A **223** 521

Monchalin J-P, Heon R, Bussiere J F and Farabakhsh B 1987 *Nondestructive Characterisation of Materials* vol 2, eds J F Bussiere, J-P Monchalin, C O Ruud and R E Green Jr (New York: Plenum) p 725

Morse P M 1936 *Vibration and Sound* (New York: McGraw-Hill) p 175

Moss B C and Scruby C B 1988 *Ultrasonics* **26** 179

Newton K 1987 *Proc. Offshore Europe 87 Aberdeen*

Ogilvy J A and Temple J A G 1983 *Ultrasonics* **21** 259

Pao Y H, Gajewski R and Ceranoglu A 1979 *J. Acoust. Soc. Am.* **65** 96

Pekeris C L and Lifson H 1957 *J. Acoust. Soc. Am.* **29** 1233

Ravenscroft F A, Newton K and Scruby C B 1989 *Harwell Laboratory Report* AERE-R13334

Reynolds W N and Smith R L 1984 *J. Phys. D: Appl. Phys.* **17** 109

—— 1985 *Br. J. NDT* September 291

Rosen M, Wadley H N G and Mehrabian R 1981 *Scripta Metall.* **15** 1231

Sachse W and Hsu N N 1979 *Physical Acoustics* vol XIV, eds W P Mason and R N Thurston (New York: Academic) p 277

Sayers C M, Allen D R, Haines G E and Proudfoot G G 1986 *Phil. Trans. R. Soc.* A **320** 187

Scruby C B 1985 *J. Acoust. Emission* **4** 9

—— 1989 *Ultrasonics* **27** 195

Scruby C B, Beesley M J, Stacey K A, Bentley P G, Daniels W and Buttle D J 1990a *Nucl. Energ.* **29** 247

Scruby C B, Brocklehurst F K, Moss B C and Buttle D J 1990b *Nondestruct. Testing Eval.* **5** 97

Scruby C B, Jones K R and Antoniazzi L 1987 *J. NDE* **5** (3–4)

Scruby C B, Smith R L and Moss B C 1986 *NDT Int.* **19** 307

Scruby C B and Stacey K A 1988 *J. Acoust. Emission* **7** 81

Scruby C B, Stacey K A and Baldwin G R 1986 *J. Phys. D: Appl. Phys.* **19** 1597

Scruby C B, Wadley H N G, Dewhurst R J, Hutchins D A and Palmer S B 1981 *Mater. Eval.* **39** 1250

Scruby C B and Wedgwood F A 1986 *UK Patent* GB2185816A

Silk M G 1977 *Research Techniques in Nondestructive Testing* vol 3, ed R S Sharpe (New York: Academic)

Szilard J 1982 *Ultrasonic Testing* (New York: Wiley)

Tam A C 1984 *Appl. Phys. Lett.* **45** 510

Tam A C and Leung W P 1984 *Appl. Phys. Lett.* **45** 1040

Vogel J A and Bruinsma A J A 1988 *Non-destructive Testing* vol 4, eds J M Farley and R W Nichols (Oxford: Pergamon) p 2267

von Gutfeld R J 1980 *Ultrasonics* **18** 175

Wadley H N G, Norton S J, Mauer F and Droney B 1986 *Phil. Trans. R. Soc.* A **320** 341

Wadley H N G, Scruby C B and Shrimpton G 1981 *Acta Metall.* **29** 399

Ward I M 1962 *Proc. Phys. Soc.* **80** 1176

Weaver R L and Pao Y H 1982 *J. Appl. Mech.* **49** 821

Wellman R L 1980 *Harry Diamond Laboratory Report* HDL-TR-1902

Wells Krautkramer 1984 *Laser Generated Ultrasound* unpublished report

Wickramasinghe H K, Bray R C, Jipson V, Quate C F and Salcedo J R 1978 *Appl. Phys. Lett.* **33** 923

Wilcox W W and Calder C A 1978 *Instrum. Techn.* **25** 63

Wilson D M, Cole P T and Whittington R 1977 *Proc. Ultrasonics International 77, Brighton* (Guildford: IPC) p 378

7 Conclusions and Future Prospects for Laser Techniques in Ultrasonics

The purpose of this book has been to introduce the reader to the combination of lasers with ultrasonics. Most of the chapters have been concerned with the use of laser techniques within the field of ultrasonics, notably with the principles of laser interferometric reception (Chapter 3) and laser generation (Chapter 5) of ultrasound, followed by applications (Chapters 4 and 6). Optics and ultrasound also interact in other ways, e.g. in Bragg diffraction, Brillouin scattering and in the knife-edge technique. These interactions are discussed in Chapter 2, together with practical examples of their application.

This final chapter will first summarize the main conclusions regarding the optical techniques that have been the subject of the previous chapters. In §7.2 we discuss possible future developments of the techniques themselves, and also suggest some recommendations for further research. In the final section we discuss future prospects for these techniques in terms of their practical application to industrial problems. By definition, this last section is more speculative than the rest of the book: we hope nevertheless that readers find it both interesting and stimulating.

7.1 SUMMARY AND CONCLUSIONS

7.1.1 Acousto-optics

Acousto-optics is a well-established field where there are already many devices commercially available and in common use. Lasers have greatly enhanced this field, because acousto-optic devices such as Bragg cells are much better suited to the deflection and modulation of the narrow, parallel beams that lasers produce. Chapter 2 demonstrated how various techniques based upon diffraction can be used to characterize ultrasonic fields in transparent fluids. There is some scope for the application of these techniques to transducer calibration, but the authors consider that in most cases laser interferometry is to be preferred because it is capable of giving more accurate and detailed information about ultrasonic fields.

Full field visualization techniques based on diffraction have been used quite extensively to study surface waves. Chapter 2 noted that one particularly important approach involves the use of a knife edge. It was shown that the knife-edge and interferometric techniques have virtually the same maximum theoretical sensitivity for the same bandwidth, highly reflective surface finish and using silicon diode detectors. Thus for both techniques the theoretical sensitivity limit using a 2 mW helium–neon laser (§§2.6.3 and 3.2.2) is approximately 1 pm for a 1 MHz bandwidth, i.e. 10^{-15} m Hz$^{-1/2}$. It was noted, however, that, unlike the case of the interferometer, the sensitivity of the knife-edge technique depends on the beam diameter at the surface. The knife-edge technique forms the basis of a number of methods for visualizing surface waves, which are discussed in §2.6.4, including the scanning laser acoustic microscope. However, although the knife-edge technique will continue to be important, the authors consider that interferometry is more versatile for the general measurement of surface acoustic wave fields, especially in the context of non-destructive testing.

Acousto-optic devices were discussed in §2.4. By far the most common type of acousto-optic device relies upon Bragg diffraction of the light beam by ultrasonic waves. These devices are therefore often called Bragg cells. Their principal uses include the modulation of light beams for applications such as rapid printing (§2.5.1). Response times as short as 1 ns are attainable using gigahertz ultrasonics. Bragg cells can also be used to deflect or scan a laser beam at speeds beyond those obtainable with mechanical scanners (§2.5.2), and also to shift the optical frequency (§2.5.3). Acousto-optic cells can also be used within the laser cavity for Q-switching, or for cavity dumping or for mode locking (§2.5.4). They can also be used for a whole range of spectral analysis and signal processing devices (§2.5.5), which is an important part of the growing area of optical signal processing.

7.1.2 Laser interferometry

The benefits of laser interferometry as a method for measuring ultrasonic wave fields are discussed at various points in Chapters 3 and 4. The main advantages are as follows.

(a) It is non-contact. Thus it does not adversely affect the dynamics of the system, e.g. by loading the surface. This is of more importance in calibration studies than for most non-destructive testing.

(b) It is remote. This is a particularly useful feature when the attachment or close proximity of a transducer to the specimen is not practical due to heat, distance, radioactivity, etc. It is also a major advantage when measurements have to be made quickly (e.g. on a production line), or at a large number of points.

(c) It has exceptionally high resolution, both spatially and temporally. Thus a diffraction-limited spot of a few micrometres diameter can be used to probe a surface with a time resolution of the order of nanoseconds (corresponding to a bandwidth of several hundred megahertz).

(d) It can be used to provide absolute calibration in terms of the wavelength of light.

The only significant technical drawback is in terms of sensitivity, especially on rough or poorly reflecting surfaces, compared with conventional transducers but, as Chapter 3 showed, considerable efforts are being made to develop interferometers that can perform satisfactorily on such surfaces. Other potential drawbacks, compared with a piezoelectric transducer, concern its relatively high cost, complexity and bulk, and the need for safety precautions.

The laser has made a major impact on interferometry. Without the laser, interferometry would not be a practical technique for ultrasonic detection. There are two main properties of the laser that make it so crucial to practical interferometry. Firstly, the light is monochromatic and hence has good temporal coherence. It also has good spatial coherence. In an interferometer based on a pseudo-monochromatic light source such as a discharge lamp, the short coherence length means that the reference and signal paths have to be carefully matched in length. The large coherence length of, for instance, a single-mode helium–neon laser means that a relatively large and variable distance between interferometer and sample can be accommodated for a much smaller reference path. The laser interferometer is thus much easier to apply to industrial measurement problems than one employing a conventional source.

The second relevant laser property is high intensity in a given direction (i.e. within a narrow solid angle). As was shown in §3.2.2, the sensitivity is proportional to the square root of the light intensity falling on the detectors. It is only possible to direct a minute fraction of the intensity from a conventional light source through the interferometer onto the detectors, so that the sensitivity (already a weak point of interferometers in general) would be so much less than the laser system as to make it virtually useless for determining the small displacements that characterize ultrasonic fields.

The reference-beam type of interferometer is likely to continue to dominate ultrasonic detection on good surfaces at low frequencies and for microscopic work. The most common problem with the reference-beam interferometer is its sensitivity to low-frequency background vibration. Various methods for operating in the presence of vibration were discussed in Chapter 3. The three main types of instrument, i.e. compensated, quadrature and heterodyne interferometers, all have their applications, and this situation will continue. Although the electro-optic method of compensation using Pockels cells is the most satisfactory, its widespread use is restricted by the poor availability of suitable cells.

One advantage of the reference-beam laser interferometer is that it can be made extremely compact, either by using one of the new high-performance diode lasers or by using a fibre-optic input from a gas or solid state laser. Such units could be used in relatively inaccessible locations.

As discussed in §3.11, a Fabry–Perot interferometer, in which the specimen surface forms one mirror of the étalon, offers significant sensitivity advantages (e.g. as much as 40 dB for 100 passes through the etalon) over a Michelson instrument for highly reflective surfaces. However, the requirements for flatness, reflectivity and alignment of the specimen surface are so high that it is only a practical proposition for a few specialized applications where these conditions can be met. It was shown in §3.10.3 that the velocity interferometer offers no advantages over the reference-beam instrument in the normal ultrasonic frequency range on good surfaces. The light retains its spatial coherence after reflection from a polished surface, so that a reference-beam interferometer can be used at full efficiency. The sensitivity of the instrument is limited by the threshold for saturation of the photo-detectors, rather than the diffusion of light from the specimen surface. The long-path interferometer is also likely to be somewhat more bulky.

Considerable attention was paid in Chapter 3 to the performance of different types of interferometer on rough surfaces. This was because, although polished surfaces can be expected for applications such as transducer calibration, they are definitely a luxury for most non-destructive testing. The sensitivity of the reference-beam laser interferometer deteriorates seriously on a rough surface. Thus it was calculated in §3.3.5 that there might be about a 40 dB loss of sensitivity compared with a good reflecting surface. This could be partly ameliorated by using a highly convergent beam to increase speckle size and hence the fraction of light returning to the interferometer.

The challenge of non-contact ultrasonic measurement on rough surfaces was probably the main motivation for the development of alternatives to the reference-beam interferometer. Those discussed in Chapter 3 were the long-path-difference (LPD) interferometer and the confocal Fabry–Perot (CFP) interferometer, both being velocity rather than displacement measurement devices. It was shown (figure 3.40 and table 4.2) that the sensitivity of the LPD instrument on an ideal rough surface was about 15 pm for a 1 W laser, measuring ultrasonic waves with a centre frequency of 10 MHz and a bandwidth of 10 MHz. This is comparable with the sensitivity of a reference-beam interferometer employing a 1 W laser. Because of its velocity sensitivity, the LPD instrument is to be preferred for frequencies ≫ 10 MHz. The reference-beam instrument is, however, more suitable for frequencies ≪ 10 MHz.

The confocal Fabry–Perot interferometer (§§3.11.3–6) shows particular promise for ultrasonic reception on rough surfaces. Thus, as figure 3.40 and table 4.2 show, a CFP instrument can be constructed to give slightly better sensitivity than the reference-beam interferometer at typical ultrasonic frequencies. An exciting recent development is the addition of 'sideband stripping' (Monchalin *et al* 1989) since it gives the instrument a much better broadband frequency response.

However, there is an important advantage of both velocity types of interferometer. Whereas the signal beam must be focused to as small a spot as possible in the reference-beam interferometer (RBI) to attain optimum performance on rough surfaces, the LDP and CFP instruments can operate with a much wider probing beam so that they do not suffer from the large fluctuations in sensitivity caused by speckle that affect the RBI. However, under certain conditions the signals may be less easy to interpret and calibrate. Insensitivity to low-frequency vibration of the surface (a problem in the reference-beam interferometer) is offset in the LDP interferometer by sensitivity to vibration in the long optical path, for which stabilization is likely to be required.

At the end of Chapter 4 an attempt was made to compare the sensitivity of interferometric reception of ultrasound with that of other types of transducer. It was shown that the sensitivity of a typical capacitive transducer was of the same order of magnitude as a typical laser interferometer under photon noise limited conditions. It was more difficult to make a direct comparison with the EMAT, the alternative non-contact transducer, but experimental data suggested that a typical EMAT was perhaps 10 dB less sensitive than the interferometer. A piezoelectric transducer was, however, considerably more sensitive than an interferometer: a resonant piezoelectric device being perhaps 50 dB more sensitive, but with a much reduced bandwidth (say 100 kHz rather than 10 MHz). A broadband piezoelectric transducer with a bandwidth of 10 MHz would thus be about 30 dB more sensitive than a laser interferometer covering a comparable bandwidth. Also highlighted were important differences in effective probe area between the interferometer and other types of probe.

These comparisons of sensitivity were all made with the interferometer working under ideal conditions on a polished surface. The EMAT and piezoelectric transducer function equally well on a rough surface, whereas the sensitivity of a reference-beam interferometer could be reduced by as much as 40 dB. Whether a reference-beam instrument is employed or is replaced by a velocity interferometer, two courses of action were discussed in Chapters 3 and 4 to improve the sensitivity of ultrasonic detection on rough surfaces. The first was to use higher power lasers, but these could bring problems of their own, such as laser noise and the difficulties that photodetectors can experience in coping with small fluctuations in high light intensities.

The second option was to use signal averaging. With modern digital instrumentation, this is a standard procedure. Although signal averaging can now be carried out rapidly, it must by definition always slow down the recording process. n^2 signals must be averaged in order to reduce the random noise level by a factor n relative to the signal, and there are usually practical limitations to improving the signal to noise ratio by this method. Thus to reduce noise by 30 dB requires 1000 signals to be summed, which would take several seconds at the fastest rates presently available. Other problems may also arise if n is too large, because of lack of longer term stability in the apparatus. Non-random noise may also be present, which by definition cannot be reduced by averaging. The digitizer must have sufficient dynamic range both to accommodate the high noise levels prior to averaging and to record the much smaller signal with adequate precision, otherwise the final result will be dominated by digitization noise.

Chapter 4 briefly discussed holographic interferometry and speckle pattern interferometry as examples of full-field visualization techniques with application to ultrasonic displacement measurement. It was shown that these techniques work best when the displacements are greater than $\frac{1}{2}\lambda$. Thus for the large-amplitude surface motion that tends to characterize vibration at the lower end of the ultrasonic frequency range they are ideal, since they enable a complete image of the pattern of surface motion to be obtained in a single exposure. The techniques are more difficult to apply to the megahertz range where displacements are small fractions of the optical wavelength. Here it is probably as easy to scan the sample using a reference beam interferometer to make point by point measurements.

Randomly occurring signals, such as those generated in most applications of acoustic emission, were shown in Chapter 4 to present the greatest challenge to laser interferometry, since the option of improving sensitivity by signal averaging was not available. It was concluded that laser interferometry was unlikely to make a significant contribution in this field except for a small number of special applications, for instance where the ultrasonic amplitude was very large, and/or where a high-power laser could be employed.

However, Chapter 4 showed that laser interferometry was particularly well suited to calibration studies, principally of transducers, either in fluids or in contact with solids. In this context the development of parallel scanning was particularly important. Calibration applications make good use of all the main features of optical detection listed at the beginning of this section. Although transverse and vector displacement measurements were discussed in Chapter 4, very few applications could be reported.

7.1.3 Laser generation of ultrasound

Most of the benefits of using lasers for generating ultrasound compared with conventional contact transducers are similar to those for laser reception. Thus laser generation offers a technique which is: (a) non-contact, (b) remote and (c) of high resolution in space and time. Although laser generation can be used as a standard source of ultrasonic waves, unlike interferometry it is not inherently an absolute technique. Some of the factors (such as surface reflectivity) which determine the relationship between incident optical energy and ultrasonic amplitude are difficult to quantify, so that laser generation is really only of use for relative calibration. However, unlike interferometry, laser generation does not suffer from problems with rough surfaces; indeed they often enhance the generation process. Nor is careful alignment of the laser relative to the specimen required for the vast majority of measurements, making laser generation easier to apply in practice than laser reception.

Lasers have made a unique impact on the generation of ultrasound, having opened up a field that previously did not exist. Thus it was shown in Chapter 5 that an instantaneous optical power approaching 10^5 W must be deposited over a very small area (~ 1 mm^2) in order to generate a typical ultrasonic amplitude of 10^{-10} m (1 Å) at a range of 100 mm in aluminium. Furthermore, the energy must be deposited in a very short pulse (< 100 ns say) in order to generate at typical ultrasonic frequencies in the megahertz range. The energy from cw and longer pulsed lasers is mainly dissipated as heat, although lower frequency acoustic vibrations will accompany the thermal waves generated. These conditions of peak power density and duration cannot be met by non-laser light sources. Only a short pulse (e.g. Q-switched) laser can deposit enough energy in a sufficiently short time to set up the large thermoelastic stresses required to generate ultrasonic signals of comparable magnitude and bandwidth to those from piezoelectric transducers. A high-power cw laser modulated by, for instance, an acousto-optic cell could be used to generate ultrasonic signals, but the amplitudes would be so low that a very narrow-band detection system (e.g. using a lock-in amplifier) would be needed.

The coherence properties of laser light are largely irrelevant to the generation of ultrasound by laser, in sharp contrast to laser reception. The properties that matter most are the duration and energy of pulsed radiation produced by the laser. The directionality of the beam is also important since it permits high power densities to be attained. Wavelength is only important in that it affects the absorptivity of the radiation, and ease of alignment.

Three main mechanisms of generation were considered in Chapter 5 to have relevance to ultrasonics: these are the thermoelastic, ablation and constrained surface regimes. The last of these is generally capable of generating the strongest ultrasonic fields. It is, however, unlikely to be of major importance for industrial applications because it requires the surface to be covered either by a liquid or a transparent solid, and thereby eliminates one of the most important features of laser ultrasound, i.e. its non-contact nature.

The next most powerful mechanism for generating ultrasound was ablation. It was noted however that this depends upon the expulsion of small amounts of material from the surface of the specimen, and therefore cannot be called non-destructive. Nevertheless, because of its efficiency and because a small level of surface damage is acceptable for some inspection applications, it maintains a place of importance. In §5.8.4 it was shown that ablative forces in the range 5–50 N (such as would be generated by a focused 100 mJ Q-switched Nd:YAG laser) generate peak compression-wave displacements on the opposite side of a 100 mm aluminium plate in the range 0.3–3 nm. At a distance from the source of 10 mm the displacement range would be 3–30 nm. This is the same order of magnitude as the displacements generated by piezoelectric transducers. Figure 4.35 for instance shows that the peak displacement amplitude from a 5 MHz probe is 10 nm at a distance of 10 mm.

Larger ultrasonic displacements can be generated in the ablation regime by using a higher energy laser pulse, the same power density at the surface, but a correspondingly larger irradiated area.

The most interesting regime from the point of view of most practical applications is thermoelastic, since it is both non-contact and non-destructive. As discussed in Chapter 5, care must be taken to ensure that the instantaneous temperature rises caused by the absorption of laser energy do not exceed the threshold for damage, or other thermal modification of the material. It was shown in §5.3 that the thermoelastic stresses can be quite high, e.g. of the order of 100 MPa per mJ absorbed energy in aluminium and its alloys. Thus care must be taken not to exceed the local yield stress (300–1000 MPa in most structural metals) thereby causing permanent mechanical damage. Interestingly, it was shown that these stresses are of the same order of magnitude as those produced by ablation.

In §5.8.4 the magnitude of the ultrasonic wave field generated by the thermoelastic source was estimated. The maximum in the compression-wave field was 0.9 nm for a 100 mJ pulse incident on polished aluminium. The maximum is at an oblique angle to the surface rather than normal because of the unusual directivity pattern of this source. This estimate did not take into account any pulse broadening due, for instance, to source aperture, losses in the material or receiver response. Practical thermoelastic amplitudes are likely to be slightly smaller than ablation amplitudes, and therefore towards the lower end of the range of amplitudes produced by piezoelectric transducers. However, the oblique orientation of the maximum might not always be convenient, and the thermoelastic source is considerably weaker than ablation at more common orientations close to the surface normal. There is some scope for intensifying the thermoelastic source by using a higher power laser, but this needs great care if the power density is to be kept below the threshold for ablation. Increasing the area of the source may also be problematical since it may broaden the ultrasonic pulses.

Chapter 5 shows that laser generation is capable of producing excellent surface and other guided waveforms, in addition to compression and shear bulk waveforms. In all cases the surface wave amplitudes produced by ablation are of the same order of magnitude as those produced by conventional transducers for comparable bandwidth. The situation is slightly more complicated for the thermoelastic source because there is a conflict between keeping the power density below the threshold for ablation yet needing to focus the energy into a small enough area to reduce source aperture effects and give a short, broadband ultrasonic pulse. One of the best and simplest methods of achieving this compromise was to focus the laser light into a line perpendicular to the proposed direction of ultrasonic propagation. It is nevertheless still difficult in the thermoelastic regime to obtain the same ultrasonic amplitudes as with piezoelectric transducers.

Considerable space was given in Chapter 5 to describing the radiation patterns and temporal waveforms generated by the absorption of pulsed laser radiation. Of particular note was the difference between the radiation patterns of the thermoelastic and ablation sources. While the latter produced a radiation pattern identical to a very small compression-wave probe, with the bulk of the compression-wave energy directed about the surface normal, and could therefore be used as a direct replacement for such a contact probe, the directivity of the former was entirely new. The compression-wave energy was directed in broad lobes centred around 60°, so that the thermoelastic source demands different bulk-wave applications from the ablation source; for instance, those which call for an angled beam. In both cases the shear-wave directivity was more complex, with multiple lobes and changes of phase. These sources (and also the constrained surface source) generated extremely broadband ultrasonic pulses with ideal characteristics for many ultrasonic investigations.

Perhaps somewhat surprisingly, thermal conductivity effects only appeared as second-order in the waveforms from metallic materials. This was because the thermal-wave field had only progressed a distance of a few micrometres during the extremely short lifetime of the optical pulse. This is very much smaller than either the distance propagated by the ultrasound, the lateral dimensions of the source, or the ultrasonic wavelength (~ 1 mm at 5 MHz).

Most of Chapter 5 was devoted to generation in metals. Generation in insulators was only discussed briefly, mainly because less work has been done on these materials, although some applications to ceramics and composites were reported in Chapter 6. Although the absorption processes and the extent of the source are different for these materials, the data suggested that the laser technique is just as effective a method for non-contact generation as in metals.

7.1.4 Applications of laser ultrasonics

Some applications of laser generation alone were discussed in Chapter 6, notably acoustic emission transducer calibration, and measurements in conjunction with either contact ultrasonic receivers '(e.g. for acoustic microscopy and medical ultrasonics) or non-contact EMATS. However, the majority of the significant potential applications required a combination of laser generation and laser reception of ultrasound. Where a receiving transducer other than laser interferometer was used, it was rarely a choice based entirely on technical performance. Factors such as expense or difficulty in obtaining a suitable laser system were also significant.

Most of the applications of ultrasound that have so far been explored are in non-destructive testing and materials characterization. Thus it has been demonstrated that a wide variety of bulk and surface defects could be detected and characterized during non-contact laser ultrasonic inspection. Surface defects have been extensively studied recently, and it has been shown that laser techniques perform at least as well as piezoelectric transducers, in many cases with higher spatial resolution. However, the results to date for defects within the bulk of specimens have been less comprehensive and regrettably less convincing. Nevertheless, preliminary data suggest that laser ultrasonics is well suited to time-of-flight diffraction measurements. Another useful application is to the detection of near-surface defects, a difficult task with contact probes because of near-field effects.

A wide range of materials property measurements has been attempted using laser ultrasonics. As §6.2 showed, these include the determination of such parameters as specimen thickness, elastic constants, grain texture, microstructure, the progress of phase transformations, coating quality, bond quality and internal temperature distribution, in a wide range of monolithic and composite materials. In most cases the potential of the laser technique has only been demonstrated, with very few instances of studies leading to commercial exploitation.

It was recognized that optical techniques were only likely to displace the relatively cheap and robust piezoelectric transducer in areas where the latter fell significantly short in performance. There also needed to be quantifiable economic or safety benefits for making the measurements in the first place. Both the defect detection and materials characterization studies were undertaken with a view to eventual application in environments that are especially hostile to contact transducers, such as the elevated temperatures associated with the production lines in the metals fabrication industries. It was however only possible to report a small number of high-temperature studies. Two of these were of phase transformations, one of which made further use of the remote nature of this technique to isolate the laser transducers from a radioactive material by means of a sealed container with transparent windows. Both studies fully demonstrated the capability of laser ultrasonic technology to make critical measurements under conditions where piezoelectric transducers are difficult or impossible to deploy. The third main study was to internal temperature measurement, but this used EMAT receivers rather than lasers. The final study is probably the most important in the sense that it is the closest to industrial exploitation. It involved the application of laser ultrasound to wall thickness measurement on hot (i.e. at 1230 °C) steel tubes in a plant environment.

7.2   FUTURE RESEARCH AND DEVELOPMENT

In this section we discuss possible future research and development of the laser techniques and their application to ultrasonics which form the subject of this book. We shall attempt to highlight areas where there are gaps in understanding, or where further experiments are required to resolve certain issues. We shall also try to recommend developments to the technology that are desirable and/or necessary in order to meet the needs of various industrial applications. Finally we shall attempt to assess the impact of developments in various technologies on which laser ultrasonics depends, such as laser design, optical components and computing, that have either become recently available or can be predicted for the 1990s. Because it is already a relatively mature field, we shall not explicitly discuss future developments of acoustic-optics. However, many of the general comments regarding, for instance, the impact of new technologies that will be made in the context of laser interferometry, will also apply to the techniques reviewed in Chapter 2.

7.2.1   Laser reception of ultrasound

There needs to be further work to improve the sensitivity of laser techniques for measuring ultrasonic fields. In the case of good reflective surfaces, there are two main avenues to explore when sensitivity is photon-noise limited. The first is higher power lasers, using the fact that the signal to noise ratio increases with the square root of the laser power. Care will have to be taken with the choice of laser to avoid noisy systems. However, the scope for increasing laser power will be restricted because of the limited capacity of photodetectors for high light levels. Drastic measures, such as splitting the light up and using multiple photodetectors, are worth serious consideration. The second option is to develop and/or use faster, higher capacity signal averaging. The disadvantage here is the time taken to collect the data; even with extremely fast signal averaging, the finite ultrasonic reverberation time will impose a lower limit on the interval between consecutive pulses.

It is, however, in respect of rough surfaces that the greatest effort needs to be made. Here there needs to be further development on the design and testing of laser interferometer systems that will perform satisfactorily on these poorly reflecting surfaces. At present the two best techniques for further development for high-frequency use appear to be velocity interferometers, i.e. the long-path-difference (LPD) instrument and the confocal Fabry–Perot (CFP) instrument. Currently the latter offers a more compact system, although there is no reason why the LPD instrument could not be made smaller by using some ingenious multiple folding of the interferometer arms. It will still be important to explore the use of higher power lasers for these velocity interferometers, and rapid on-line signal averaging.

Having established systems that perform satisfactorily on static, rough surfaces, there will then need to be research and development into the application of laser interferometry to moving surfaces, since this capability will be needed for some of the major potential applications of laser ultrasonics. This problem need not arise when scanning a stationary object, since the interferometer beam can be stepped along so that it is momentarily stationary during the recording of the measurement. The authors are only aware of one set of published measurements on moving specimens (Keck *et al* 1987), but the practical problems were not discussed in detail. There may need to be major modifications to the design of some types of interferometer to accommodate this new situation.

The measurement of transverse and vector displacement by means of the laser interferometer has not yet been seriously tackled, and further work here is recommended by the authors.

Although techniques for stabilizing interferometers against vibration are now well established, there is still a need for further development. Again, conditions in industrial plant are more hostile as regards acoustic interference than typical laboratories, making this a relatively serious matter. We note again that, although velocity interferometers based for instance on the LPD design, do not need to compensate for fluctuations in the distance between interferometer and specimen, they do need to stabilize the magnitude of the long path difference. Electro-optic cells perform well as compensators in reference-beam interferometers, but they are expensive and sometimes difficult to obtain with adequate performance. Some improvement in this situation is desirable, or alternatively some further development of mechanical (e.g. piezoelectric) compensation systems for both the reference-beam and LPD instruments should be undertaken.

At the time of writing, there is still a number of interferometer designs in the running for various potential applications. Some extensive critical studies could be very valuable in order to demonstrate not only their comparative performance, but also to assist in matching instruments to applications. There is also a general need to make these laser interferometer systems more compact, simpler to operate and less expensive. Many current systems require substantial setting up and fine tuning, and intermittent adjustments that require an expert always to be available. If these instruments are to find wide industrial acceptance, they will need to be more robust and maintenance free. To this end, once the best type of system is chosen for a given measurement, further engineering design work will be required to achieve compactness, robustness, self-containment and hopefully also lower cost.

Irrespective of which design is chosen, the laser interferometer is dependent on a number of other technologies, mainly laser design, optical devices, electronics and computing. Major advances in any of these fields will most certainly have some impact on the interferometric detection of ultrasound, which we now consider.

Firstly, there are continual improvements in the design and performance of lasers. The general trend is towards higher powers, greater stability, longer coherence, smaller size and a larger choice of wavelengths and specifications. There have recently been particularly impressive advances in laser diodes. These are by far the smallest sources of laser radiation, and could be used directly to make miniature interferometers (i.e. of comparable size to piezoelectric transducers). Although beam quality and coherence are still inferior to gas lasers, there are improvements all the time. Although there is a slow but steady move towards visible wavelengths, laser diodes are unlikely entirely to displace the helium–neon laser in the foreseeable future because the latter is so well suited to general interferometric applications, being both inexpensive and rugged.

In the search for higher power lasers to improve sensitivity on rough surfaces, the researcher may try replacing the rather inefficient helium–neon system with the more powerful argon ion laser. However, these lasers have certain disadvantages, including high noise and low efficiency, hence the need for some form of forced cooling. A more exciting, recent alternative is to use laser diodes indirectly to pump a cw Nd:YAG laser. In this mode of operation the poor beam quality and coherence of the diode are of little consequence. The diode-pumped YAG laser offers a compact and efficient system which only requires a low-voltage power supply (in contrast to the cumbersome high-voltage supplies required for flash-tube pumping and gaseous discharge), but which generates a high-coherence beam at intermediate powers (~ 100 mW at the time of writing). A particularly attractive option is the frequency-doubled version at a wavelength of 0.532 nm which competes directly with the argon laser. Laser diode development could thus be a key to the design of compact optical probes for the reception of ultrasound, whether used directly, or indirectly as a pump for a YAG system.

The highest power cw lasers, based on carbon dioxide, are at too long a wavelength for the type of interferometry discussed in this book. Because it is difficult to maintain a high rating for other cw lasers without a substantial cooling system, and because of problems with heating the specimen, the further development of long-pulse lasers (e.g. Nd:YAG without Q-switching) is awaited with interest. These could give high powers over a period of a fraction of a millisecond during which the measurement would be made, yet with an average power one-thousandth or less of the peak power. If such lasers were to become generally available with good long-pulse characteristics, then they could be of importance to the development of velocity interferometers for use on poorly reflecting surfaces (this idea has been tested out by Keck *et al* (1987)).

Optical fibres have rapidly assumed an important role in modern optics. It is interesting to speculate on their long-term impact on the laser reception of ultrasound. Indeed they have already been tested for a number of functions in a laser interferometer. Thus a single-mode optical fibre can be used to link a very small probe head, which contains the optical components that constitute the interferometer, to a laser which would otherwise make the head too bulky for the proposed applications. In the system designed at Harwell (Moss *et al* 1987) the quadrature interferometer head was 21 mm in diameter by 100 mm long and linked by fibre to a separate helium–neon laser for trial purposes, but this latter could readily be replaced by a more powerful laser such as argon ion, or diode-pumped Nd:YAG.

Alternatively, the optical fibre could be used to transmit the light from the interferometer to the surface of the specimen, enabling the interferometer to be entirely remote for applications where access is difficult. In the case of the reference-beam instrument it would need to be a single-mode fibre since it would constitute part of one arm of the interferometer. Variations in phase caused by vibrations, etc, of the fibre would moreover require compensation. However, a multimode fibre could be used for either the LPD or CFP interferometers in order to collect more light from a rough surface and thus improve sensitivity, provided the interferometer has a sufficiently high angular tolerance (étendue). In a demonstration instrument Vogel and Bruinsma (1988) use a fibre for this purpose, although they do not specify which type. Optical fibres have a further potential use in the LPD interferometer in enabling a long delay path to be produced reasonably compactly (Vogel and Bruinsma 1988). However, because a single-mode fibre must be used here in order to maintain coherence, the instrument loses the benefit of a wide acceptance angle, the feature which gives velocity interferometers their edge over reference-beam interferometers for rough surfaces.

There may be other possibilities for the application of optical fibres to the reception of ultrasound, e.g. transmitting two beams of different polarization through a polarization-preserving fibre onto the specimen as part of a heterodyne system in which one beam is frequency shifted.

Another rapidly developing technology which could have a major impact on laser interferometry in the near future is in optical devices, especially when incorporated into an integrated system. Further development of existing electro-optic devices or entirely new non-linear devices could provide opportunities for more compact and/or more robust interferometer designs. Advances in integrated optics could facilitate the manufacture of a full interferometer in a single optical component.

Photodiode detectors are already well developed to have efficiencies close to the theoretical maximum. A reduction in price of avalanche photodiodes could however be beneficial. With regard to electronics, the trends towards miniaturization and integration would doubtless aid the development of compact interferometer systems. One advantage of laser interferometers over piezoelectric transducers is their high bandwidth. The development of inexpensive, broadband, low-noise analogue electronics would also be beneficial. The continual small improvements in electric motors suitable for scanning should aid ultrasonic beam plotting and similar applications of interferometry.

Digital electronics and computing continue to advance rapidly, with constant improvements and price reductions for existing systems and the regular introduction of new technologies. Advances that enable cheaper and more rapid on-line signal averaging and thereby improve interferometer sensitivity will be important. Also of value to many applications of laser interferometry will be faster techniques for data analysis and interpretation, especially those which are rapid enough to permit on-line analysis. There is more information in the broadband data from laser interferometry than from conventional ultrasonics, and much of this is ignored in simple analysis methods such as rectification and thresholding. More advanced signal processing techniques, such as synthetic aperture processing and pattern recognition, would make better use of the broadband information, but they currently take too much computer time to be used on line. Further improvements in speed and capability could be particularly important because some predicted applications of laser ultrasonics are in process monitoring and control, where rapid feedback is essential.

7.2.2 Laser generation of ultrasound

Although the study of laser-generated ultrasound has only been under way since 1976 (apart from a few important pioneering studies) most of the basic work has been completed, and the fundamental principles underlying the various source mechanisms identified. There is, however, still a need for more rigorous theoretical derivations and/or experimental measurements in a few areas. For instance, studies of laser generation processes in hot specimens could provide valuable background information for industrial applications, to include investigations of thresholds for melting and vaporization, changes in reflectivity, elastic properties and skin depth.

Some considerable effort has already been invested in the design of source geometries to increase ultrasonic amplitudes in the thermoelastic regime without exceeding any threshold for damage, and also to generate directional beams. The simplest and most effective has been the line source (e.g. Aindow *et al* 1980), but an annular source (Cielo 1985) and phased arrays (Vogel and Bruinsma 1988) have also been reported. There is scope for further development of array sources, and for research into other methods for controlling the spatial variation of deposited laser energy using lenses, apertures, zone plates, optical fibres, etc. It may also be valuable to investigate new source configurations as means for selectively generating specific ultrasonic modes.

Most published laser generation studies have used nanosecond optical pulses, when thermal conductivity effects are sufficiently small to be generally second-order, even in good conductors such as aluminium. It would be interesting to see more studies extending the field into the regime where thermal diffusion effects become more significant, e.g. by using longer pulse lasers, or high-frequency modulated cw lasers. There could be interesting effects in the combination of elastic-wave and thermal-wave imaging that have not already been explored in the fields of photoacoustics and thermography.

At the other extreme, further studies with mode-locked picosecond pulse lasers are recommended, following the work of Dewhurst and Al'Rubai

(1989). Not only do these lasers generate much shorter ultrasonic pulses in metals, but they could also be used to study new materials effects. Reduced pulse lengths take the ultrasonic bandwidth into the > 100 MHz regime where important microstructural variations can be studied, and small defects in high-strength materials detected. Further studies of the air-plasma source of Edwards *et al* (1989) are also of interest since the carbon dioxide laser is already an accepted modern manufacturing tool. Research into other laser-induced effects, e.g. at ultraviolet wavelengths, is indicated on the basis that it might provide a further alternative method for generating ultrasound.

The majority of the studies to date have been carried out on metallic materials. This is justified by the fact that many of the in-process measurements for which laser ultrasonics may be required are during the manufacture of structural steels and aluminium alloys. There is nevertheless a need for more extensive and systematic investigations of laser-generation processes in non-metallic materials, including ceramics, plastics and composites. The generation processes may be particularly complex in composite materials which comprise metallic and insulating phases, and the results should prove interesting.

Advances in laser technology will clearly continue to have an impact on the laser generation of ultrasound as they will on laser interferometry. External developments which could be of particular significance here include the availability of inexpensive lasers; firstly to make it easier to operate in the visible and ultraviolet, and secondly to generate sub-nanosecond pulses. A further development of importance for a number of applications is the availability of pulsed lasers with high repetition rates, i.e. comparable with those of conventional ultrasonic sets driving piezoelectric transducers, but without too much reduction in pulse energy. The further development of metal vapour lasers and slab lasers may be significant in this respect.

Many pulsed laser systems are still somewhat bulky, either because of cooling requirements, or because of the large power supplies which are needed to energize gaseous discharges. Compact YAG lasers have been manufactured (e.g. Dewhurst 1983) but, although they can deliver in excess of 10 mJ per pulse, the average power they can handle is small so that repetition rates remain well below 1 Hz, making them ideal for one-off measurements (e.g. *in situ* transducer calibration) but unsuitable for continuous ultrasonic inspection. An obvious first alternative would be a system based on laser diodes, but these cannot store sufficient energy within the laser medium to generate high-amplitude short pulses of light. However, it is interesting to speculate whether laser diodes could emit enough energy to pump a YAG laser, which could then be Q-switched to generate energetic short pulses. This would be more compact than the conventional discharge-tube pumped YAG, and only slightly larger than the smallest YAG laser already referred to, but capable of more realistic repetition rates.

It is not yet clear how wide an impact optical fibres will have on laser generation of ultrasound. Certainly, fibre transmission enables dangerous laser beams to be taken safely across the laboratory or factory so that a bulky laser can be situated some distance from the specimen. The capacity of optical fibres to handle high instantaneous power levels could be a difficulty. Fortunately, however, since very small spot sizes on the specimen surface are not usually required, the use of one or more multimode fibres is acceptable. These have a much larger core cross-sectional area than single-mode fibres and can therefore handle much greater power levels. The transmission of high-energy pulses through such fibres to generate ultrasound has already been demonstrated (Dewhurst *et al* 1988, Vogel and Bruinsma 1988).

There are constant developments in the theory of the interaction of elastic waves with materials and defects, which could encourage the further use of laser ultrasound. This is because the pulsed laser generates an ideal ultrasonic pulse for the experimental confirmation and further investigation of such interactions. It is likely therefore that advances in the general field of ultrasonics will encourage further specific research and development of laser-generated ultrasound.

7.2.3   Non-contact ultrasonic measurements

In addition to the suggestions for work outlined in §§7.2.1 and 7.2.2 above, we would like to suggest some areas for future research and development in respect of non-contact ultrasonic systems based on a combination of laser generation and reception.

Firstly, it would be good to see complete laser generation and reception systems being used to make ultrasonic measurements over a wide range of temperatures to include those routinely encountered during the production of structural materials such as steel. Irrespective of whether these investigations are carried out in the laboratory or on industrial pilot plant, the influence of various environmental effects, such as convection currents, steam and oxide surface scale formation, ought to be included.

Again with industrial applications in mind, there needs to be further work to develop scanned laser ultrasonic systems for the geometries and levels of access that might be encountered on line. In some cases the inspection system must carry out a raster scan on a temporarily stationary specimen, while in others the specimen provides one scan direction as it moves past the laser beams.

Laboratory studies could usefully be widened to a larger range of materials, to include not only the most common constructional materials such as ferritic and stainless steels and aluminium alloys, but also special steels, nickel alloys, titanium alloys, ceramics, composites, etc. A wider range of specimen geometries would also be valuable, bearing in mind that steel, for instance, is produced with a range of cross sections and not only as plate. Variations in accuracy and resolution in these widely different materials should be quantified.

The review of applications of laser ultrasonic systems earlier in this chapter noted that most recent work had concentrated on surface defects. It is therefore suggested that further work is carried out on the detection and characterization of bulk defects using laser generation and reception, noting that the early identification of such defects is most important to the production of steel plate, for instance. It was shown that laser ultrasonics is well suited for the time-of-flight diffraction technique. Further work to explore this combination would be valuable. We noted that laser ultrasonics had a wide range of potential applications to materials characterization, many of which deserve further work, including the monitoring of microstructure and texture, the measurement of internal temperature and the monitoring of coatings and surface treatments. We also suggest that a research programme is undertaken to assess whether laser ultrasonic techniques can be applied to residual stress measurement.

It would be interesting to see laser ultrasonic techniques more widely applied to the measurement of elastic constants. This is because they offer good accuracy on small specimens over a very wide temperature range. The broadband form of the data means that, although a single frequency can be selected if desirable, the measurements need not be restricted to a limited frequency range as in the case of piezoelectric transducers. If a scanning system is incorporated for measurements on single crystals, then point-by-point laser ultrasonic data should enable all the velocities (and hence elastic constants) to be plotted out as a function of orientation in the unit cell. This would ideally incorporate measurements of transverse and vector displacement.

A further area for research would be to explore the high resolution potential of a miniaturized laser ultrasonics system. Thus picosecond-pulse laser generators and high-bandwidth interferometers could be used to carry out non-contact defect detection in electronic materials or in high-strength, low-ductility engineering materials such as ceramics. The critical size range may be as small as 10–100 μm, and one of the few techniques that can detect such small defects is acoustic microscopy. The drawback here is that acoustic microscopy requires water to couple the ultrasound into the specimens, some of which may undergo slight degradation as a result. An alternative approach to be explored is the use of a high-frequency modulated cw laser to generate the ultrasound, because it avoids the possibility of damaging the specimen with a high-power single pulse. This is a particularly important consideration for applications within the microelectronics industry.

7.3 FUTURE PROSPECTS FOR APPLICATIONS

Most of the essential fundamental work on laser generation and reception of ultrasound has been carried out. There are still some gaps in our understanding, as discussed in the previous section, but most of these are likely to be filled over the next few years. Laser ultrasonics is therefore ready, or at least close to being ready, for exploitation by industry ('industry' will for simplicity be taken to include any recipient of laser ultrasonic technology). In this last section of Chapter 7 we attempt to assess prospects for the exploitation of laser ultrasonic techniques. We shall, however, first list some of the general principles that govern the transfer of new technologies to industry, in order to lay a foundation for our discussion of these future applications.

Sharpe (1975) has studied the development of NDT techniques in general, and has found that many of them follow a somewhat similar pattern to figure 7.1. In the early years of the research phase, the main motivation for research and development is technical (scientific). There is some pioneering work, which in the case of laser generation of ultrasound would for instance include the work of White (1963) and Ready (1965). This is followed by a growing number of fundamental laboratory studies during an optimistic period in which the technology is presented with the capability of solving many practical problems (especially in research proposals!). For laser ultrasound this was the late 1970s and 1980s. Sometimes this gives way to a fall-off in interest as funds for basic research are spent, but before industrial support

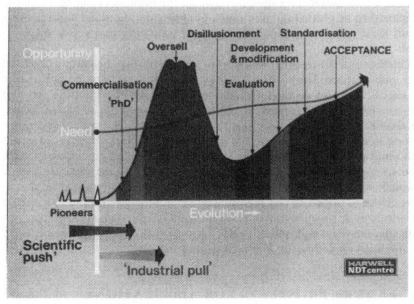

Figure 7.1 Illustration of how interest in a new technology initially develops as a result of 'scientific push', which may be followed by a period of disillusionment before (hopefully) industrial pull promotes transfer into application (Sharpe 1975).

is forthcoming in a period of realism (or even slight pessimism) when it is understood that (inevitably) the technology does not fulfil all that was promised on its behalf earlier on. In the case of laser ultrasonics, although its limitations are generally recognized, there has been no really noticeable dip in interest as yet.

In the final phase, industrial 'pull' takes over from scientific 'push' as the technology begins to be transferred to industry for specific applications. Although some industrial trials have been undertaken, laser ultrasonics is probably somewhere in the intermediate region prior to exploitation. This makes it especially timely for us to consider briefly the steps necessary to ensure successful technology transfer to industry. Some of the points that follow may seem self-evident to our readers, but it is often the most obvious that get forgotten by researchers.

(1) Firstly, industry must have problems that affect efficiency, economics, cash-flow, international competitiveness, profitability, safety or image in the market-place, which it is imperative to solve.

(2) It must be either impossible, very difficult, very costly, very time consuming or very unreliable to solve these problems by other means, e.g. changing the production route, designing out the problem, or modifying existing technology (piezoelectric transducers in the case of ultrasonic inspection).

(3) It must be demonstrated by means of laboratory, and preferably also industrial trials, that the new technology is physically capable of solving the problems that industry has identified.

(4) It must be possible to engineer the laboratory apparatus into a system suitable for industrial installation, i.e. robust, capable of running for long periods unattended, fulfilling industrial safety procedures, and not interfering with the efficient running of the plant or process.

(5) The price (i.e. including running costs as well as purchase and installation costs) must be low enough to repay the investment within a suitably short period of time. If the application is to a production line, the investment may, for instance, be recovered from savings in scrap, re-worked material, reheat energy, off-line NDT and inspection, and time (which equates to lost production). Following a cost–benefit analysis, there must be a sufficiently large and rapid return on investment to persuade financial management to commit funds to what they will inevitably perceive as a 'risk'.

(6) The route for exploitation of the technology (manufacture, marketing, etc) must be well defined, with collaboration between suitable parties, protection of interests (intellectual property, investment etc), and long-term commitment assured. If the basic idea is provided by a university or research institute and the end-user is a materials fabricator, for instance, then a manufacturer of plant instrumentation may need to join the collaboration to ensure fully effective exploitation.

(7)   There needs to be good communication throughout this process. Industry must know as soon as possible what technology is available for exploitation. Although important for other reasons, publication in learned journals or academic conferences may be a less effective way of communicating with industry than publication in trade journals and presentations at trade fairs. Similarly, the research community needs to be aware of major industrial problems so that it can direct at least some of its effort in this direction. Personal contact is undoubtedly the most effective means of ensuring communication in both directions.

Unless these criteria (starting with a major unfulfilled need as described in (1)) can be fulfilled, there will be no industrial 'pull' to get advanced techniques such as laser ultrasound into industry. With scientific push alone, there may well be interest in the technology but it will lack financial commitment from industrial managers.

Turning directly now to laser techniques, it is clear that unless there are some very major reductions in the cost of lasers and optical components, the price of a non-contact laser ultrasonic inspection or measurement system is going to be far in excess of the price of a comparable contact system based on piezoelectric transducers. As a new technology it may meet with additional resistance because it will require the involvement of specialized staff (who may have to be recruited specially) for supervision and maintenance. Extra safety precautions are likely to be necessary, even if some of the less hazardous laser systems are used. Thus there must be problems that industry cannot solve by other means, and where the economic advantage of finding a solution strongly outweighs the cost of the laser ultrasonic measurement system.

This is the main reason why the authors consider that the most likely area for this technology to make a major impact is in ultrasonic inspection and measurement under extreme environmental conditions. The most important of these is elevated temperatures above $\sim 650\,°C$, where piezoelectric transducers can no longer be used. Here the main opportunities lie within a well established industry: steel production. Because of demands for higher quality and lower price steel, a number of measurement and inspection needs have been identified. These include the following:

(1)   locating the solid–liquid interface during e.g. welding and casting (figure 7.2(a)),
(2)   detecting major defects (laminations, porosity) during forging and hot-rolling (figure 7.2(b)),
(3)   measuring internal temperature distributions during continuous casting, reheating and forging (cf figure 6.30),
(4)   dimensioning hot products, during processes such as hot-rolling and forging (figure 7.3(a), (b)),
(5)   monitoring microstructural parameters (e.g. grain size, texture) during hot rolling (figure 7.4(a)),
(6)   detecting surface and bulk defects in hot products (figure 7.4(b)).

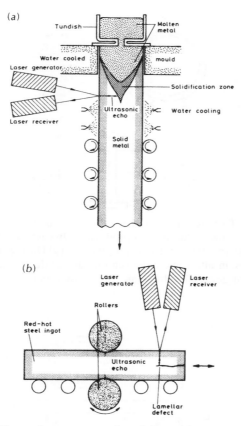

Figure 7.2 Illustrating two examples of the application of laser ultrasonic techniques to steel-making. In each case ultrasonic echoes are detected and interpreted. (*a*) Detection of the solid–liquid interface during continuous casting, (*b*) detection of lamellar defects during the rolling of hot steel ingots.

The steel industry worldwide has a huge investment in plant (i.e. steel mills), worth billions of pounds. The annual value of manufactured steel product is equally impressive—hundreds of millions of tonnes, worth tens of billions of pounds. A laser ultrasonic system would not be cheap by the time it has been engineered into a suitable system for installation on a steel line. However, with such a high volume product, only a very small percentage saving is necessary to justify investment in new technology, although it must be remembered that profit margins may be equally slender.

High temperatures are also used in the manufacture of other structural materials, although they are less likely to be combined with such a large volume product as the steel industry which gives such scope for potential savings to offset the initial investment. Thus there may also be opportunities

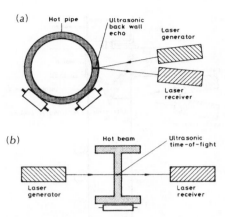

Figure 7.3 Two examples of the application of laser ultrasonics to non-contact gauging during steel production. In each case the thickness is determined from the ultrasonic time of flight. (a) During the hot rolling of steel tubes, in ultrasonic pulse-echo mode, (b) during the hot rolling of beams and girders, in ultrasonic transmission mode.

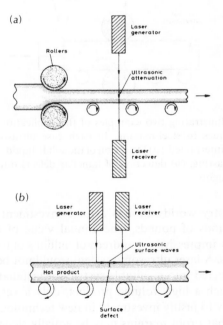

Figure 7.4 Two examples of the application of laser ultrasonics to on-line NDT of hot materials: (a) monitoring microstructure by means of ultrasonic attenuation and scattering following hot rolling, (b) detection of surface defects in hot products.

for exploiting laser ultrasonics in the aluminium industry, although temperatures are not as high there. In the case of high-specification materials such as nickel alloys, forging temperatures are again very high. Although the volume of the product is orders of magnitude smaller than steel, its specific value is much higher, so that an economic case for laser ultrasonic inspection might still be feasible.

Environmental extremes are likely to be encountered during the service of certain types of plant and machinery. Nuclear plant often combines high temperatures with radioactivity (which degrades most piezoelectric materials) so that there is an added motivation to employ remote measurement techniques. Not only would laser ultrasonics eliminate the need to identify a suitable non-corrosive couplant for high-temperature components, but it might also enable inspection to be carried out at much higher temperatures than otherwise, so that the plant would not have to be taken off line. Gas and steam turbines are further examples of high-temperature plant. Here laser ultrasonic techniques could possibly be applied to rotating components at normal running temperatures, as part of regular in-service inspections to detect defects and monitor their growth (figure 7.5). Some system of optical de-rotation would be required.

The microelectronics industry relies upon using well-characterized defect-free materials and devices. There could therefore be some interesting opportunities for non-contact ultrasonic sensors to monitor quality. This industry is already exploring high-resolution non-destructive evaluation techniques such as thermal-wave microscopy (which also has the benefit of being non-contact). Indeed, some of the development work on laser interferometry was directed at the characterization of semiconductors. High-resolution ultrasonics and acoustic microscopy would be much more attractive if the requirement for a coupling fluid could be removed.

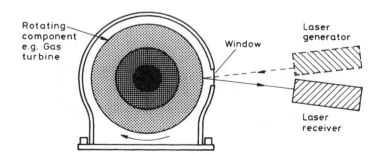

Figure 7.5 Schematic to illustrate possible applications of laser ultrasonic generation and reception to defect detection in hot rotating components such as gas turbines. A similar arrangement could be used to monitor plant condition and vibration using a different design of interferometer, and without the laser generator.

Turning from established industries, it is interesting (but much more difficult!) to speculate about possible applications in new areas of industry. One field that is worth serious consideration is in the characterization and inspection of advanced materials such as ceramics and composites. Most advanced ceramic materials are designed for high-strength (and high-temperature) applications, and tend to have low fracture toughness, so that critical defect sizes are below 100 μm, making them virtually undetectable by conventional ultrasonic NDT. Thus high-frequency ultrasonic techniques are under development. However, there is evidence that couplants such as water may cause slight degradation of some ceramics, making a non-contact technique desirable. As in the case of microelectronic materials, laser ultrasonic techniques could be used to give the desired resolution if, for instance, a mode-locked laser were used in combination with a high-bandwidth interferometer. Another field could be in association with robotic manufacturing systems, e.g. for on-line weld quality monitoring and feedback control (figure 7.6).

Turning now to smaller specialized application areas, laser ultrasonics has considerable potential for the characterization of materials during their development and testing. It is expected that materials scientists will increasingly appreciate the non-contact and high-resolution features of laser ultrasound for studies of new materials such as advanced alloys, ceramics and metal matrix composites. No other technique can give point-by-point measurements of elastic constants over an almost unlimited range of temperature, and this feature is surely asking to be more widely exploited.

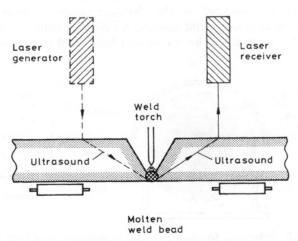

Figure 7.6 Schematic to demonstrate possible application of laser ultrasonic generation and reception to the on-line monitoring and control of welding. The laser interferometer receiver could be used alone to detect acoustic emission and other process noise.

Whether of individual ultrasonic transducers, complete ultrasonic systems or simulated testing geometries, calibration applications are likely to undergo a modest amount of growth, and remain a small but important field for laser ultrasonics. It is unlikely that there will be sufficient demand for absolute transducer calibration by laser interferometry for its use to spread beyond a small number of industrial and standards laboratories who can offer a specialized calibration service.

There is room for substantial growth in the use of laser measurement systems based on interferometry and some of the techniques described in Chapter 2, for the characterization of vibration and condition monitoring. While most work is traditionally carried out in the sonic and low ultrasonic frequency range for practical reasons, the use of laser sensors could extend this further into the ultrasonic frequency range for certain applications where it is known that more information is available. For instance, in the 'machine health monitoring' of a motor or pump, lower frequency measurement tends to be sensitive to changes in the normal modes of vibration of the whole assembly. Higher frequency information, as supplied by a broadband laser interferometer, would be more directly related to the performance and wear of a specific bearing, or even one ball in a ball-race. Provided optical de-rotation is included, there is no reason why laser sensing could be used for condition monitoring of spinning components (compare figure 7.5). There would be major benefits in such an application since conventional transducers or strain gauges must inevitably change the moment of inertia of the rotor.

To conclude, laser ultrasonics is an area of research that has grown rapidly during the past ten years. This has been partly possible because it has combined known areas of physics such as thermal effects, elastic-wave propagation, ultrasonics and optical interferometry, with small but crucial pieces of new work. It has also been greatly helped by the extensive ongoing development of new optical devices and high-performance lasers. Although non-contact laser ultrasonics cannot be viewed as a serious competitor to conventional contact probes for the vast majority of ultrasonic testing, it appears to be the best long-term alternative for inspection under great extremes of environment. It may therefore be significant that its development has coincided with the growing industrial interest in process monitoring and on-line quality control, with the development of new high-performance materials, and also with fresh requirements to inspect products at elevated temperatures.

Earlier in its development, laser ultrasonics was often described as being of great scientific interest and useful for laboratory studies, but of questionable industrial relevance. However, its timeliness in offering much-needed solutions to, for instance, some major hot inspection problems gives this new technology its first serious opportunity to make an industrial impact. Although there is clearly still a great deal of work to be done in terms of the design and manufacture of engineered laser ultrasonic systems for routine plant application, we believe we can now look forward to seeing this technology take its place as a cost-effective industrial tool as well as a valuable research and calibration technique.

REFERENCES

Aindow A M, Dewhurst R J, Hutchins D A and Palmer S B 1980 *Proc. Acoustics 80 Conf.* (Edinburgh: Institute of Acoustics) p 277
Cielo P 1985 *International Advances in NDT* vol 11, ed W J McGonnagle p 175
Dewhurst R J 1983 *NDT. Commun.* **1** 93
Dewhurst R J and Al'Rubai W S A R 1989 *Ultrasonics* **27** 262
Dewhurst R J, Nurse A G and Palmer S B 1988 *Ultrasonics* **26** 307
Edwards C, Taylor G S and Palmer S B 1989 *J. Phys. D: Appl. Phys.* **22** 1266
Keck R, Krüger B, Coen G and Häsing W 1987 *Stahl und Eisen* **107** 1057
Monchalin J-P, Héon R, Bouchard P and Padioleau C 1989 *Appl. Phys. Lett.* **55** 1612
Moss B C, Drain L E and Brocklehurst F K 1987 Unpublished
Ready J F 1965 *J. Appl. Phys.* **36** 462
Sharpe R S 1975 Unpublished
Vogel J A and Bruinsma A J A 1988 *Non-destructive Testing* vol 4, eds J M Farley and R W Nichols (Oxford: Pergamon) p 2267
White R M 1963 *J. Appl. Phys.* **34** 3559

Index